ST. MARY'S CITY, MARYLAND 20686

SOLITON EQUATIONS
AND
HAMILTONIAN SYSTEMS

ADVANCED SERIES IN MATHEMATICAL PHYSICS

Editors-in-Charge

V G Kac (*Massachusetts Institute of Technology*)
D H Phong (*Columbia University*)
S-T Yau (*Harvard University*)

Associate Editors

L Alvarez-Gaumé (*CERN*)
J P Bourguignon (*Ecole Polytechnique, Palaiseau*)
T Eguchi (*University of Tokyo*)
B Julia (*CNRS, Paris*)
F Wilczek (*Institute for Advanced Study, Princeton*)

Published

Vol. 1: Mathematical Aspects of String Theory
         edited by S-T Yau

Vol. 2: Bombay Lectures on Highest Weight Representations
         of Infinite Dimensional Lie Algebras
         by V G Kac and A K Raina

Vol. 3: Kac-Moody and Virasoro Algebras: A Reprint Volume for Physicists
         edited by P Goddard and D Olive

Vol. 4: Harmonic Mappings, Twistors and σ-Models
         edited by P Gauduchon

Vol. 5: Geometric Phases in Physics
         edited by A Shapere and F Wilczek

Vol. 7: Infinite Dimensional Lie Algebras and Groups
         edited by V Kac

Vol. 8: Introduction to String Field Theory
         by W Siegel

Vol. 9: Braid Group, Knot Theory and Statistical Mechanics
         edited by C N Yang and M L Ge

Vol. 10: Yang-Baxter Equations in Integrable Systems
         edited by M Jimbo

Vol. 11: New Developments in the Theory of Knots
         edited by T Kohno

Advanced Series in Mathematical Physics
Vol. 12

SOLITON EQUATIONS

AND

HAMILTONIAN SYSTEMS

L. A. Dickey
Department of Mathematics
University of Oklahoma

World Scientific
Singapore • New Jersey • London • Hong Kong

Published by

World Scientific Publishing Co. Pte. Ltd.
P O Box 128, Farrer Road, Singapore 9128
USA office: Suite 1B, 1060 Main Street, River Edge, NJ 07661
UK office: 73 Lynton Mead, Totteridge, London N20 8DH

SOLITON EQUATIONS AND HAMILTONIAN SYSTEMS
— Advanced Series in Mathematical Physics Vol. 12

Copyright © 1991 by World Scientific Publishing Co. Pte. Ltd.

All rights reserved. This book, or parts thereof, may not be reproduced in any form or by any means, electronic or mechanical, including photocopying, recording or any information storage and retrieval system now known or to be invented, without written permission from the Publisher.

ISBN 981-02-0215-6

Printed in Singapore by JBW Printers and Binders Pte. Ltd.

CONTENTS

0. Introduction   1

Chapter 1. Integrable Systems Generated by Linear Differential nth Order Operators   7

1.1. Differential algebra $\mathcal{A}$ 7
1.2. Space of functionals $\tilde{\mathcal{A}}$ 8
1.3. Ring R of pseudodifferential operators   9
1.4. Lax pairs. KdV-type hierarchies of equations   12
1.5. First integrals (constants of motion)   14
1.6. Soliton solutions   15
1.7. Resolvent. Adler mapping H 17

Chapter 2. Hamiltonian Structures   21

2.1. Finite-dimensional case   21
2.2. Hamilton mapping   26
2.3. Variational principles   27
2.4. Symplectic form on an orbit of the coadjoint representation of a Lie group   31
2.5. Purely algebraic treatment of the Hamiltonian structure   34

Chapter 3. Hamiltonian Structure of the KdV-Hierarchies — 40

3.1. Lie algebra $\mathfrak{E}$ and complex Ω — 40
3.2. Proof of the Hamiltonian property of the Adler mapping — 42
3.3. Poisson bracket — 46
3.4. Reduction to the submanifold $u_{n-1} = 0$ — 48
3.5. Variation of the resolvent — 50
3.6. Hamiltonians of the KdV-hierarchies — 51
3.7. Virasoro algebra — 52

Chapter 4. The Kupershmidt-Wilson Theorem — 55

4.1. A generalization of the Gardner-Zakharov-Faddeev bracket — 55
4.2. Miura transformation. Kupershmidt-Wilson theorem — 57
4.3. Modified KdV equation. Lie-Bäcklund transformations — 59

Chapter 5. The KP-Hierarchy — 62

5.1. Introductory notes — 62
5.2. Definition of the KP-hierarchy — 63
5.3. Reduction of the KP-hierarchy to the KdV — 65
5.4. First integrals — 66
5.5. Soliton solutions — 67

Chapter 6. Hamiltonian Structure of the KP-Hierarchy — 70

6.1. Hamiltonian structure — 70
6.2. Resolvent — 73
6.3. Hamiltonians of the KP-hierarchies — 76

Chapter 7. Baker Function, τ-Function — 79

7.1. Dressing — 79
7.2. Baker function — 81
7.3. Bilinear identity — 82
7.4. Riccati equation — 85
7.5. List of useful formulas for the Faà di Bruno polynomials — 91
7.6. Resolvent and Baker function — 92
7.7. τ-function — 94
7.8. Additional symmetries — 100
7.9. τ-function and Fock representation — 104

Chapter 8. Grassmannian. τ-Function and Baker Function after Segal and Wilson. Algebraic-Geometrical Krichever's Solutions   **109**
8.1. Grassmannian after Segal and Wilson   109
8.2. Algebraic-geometrical solutions of Krichever   115
8.A. Appendix: Abel mapping and the θ-function   119

Chapter 9. Matrix First-Order Operators   **123**
9.1. Hierarchy of equations generated by a first-order matrix differential operator   123
9.2. Hamiltonian structure   130
9.3. Hamiltonians of the matrix hierarchy   133
9.4. Segal-Wilson style theory, soliton solutions   137
9.A. Appendix: Extension of the algebra $\mathcal{A}$ to an algebra closed with respect to indefinite integration   138

Chapter 10. KdV-Hierarchies as Reductions of Matrix Hierarchies   **140**
10.1. Reduction of matrix manifolds   140
10.2. Hamiltonian structures. Resolvent   144
10.3. Reduction to a factor-manifold   148
10.4. Kupershmidt-Wilson theorem   151
10.5. The general Drinfeld-Sokolov scheme   154

Chapter 11. Stationary Equations   **156**
11.1. The ring of functions on the equation phase space   156
11.2. Characteristics of first integrals   159
11.3. Hamilton structure   160

Chapter 12. Stationary Equations of the KdV-Hierarchy in the Narrow Sense ($n = 2$)   **165**
12.1. Theory of the KdV-hierarchy ($n = 2$) independent of Chap. 1   165
12.2. Stationary equations   170
12.3. Integration after Liouville   176
12.4. Return to original variables   181

Chapter 13. Stationary Equations of the Matrix Hierarchy — 186

13.1. Hamilton structure. First integrals — 186
13.2. Hamilton structure of stationary equations — 193
13.3. Action-angle variables — 198

Chapter 14. Stationary Equations of the Matrix Hierarchy (Continuation) — 202

14.1. Baker function. Return to original variables — 202
14.2. Rotation of an n-dimensional rigid body — 208

Chaper 15. Stationary Equations of the KdV-Hierarchies — 213

15.1. Reduction to stationary equations of the matrix hierarchy — 213
15.2. First integrals — 220
15.3. Hamilton structure. Integration — 226

Chapter 16. Matrix Differential Operators Polynomially Depending on a Parameter — 229

16.1. Resolvent — 229
16.2. Another hierarchy of equations — 232
16.3. Two Hamiltonian structures — 233
16.4. Piosson-Lie-Berezin-Kirillov-Costant bracket, central extension, and Kac-Moody algebras — 237
16.5. Hamiltonians in involution — 238
16.6. Still more general equation — 240
16.7. Principal chiral field equation — 241

Chapter 17. Multi-time Lagrangian and Hamiltonian Formalism — 245

17.1. Introduction — 245
17.2. Variational bi-complex — 247
17.3. Exactness of the bi-complex — 251
17.4. Variational derivative — 256
17.5. Lagrangian-Hamiltonian formalism — 260
17.6. Variational bi-complex of a differential equation. First integrals — 264
17.7. Poisson bracket — 268

17.8. Connection with the single-time formalism 269

Chapter 18. Further Examples and Applications 274

18.1. KP-hierarchy 274
18.2. Polynomial matrix family 279
18.3. Principal chiral field 288
18.4. Integrable systems related to nth-order differential operators 297

References 303

0. INTRODUCTION

0.1. The author ventures to offer one more monograph to the reader's atttention despite a rather great amount of books in this field which recently appeared. The first of them was: Novikov (ed.) [1] followed by many others, e.g. Ablowitz and Segur [2], Calogero and Degasperis [3], Newell [5], Takhtadjan and Faddeev [4], Leznov and Saveljev [111], to say nothing about a much older survey by Dubrovin, Matveev, and Novikov [6]. We do so because we believe that nowadays no one can claim to have written a book which can be regarded as a standard manual in this science as a whole. Neither do we. Takhtadjan and Faddeev write in the foreword: "So we picture this science to ourselves here in Leningrad". We think that even this statement of theirs is too strong. Their book does not contain, for example, algebraic-geometrical methods originated in the works of the Leningrad authors Its and Matveev. We are not intended to belittle great merits of this book, we only want to underline that all the books (ours included) reflect approaches of individual schools. We, in our turn, could say: "So I. M. Gelfand and the author, picture this science to themselves". (We have already published a short survey of our joint works with Gelfand in [7]).

For a long time books had not been written but the flood of papers was overwhelming: many hundreds, maybe thousands of them. All this followed one single work by Gardner, Green, Kruskal, and Miura [8] about

1

the Korteweg-de Vries equation (KdV):

(0.2) $$u_t = 6uu_x + u_{xxx}$$

which, before that work, seemed to be merely an unassuming equation of mathematical physics describing waves in shallow water.

We have written "this field", "this science", but have not said yet which science. This has its reason and not an unimportant one. The matter is that "this science" has not found its proper name yet. Looking through "Mathematical Abstracts" or "Reviews" one can expect to meet papers in this science most probably somewhere in "miscellaneous topics". One conventionally calls this science "the theory of completely integrable systems", but this does not reflect the main point exactly. There are completely integrable systems having no relation to this theory. On the other hand the systems studied here are completely integrable only in a relative sense. The name often used, "the inverse scattering method", may sound better because it makes clear at once that one deals with "this science". But this name is also very relative. It reflects only one of the methods used, albeit the first one.

However, what enables one to speak about one branch of science? What part of the theory of differential equations this represents? The most conspicuous is that non-linear equations, e.g. KdV, are here merely by-products of the theory of linear differential operators. The brilliant and surprising idea of the authors of the first article was that the unknown function $u(x,t)$ in KdV must be regarded not simply as a function of x and t, but as a coefficient (potential energy) in the Schrödinger equation $-\varphi'' + u\varphi = \lambda\varphi$ depending on some evolutionary parameter t.

This origin of the equations of this theory implies that they are very rare and refined. For example, the next equation of the "KdV-hierarchy" is

(0.3) $$64\dot{u} = -(10u^3 + 10uu'' + 5(u')^2 + u^{IV})', (\dot{u} = u_t, u' = u_x).$$

Three of the coefficients can be made arbitrary by a scaling transformation $t \to at, x \to bx, u \to cu$, two remaining coefficients being uniquely determined. Any small perturbation of coefficients suffices to take the equation out of the theory. For this, equations which happen to belong to the theory are aristocrats among all the others: they have infinitely many first integrals effectively written, they can be presented in the Hamilton form, and, in a sense, variables "action-angle" can be found in which they can be integrated without difficulty (Zakharov-Faddeev [9]). Finally, they have classes of exact analytical solutions, the most striking of which are soliton solutions. A soliton is a solitary wave moving with a constant velocity. This is not surprising by itself since many equations have such solutions. What is remarkable is that there exist solutions having the form of a system of any number of interacting solitary waves which come from the infinity, then in a way interact among themselves and return into infinity. The existence of such solutions (if they are understood in the broad sense, not necessarily localized in a finite part of the space) is the most noticeable feature of these equations; therefore, the name "soliton equations" fits them very well.

Thus, the theory has two ends: non-linear partial differential equations and the theory of linear differential operators. Now, in retrospective, it is clear that the way to this theory was shorter from the side of linear operators. This, however, was not done (although in some old works one can find all the first integrals of KdV and also, for example, right-hand sides of Eqs. (0.2) and (0.3), see Refs. [10,11]). It is astonishing that the authors of Ref. [8] dug from the opposite end. It is hard to understand how to guess regarding Eq. (0.2) that this equation is related to the Schrödinger equation, and that a remarkable algebra stands behind it. It is also surprising that the first idea about the exceptional properties of this equation was generated by computer calculations. This story was captivatingly narrated by one of the authors M. Kruskal [12].

0.4. The relation of non-linear equations to linear ones was explained best by Lax [13]. It is easy to verify that Eq. (0.2) is equivalent to the operator equation

$$(0.5) \qquad dL/dt = [P, L]$$

where $L = d^2/dx^2 + u$ and $P = d^3/dx^3 + (3/4)(ud/dx + d/dx \cdot u)$. This implies that if u is a solution of the KdV-equation then the dependence of the operator L on time is $L(t) = T(t)L(0)T(t)^{-1}$, where $T(t)$ is an operator. This, in its turn, implies that the spectrum of the Schrödinger operator $L(t)$ is independent of t. Thus, the change of L in time is an isospectral deformation. The spectrum gives the constants of motion of the KdV-equation, very inconvenient ones however, since they are expressed in terms of the function u only in an implicit form. Other quantities can be chosen which are expressed in terms of the spectrum but without this disadvantage. The simplest is the trace of the operator $(L+\zeta)^{-1}$ i.e. of the resolvent of the operator L: $tr(L+\zeta)^{-1} = \int R(x,x,\zeta)dx$ where $R(x,y,\zeta)$ is the Green function. A noticeable property of the kernel $R(x,y,\zeta)$ is that its diagonal $R(x,x,\zeta)$ can be expanded in an asymptotical series in $\zeta^{-1/2}(\zeta \to \infty)$: $R(x,x,\zeta) = \sum_{0}^{\infty} R_k \zeta^{-k+\frac{1}{2}}$ with local coefficients R_k, which means that they are polynomials in $u, u', u'', \ldots$. The integrals of these coefficients $\int R_k dx$ are first integrals of the KdV-equation.

As a matter of fact, instead of the resolvent, another function, the so-called "zeta-function of the operator L" can be considered. This is $tr L^{-s}$. It is connected with the resolvent by the Mellin transformation. The above coefficients R_k are equal to the residues of the diagonal of the kernel of the operator L^{-s} in the points $s = -1/2, 1/2, 3/2, \ldots$

An explicit expression for the coefficients R_k can be obtained as follows. The Green function $R(x,y,\zeta)$ is a product of two solutions of the equation $(L+\zeta)\varphi = 0$: $R(x,y,\zeta) = \varphi(x,\zeta)\psi(y,\zeta)$. It is easy to see that $R(x,x,\zeta) = \varphi(x,\zeta)\psi(x,\zeta)$ satisfies the third-order equation

$$(0.6) \qquad R''' + 4uR' + 2u'R + 4\zeta R' = 0.$$

After multiplication by R it can be integrated once

$$(0.7) \qquad 2RR'' - R'^2 + 4(u+\zeta)R^2 = c(\zeta)$$

where $c(\zeta)$ is an arbitrary constant formal series. Taking $c(\zeta) \equiv 1$ and substituting the formal series $R = (1/2)\sum_{0}^{\infty} R_k \zeta^{-k-1/2}$ into Eq. (0.7), one obtains a recurrence formula which yields in succession all R_k, e.g. $R_0 = 1, R_1 = -u/2, R_2 = (3u^2 + u'')/8$. (This procedure is described in detail in Chapter 12, the first part of which can be read independently).

A problem arises as to how to find all the operators P for which Lax equation (0.5) can be written. This means the following. The left-hand side involves a zero-order operator (namely, that of multiplication by the function $\dot{u}$). Hence the right-hand side must be also a zero-order operator. Thus, all possible operators P must be found such that $[P, L]$ are zero-order operators. Such two operators P and L are said to form a $P - L$ Lax pair. (More often it is called $L - A$ pair but we prefer to call it $P - L$ pair in honour of P. Lax). There are various methods to construct the P-operators. One can use the same resolvent. Let us take the first-order operator $\tilde{P} =$

$-R\partial/\partial x + R'/2$. It is easy to see that $[\tilde{P}, L] = +2R'(L+\zeta)$, whence $[\tilde{P}(L+\zeta)^{-1}, L] = +2R'$. The right-hand side is a zero-order operator which can be expanded in $\zeta^{-1/2}$. The operator $\tilde{P}(L+\zeta)^{-1} = \tilde{P}\sum_{0}^{\infty}(-1)^k L^k \zeta^{-k-1}$ can be also expanded in $\zeta^{-1/2}$, the coefficients being differential operators of growing orders. Thus, $\tilde{P}(L+\zeta)^{-1}$ is a generator for P-operators. This generator was found by Dubrovin [6]. The Lax equations are

$$(0.8) \qquad \dot{u} = 2R'_k$$

If $k = 2$ this is the KdV-equation.

0.9. We have demonstrated here only one example: the KdV-hierarchy of equations. More general hierarchies are obtained for operators L of arbitrary orders, e.g. $L = (\partial/\partial x)^3 + u_1 \partial/\partial x + u_0$, etc. They are called generalized KdV-hierarchies. Besides, matrix equations will be considered with $L = -\partial/\partial x \cdot 1 + U$ and also some others. All the generalized KdV-hierarchies can be unified into one Kadomtsev-Petviashvili (KP)-hierarchy.

0.10. As it can be seen, in the above discussion all the operators, strictly speaking, were not genuine ones: they did not act as operators in any spaces. Accordingly, neither classes of functions nor boundary conditions were involved. In fact, only the algebra of commutation relations between operators was significant. This gives rise to one feature of this book: the almost complete absence of mathematical analysis in its classical sense (except some facts about Riemann surfaces), but solely differential algebra. There are no convergence considerations, all the series are formal. Another feature of the book is a regular usage of the resolvent (or, equivalently, of fractional powers of operators).

We attach also great importance to the Hamiltonian structure of the equations. For the KdV-equation this means a possibility to write it in the form

$$(0.11) \qquad \frac{\partial u}{\partial t} = \frac{\partial}{\partial x}\frac{\delta H}{\delta u}$$

where $H = \int h\, dx$ is a functional called the Hamiltonian (which is also a local one, i.e. h is a polynomial in $u, u', u'', \ldots$), $\delta/\delta u$ is the operator of variational derivative. The resolvent $R(\zeta)$ has the property $\delta(\int R\, dx)/\delta u = \partial R/\partial \zeta$, the explanation of which is, in the end, that $(L+\zeta)^{-1} = (\partial^2/\partial x^2 + u + \zeta)^{-1}$ depends on the sum $u + \zeta$ and the differentiation with respect to ζ yields the same as variation with respect to u. Therefore $R_k = c\delta(\int R_{k+1}dx)/\delta u$, where $c = $ const, and Eq. (0.8) has the required form. It is easy to verify for Eq. (0.11) that if $F = \int f dx$ is any functional then its derivative with respect to t by virtue of the equation (0.11) is $dF/dt = \{H, F\}$, where $\{H, F\} = \int (\delta H/\delta u)'(\delta F/\delta u)dx$. This last expression has all the properties of a commutator and is called the Poisson bracket. For the first time this bracket was obtained by Gardner [14], and Zakharov and Faddeev [9]. In all the cases besides KdV it is also possible to construct a relevant Hamiltonian structure. For this, a formal algebraic definition will also be given.

0.12. An important part will be played by the study of the stationary (independent of t) solutions which satisfy stationary equations $[P, L] = 0$. The significance of these equations was firstly emphasized by Novikov. They are ordinary differential equations; hence the manifolds of their solutions are finite-dimensional. It is remarkable that if one of the solutions is taken as an initial condition for a non-stationary equation of the same hierarchy then at each moment t it remains to be a solution of the stationary equation. Thus, finite-dimensional invariant submanifolds in infinite-dimensional phase space of non-stationary equations can be obtained. This yields finite-dimensional classes of solutions, soliton and algebraic-geometrical ones. Stationary equations also are of the Hamiltonian type. They have sufficiently many first integrals in involution to be integrable in quadratures, according to the Liouville theorem. We perform this procedure of integrating explicitly.

0.13. The last part of this book is dedicated to the so-called multi-time (or the field) formalism. The matter is that sometimes variables are equal in rights, and it is unnatural to choose one of them as a time variable. We construct an algebraic variant of the multi-time variational and Hamiltonian formalism and apply it to our equations.

0.14. We have tried to make this book available to beginners in this area having only basic training in algebra and analysis. All explanations are detailed. A few computations are left to the reader as exercises, which are actually not too numerous.

This branch of science is attractive for the author because it is one of those which revive the interest to the base of mathematics: a beautiful formula.

CHAPTER 1. Integrable Systems Generated by Linear Differential nth Order Operators.

1.1. Differential algebra $\mathcal{A}$.

1.1.1. Let

(1.1.2) $$L = \partial^n + u_{n-2}\partial^{n-2} + \ldots + u_0, \partial = d/dx$$

be a linear differential operator. Further we shall associate with this operator nonlinear differential equations. The coefficients of these equations will be expressed in terms of polynomials with real or complex coefficients in $u_0, u_1, \ldots, u_{n-2}$ and their derivatives of arbitrary orders (i.e. differential polynomials in $\{u_i\}, i = 0, \ldots, n-2$). As far as we discuss construction of these nonlinear equations (and not their solution) the class of functions $\{u_i(x)\}$ under consideration is not important. Therefore we may confine ourselves to the differential algebra $\mathcal{A}_u$ (or simply $\mathcal{A}$) of polynomials in formal symbols $\{u_i^{(j)}\}$, where the operator ∂ (a differentiation) acts according to the rules: $\partial(fg) = (\partial f)g + f(\partial g), \partial u_i^{(j)} = u_i^{(j+1)}$.

If a differential polynomial has no free term we shall call it a polynomial without constants and denote the sub-algebra of such polynomial by $\mathcal{A}_0$.

Very seldom we use the more general differential algebra $\mathcal{B}$ consisting of differential polynomials whose coefficients are explicit functions of x. It is assumed that these coefficients are smooth. (A still more general algebra $\mathcal{C}$

can be introduced consisting of smooth functions of $\{u_i^{(j)}\}$; we do not touch this case at all).

Besides ∂ many other differentiations in $\mathcal{A}$ can be constructed. It is sufficient to determine their action on the generators. Let ξ be such a differentiation and $\xi u_i^{(j)} = a_{i,k}$. Then for every $f \in a$

(1.1.3) $$\xi f = \sum_{i,k} a_{i,k} \partial f / \partial u_i^{(k)}$$

holds ($\partial f / \partial u_i^{(k)}$ is defined in the natural way).

Among others, the differentiations commuting with the operator ∂ will play an exceptional role. For them $a_{i,k+1} = \xi u_i^{(k+1)} = \xi \partial u_i^{(k)} = \partial \xi u_i^{(k)} = \partial a_{i,k}$. i.e. $a_{i,k} = \partial^k a_{i,0} = \partial^k a_i$. Thus $\xi f = \sum_{i,k} a_i^{(k)} \partial / \partial u_i^{(k)}$. Let $a = (a_0, \ldots, a_{n-2})$ and let us denote

(1.1.4) $$\partial_a = \sum_{i=0}^{n-2} \sum_{k=0}^{\infty} a_i^{(k)} \partial / \partial u_i^{(k)}.$$

Obviously ∂ can be written as $\partial_{u'}$, where $u' = (u'_0, \ldots, u'_{n-2})$. We shall also call the differentiations (1.1.4) vector fields. The set of all vector fields will be denoted as $T\mathcal{A}$.

1.2. Space of functionals $\tilde{\mathcal{A}}$.

1.2.1. We consider functionals of the type $\tilde{f} = \int f(u) dx$, where $f(u) \in \mathcal{A}$. What do we understand under the notion or "integral"? Retaining the formal level, i.e. not considering the functional dependence of u on x, we need only two properties of the integral: being linear and $\int f' dx = 0$. Hence the integral is a homomorphism of the linear space $\mathcal{A}$ onto a linear space l, $\mathcal{A} \to l$, while $\partial \mathcal{A} \to 0$. A minimal mapping of the kind exists, $\mathcal{A} \to \tilde{\mathcal{A}} = \mathcal{A}/\partial \mathcal{A}$. Any other mapping can be lead through this one, i.e. a commutative diagram

$$\mathcal{A} \to l$$
$$\searrow \tilde{\mathcal{A}} \nearrow$$

can be constructed. This minimal mapping we call the formal integral and denote $f \in \mathcal{A} \to \tilde{f} = \int f dx \in \tilde{\mathcal{A}}$. Thus, the integral of $f \in \mathcal{A}$ is the equivalence class of f modulo exact derivatives. This immediately leads to the formulas $\int f' dx = 0$, $\int fg' dx = -\int f'g dx$.

If we wish to pass from the formal point of view to the substantial one, this means that we use such classes of functions that $\int f'dx$ always vanishes: e.g. the usual integral $\int_{-\infty}^{\infty} f dx$ in the class of rapidly decreasing functions or the integral over the period in the class of periodic functions etc.

1.2.2. Lemma. The vector fields ∂_a can be applied to the integrals according to the rule
$$\partial_a \tilde{f} = \int (\partial_a f) dx.$$

Proof: The differentiations ∂_a and ∂ commute; if f and g differ by an exact derivative then so do $\partial_a f$ and $\partial_a g$, i.e. $f - g = \partial h \Rightarrow \partial_a f - \partial_a g = \partial(\partial_a h)$. □

1.2.3. Proposition. The relation
$$\partial_a \tilde{f} = \int \sum_i (\delta f / \delta u_i) a_i dx$$
holds, where $\delta f / \delta u_i = \sum_{k=0}^{\infty} (-\partial)^k (\partial f / \partial u_i^{(k)})$ is the so-called variational derivative.

Proof:
$$\partial_a \tilde{f} = \int \sum_{i,k} (\partial f / \partial u_i^{(k)}) a_i^{(k)} dx = \int \sum_{i,k} ((-\partial)^k \partial f / \partial u_i^{(k)}) a_i dx$$
$$= \int \sum_i (\delta f / \delta u_i) a_i dx. \square$$

1.2.4. Remark. The meaning of the vector fields can be elucidated thus: let us substitute $u_i + \varepsilon a_i$ for u_i where ε is a small parameter. Then
$$f(u + \varepsilon a) = f(u) + \varepsilon \partial_a f + 0(\varepsilon^2)$$
Hence $\partial_a f$ describes the deformation of the function f if the argument u is shifted in the direction of a.

1.3. Ring R of pseudodifferential operators.

1.3.1. Let
$$X = \sum_{-\infty}^{m} X_i \partial^i, X_i \in a$$

be a formal series, m being arbitrary. These series can be added and multiplied together and multiplied by elements of $\mathcal{A}$ according to the commutation rule

$$\partial^k f = f\partial^k + \binom{k}{1}f'\partial^{k-1} + \ldots, k \in \mathbb{Z}, \binom{k}{i} = \frac{k(k-1)\ldots(k-i+1)}{i!}.$$

This turns the set of all such series into a module over $\mathcal{A}$. We call it the ring R of pseudodifferential operators (PDO).

1.3.2. Remark. This is not the general definition of the PDO: there are only integer powers of ∂ involved in this definition. Sometimes the name "microdifferential operators" is used.

Using commutation rules we can rewrite every PDO in the dual "left" form

$$X = \sum_{-\infty}^{m} \partial^i X_i^*$$

1.3.3. Lemma. *The above definition of multiplication is associative.*

The proof we offer to the reader as an exercise. □

We shall use the following notations: $R_+ = \{\sum_{i\geq 0} X_i \partial_i\}$ (finite number of terms) for the ring of differential operators; $R_- = \{\sum_{i<0} X_i \partial^i\}$ for the ring of integral (Volterra) operators. Further, $(\sum X_i \partial^i)_+ = \sum_{i\geq 0} X_i \partial^i, (\sum X_i \partial^i)_- = \sum_{i<0} X_i \partial^i, \operatorname{res} \sum X_i \partial^i = X_{-1}$.

Let us introduce the coupling of two operators

$$\langle X, Y \rangle = \int \operatorname{res} XY \, dx \in \tilde{\mathcal{A}}.$$

Then $\langle R_+, R_+ \rangle = \langle R_-, R_- \rangle = 0$. The dual to R_+ is R/R_+. The ring R_- is a section of R/R_+, i.e. in each class of R/R_+ there is one and only one element of R_-.

Sometimes we shall also use the notations

$$R_{(-\infty,n)} = \{\sum_{-\infty}^{n} X_i \partial^i\}, R_{(0,n)} = \{\sum_{0}^{n} X_i \partial^i\},$$

and so on. The corresponding duals are $R/R_{(-\infty,-n-2)}$ and $R_-/R_{(-\infty,-n-2)}$ with sections $R_{(-n-1,\infty)}$ and $R_{(-n-1,-1)}$.

1.3.4. Lemma. If $X = \sum X_i \partial^i$ is rewritten in the left form $X = \sum \partial^i X_i^*$, then $X_{-1}^* = X_{-1} = \operatorname{res} X$; further, $\operatorname{res} \sum X_i \partial^i Y_i = X_{-1} Y_{-1}$. The proof is obvious. □

1.3.5. Proposition. To any X and Y an element $h \in \mathcal{A}$ can be found such that $\operatorname{res}[X,Y] = \partial h$.

Proof: It is sufficient to verify this assertion for monoms $X = X_i \partial^i$ and $Y = Y_j \partial^j$ where $i \geq 0, j < 0, i+j \geq -1$. Then

$$h = \binom{i}{i+j+1} \sum_{\alpha=0}^{i+j+1} (-1)^\alpha X_i^{(\alpha)} Y_j^{(i+j+1-\alpha)}.$$

Check it! □

1.3.6. Corollary.

$$\int \operatorname{res} XY \, dx = \int \operatorname{res} YX \, dx.$$

1.3.7. Proposition. If $X = \sum_{-\infty}^{m} X_i \partial^i$ and $X_m = \operatorname{const} \neq 0$, then the PDO X^{-1} and $X^{1/m}$ exist (and unique). They commute with X.

Proof: We may put $X_m = 1$ without loss of generality. Let us write

$$X^{-1} = \partial^{-m} + Y_{-m-1} \partial^{-m-1} + Y_{-m-2} \partial^{-m-2} + \ldots.$$

with indefinite coefficients. $XX^{-1} = 1$ gives

$$1 + (X_{m-1} + Y_{-m-1})\partial^{-1} + (X_{m-2} + X_{m-1} Y_{-m-1} + Y_{-m-2} + mY'_{-m-1})\partial^{-2} + \ldots \equiv 1.$$

We obtain a sequence of recurrency equations of the form $Y_{-m-k} = -X_{m-k} + Q_k$, where Q_k are differential polynomials in $\{X_i\}$ and $\{Y_j\}, (j > -m-k)$. In the same way the right reciprocal element can be found. The right and the left reciprocals must coincide. In the same way we construct $X^{1/m} : (X^{1/m})^m = X$.

Further, $[X, X^{-1}] = 1 - 1 = 0$. From $X = X^{1/m} \ldots X^{1/m}$ we obtain, commuting the both sides with $X^{1/m}$,

$$[X, X^{1/m}] = [X, X^{1/m}] X^{1/m} \ldots X^{1/m} + X^{1/m} [X, X^{1/m}] X^{1/m}$$
$$\ldots X^{1/m} + \ldots = 0. □$$

1.3.8. Corollary. For arbitrary p thr PDO $X^{p/m}$ can be constructed. The highest term is ∂^p, and $[X, X^{p/m}] = 0$.

If $X = \sum_{0}^{n-2} X_i \partial_i \in R_{(0,n-2)}$ then ∂_X will denote the differentiation (1.1.4) corresponding to the set of coefficients $X = (X_0, \ldots, X_{n-2})$ (the same letter X denotes here a differential operator and the set of its coefficients; this does not bring about any confusion).

Let

$$\tilde{f} = \int f dx, \, \delta f / \delta L = \sum_{0}^{n-2} \partial^{-i-1} \delta f / \delta u_i \in R_{-}/R_{(-\infty, -n)}.$$

1.3.9. Proposition. If $X = \sum_{0}^{n-2} X_i \partial^i \in R_{(0,n-2)}$ then

$$\partial_X \tilde{f} = \int \mathrm{res}(X \cdot \delta f / \delta L) dx = \langle X, \delta f / \delta L \rangle.$$

Proof: This immediately follows from 1.2.2. □

1.3.10. Lemma. The relation $\partial_X L = X$ holds.

Proof: $\partial_X L = \sum_{0}^{n-2} (\partial_X u_i) \partial^i = \sum_{0}^{n-2} X_i \partial^i = X$. □

1.4. Lax pairs. KdV-type hierarchies of equations.

1.4.1. Return to the operator L (1.1.2). Let

$$P_m = (L^{m/n})_+ \in R_{(0,m)}$$

(we shall simply write $L_+^{m/n}$).

1.4.2. Prosposition. The commutator $[P_m, L]$ belongs to $R_{(0,n-2)}$.

Proof: We have

$$[P_m, L] = [L^{m/n} - L_-^{m/n}, L] = -[L_-^{m/n}, L].$$

The operator in the left-hand side is differential. In the right-hand side we have the commutator of two operators of the orders -1 and n. Its order is equal to or less than $-1 + n - 1 = n - 2$. □

We say that the differential operator P (whose coefficients belong to $\mathcal{A}$) together with L make up a Lax pair (PL-pair) if $[P, L] \in R_{(0,n-2)}$.

Thus, for any m we have constructed an operator P_m which makes up, together with L, a Lax pair.

Since $[P_m, L] \in R_{(0,n-2)}$, the differentiation $\partial_{[P_m,L]}$ has sense. According to 1.3.11 $\partial_{[P_m,L]}L = [P_m, L]$.

Let $\{u_i\}$ depend on additional parameter t. We write a differential equation

$$\partial_t L = \partial_{[P_m,L]} L$$

i.e.

(1.4.3) $$\dot{L} = [P_m, L].$$

This is equivalent to a system of partial differential equations on $\{u_i\}, i = 0, \ldots, n-2$. The system is determined by two integers, m and n.

1.4.4. Definition. The set of the systems (1.4.3) with fixed n and all m is called the nth KdV-type hierarchy of equations. The 2nd hierarchy is called the KdV-hierarchy in the narrow sense.

1.4.5. Exercise. Let $n = 2, L = \partial^2 + u$. Find P_3 and write the corresponding equation (1.4.3).

Answer: $P_3 = \partial^3 + 3/2u\partial + 3/4u' = \partial^3 + 3/4(\partial u + u\partial)$. The equation is

(1.4.6) $$4\dot{u} = u''' + 6uu'$$

which is called the Korteweg-de Vries equation (KdV).

1.4.7. Exercise. Let $n = 3, L = \partial^3 + u\partial + v$. Find P_2 and write the corresponding equation.

Answer: $L^{1/3} = \partial + 1/3u\partial^{-1} + 0(\partial^{-2}), P_2 = \partial^2 + 2/3u$; the equation is

$$\dot{u} = -u'' + 2v', \quad \dot{v} = v'' - 2/3u''' - 2/3uu'.$$

Eliminating v we obtain

(1.4.8) $$\ddot{u} = -\frac{1}{3}u^{(4)} - \frac{2}{3}(uu')'.$$

This is the Boussinesq equation.

1.5. First integrals (constants of motion).

1.5.1. Definition. The first integral is a functional $\tilde{f} = \int f dx$ which is conserved by virtue of the equation (1.4.3), i.e.

$$0 = \partial_t \tilde{f} = \int \partial_t f dx = \int \mathrm{res}[P_m, L]\delta f/\delta L dx.$$

1.5.2. Lemma. For any k, by virtue of Eqs. (1.4.4),

$$\partial_t L^{k/n} = [P_m, L^{k/n}]$$

holds.

Proof: Let $k = 1$. Then $L^{1/n} = \partial + v_{-1}\partial^{-1} + \ldots$, all the v_i being differential polynomials in $\{u_i\}$ and vice versa. If we define $\{\partial_t v_i\}$ arbitrarily and then show that $\{\partial_t u_i\}$ are correct (including $\partial_t u_i = 0$ if $i < 0$ and $i > n-2$), this means that $\{\partial_t v_i\}$ have been guessed correctly. Put $\partial_t L^{1/m} = [P_m, L^{1/n}]$. From $L = (L^{1/n})^n$ it follows that $\partial_t L = \sum_{i=0}^{n-1}(L^{1/n})^i[P_m, L^{1/n}](L^{1/n})^{n-i-1} = [P_m, (L^{1/n})^n] = [P_m, L]$ as required. Now let $k > 1$. We have

$$\partial_t L^{k/n} = \sum_{i=0}^{k-1}(L^{1/n})^i[P_m, L^{1/n}](L^{1/n})^{k-i-1} = [P_m, (L^{1/n})^k] = [P_m, L^{k/n}].$$

For negative k this follows from the fact that both ∂_t and the commutator act as differentiations in R : $\partial_t L^{-k/n} = -L^{-k/n}\partial_t(L^{k/n})L^{-k/n}$, $[P_m, L^{-k/n}] = -L^{-k/n}[P_m, L^{k/n}]L^{-k/n}$. $\square$

1.5.3. Proposition. The functionals

$$J_k = \int \mathrm{res} L^{k/n} dx, \; k = 1, 2, \ldots$$

are first integrals of all the equations of the nth hierarchy.
Proof:

$$\partial_t J_k = \int \mathrm{res} \partial_t L^{k/n} dx = \int \mathrm{res}[P_m, L^{k/n}] dx = 0$$

owing to 1.3.5. $\square$

Note that if k is a multiple of n the first integral degenerates, $J_k = 0$.

Thus, the first remarkable property of Eqs. (1.4.3) is proved: the existence of an infinite set of the first integrals.

1.5.4. Exercise. Write two or three first integrals if $n = 2$.
Answer: $J_1 = \int u dx$, $J_2 = \int u^2 dx$, $J_3 = \int [2u^3 - (u')^2] dx$.

1.6. Soliton solutions.

1.6.1. The second remarkable property of Eqs. (1.4.3) is that they possess infinitely many exact analytic solutions. The simplest are the so-called soliton solutions.

We shall construct now a differential operator L whose coefficients are genuine functions of variables x and t, which satisfy Eq. (1.4.3).

Let N be an arbitrary natural number (number of solitons). Let

$$y_k(x,t) = \exp(\alpha_k x + \alpha_k^m t) + a_k \exp(\varepsilon \alpha_k x + \varepsilon^m \alpha_k^m t), k = 1, \ldots, N$$

where $\{\alpha_k\}, \{a_k\}, k = 1, \ldots, N$ are complex numbers, $\alpha_k \neq \alpha_l, k \neq l, \varepsilon^n = 1$. Let

$$(1.6.2) \qquad \phi = \frac{1}{\Delta} \begin{vmatrix} y_1 & \cdots & y_N & 1 \\ y_1' & \cdots & y_N' & \partial \\ \cdots & & & \\ y_1^{(N-1)} & & y_N^{(N-1)} & \partial^{N-1} \\ y_1^{(N)} & & y_N^{(N)} & \partial^N \end{vmatrix},$$

where Δ is the Wronskian of $y_1, \ldots, y_N$. In the expansion of the determinant by the elements of the last column, ∂^i must be written to the right of the minors. ϕ is a differential N-th order operator with the highest coefficient 1.

1.6.3. Lemma. The functions y_k have the properties

$$\partial_t y_k = \partial^m y_k, \partial^n y_k = \alpha_k^n y_k, \phi y_k = 0.$$

The proof is obvious. □

Now we construct operator L by "dressing" the operator ∂^n with the help of the operator ϕ:

$$(1.6.4) \qquad L = \phi \partial^n \phi^{-1}.$$

1.6.5. Proposition. (i) PDO L is in fact a differential operator with the highest term ∂^n. (ii) It satisfies the equation (1.4.3).

Proof: (i) Rewrite (1.6.4) as

$$(1.6.6) \qquad L = (\phi \partial^{-N}) \partial^n (\phi \partial^{-N})^{-1}.$$

The operator $\phi\partial^{-N}$ has the form $1 + a_{-1}\partial^{-1} + \ldots + a_{-N}\partial^{-N}$. The inverse operator has the form $(\phi\partial^{-N})^{-1} = \sum_0^\infty b_{-i}\partial^{-i}, b_0 = 1$. Substituting into (1.6.6) we obtain $L = \sum_{-\infty}^n u_i \partial^i$, where $u_n = 1$ as stated. Let us expand $L = L_+ + L_-$. Then (1.6.4) can be written as

$$L_+\phi - \phi\partial^n = -L_-\phi.$$

The right-hand side is an operator of the order $< N$. Hence the left-hand side is also an operator of the order $< N$, but this operator is obviously differential. It has the property $(L_+\phi - \phi\partial^n)y_k = 0$ (see 1.6.3). If an operator of the order $< N$ sends to zero N linearly independent functions, it is identically equal to zero, $L_+\phi - \phi\partial^n = 0$. Then $L_-\phi = 0$ and $L_- = 0$ since ϕ is invertible. Thus $L = L_+$, a pure differential operator.
(ii) Eq. (1.6.4) implies

$$L^{1/n} = \phi\partial\phi^{-1}, L^{m/n} = \phi\partial^m\phi^{-1}, L_+^{m/n}\phi - \phi\partial^m = -L_-^{m/n}\phi.$$

Denote the right-hand side by Q; the order of Q is $< N$. Let $P_m = L_+^{m/n}$. We have $P_m\phi - \phi\partial^m = Q$. The same argument gives that Q is a differential operator of the order $< N$. Now we differentiate the identity $\phi y_k = 0$ with respect to t:

$$0 = \dot\phi y_k + \phi \dot y_k = \dot\phi y_k + \phi\partial^m y_k = \dot\phi y_k + P_m\phi y_k - Qy_k$$
$$= (\dot\phi - Q)y_k.$$

The differential operator $\dot\phi - Q$ of the order $< N$ sends to zero N functions y_k. Hence $\dot\phi = Q$ and $P_m\phi - \phi\partial^m = \dot\phi$. Finally,

$$\dot L = \dot\phi\partial^n\phi^{-1} - \phi\partial^n\phi^{-1}\dot\phi\phi^{-1} = (P_m\phi - \phi\partial^m)\partial^n\phi^{-1}$$
$$- \phi\partial^n\phi^{-1}(P_m\phi - \phi\partial^m)\phi^{-1} = P_m\phi\partial^n\phi^{-1} - \phi\partial^n\phi^{-1}P_m$$
$$= P_m L - LP_m = [P_m, L]. \square$$

1.6.7. *Exercise.* Write the solution for $n = 2, m = 3, N = 1, L = \partial^2 + u$. *Answer:*

$$(1.6.8) \qquad u = \frac{2\alpha^2}{\cosh^2(x\alpha + t\alpha^3 + \log a)}.$$

This solution is called "soliton". This is a solitary wave moving with the velocity $-\alpha^2$ without changing its form. The velocity depends only on the amplitude.

1.6.9. Exercise. Obtain the solution (1.6.8) directly from Eq. (1.4.6). Look for the solution in the form of a progressive wave $u = f(x + ct)$ under condition: all $f^{(i)} \to 0$ when $x \to \pm\infty$.

A formula for the N-soliton solution will appear in (7.7.13) in connection with the τ-function. It may be shown that when $t \to -\infty$ these solutions respresent v independent solitons moving with different velocities. When they come near to each other a complicated interaction proceeds. When $t \to \infty$ they diverge as exactly the same solitons, whereas during the interaction each soliton can shift.

1.7. Resolvent. Adler mapping H.

1.7.1. Let z be a complex parameter, and ε be a root of unity, i.e. $\varepsilon^n = 1$. Let

$$(1.7.2) \qquad T_\varepsilon = \frac{1}{n}\left(\sum_{r=-\infty}^{\infty} (\varepsilon z)^{-r-n} L^{r/n}\right).$$

a formal series. It can be written as a series in z, or in ∂, or as a double series:

$$T_\varepsilon = \sum_{r=-\infty}^{\infty} T_r z^{-r} = \sum_{i=1}^{\infty} S_i \partial^{-i} = \sum_{i=1}^{\infty}\sum_{r=n-i}^{\infty} P_{r,i} z^{-r}\partial^{-i}$$

The series T_ε we call the basic resolvents. The number of them is n.

1.7.3. Proposition.

$$\sum_{(\varepsilon)} T_\varepsilon = (L - z^n)^{-1}$$

where the right-hand side is understood to be a formal series in z^n.

Proof: We have $\sum_{(\varepsilon)} \varepsilon^r = n$ if $r = pn$, and 0 otherwise. This yields

$$\sum_{(\varepsilon)} T_\varepsilon = \sum_{p=-\infty}^{\infty} z^{-pn-n}(L^p)_- = \sum_{p=-\infty}^{-1} z^{-pn-n} L^p = (L-z^n)^{-1}.\ \square$$

1.7.4. Proposition. The operators $(L-z^n)T_\varepsilon$ and $T_\varepsilon(L-z^n)$ are differential.
Proof: We have

$$((L-z^n)T_\varepsilon)_- = \frac{1}{n}\left((L-z^n)\left(\sum_{-\infty}^{\infty}(\varepsilon z)^{-r-n}L^{r/n}\right)\right)_-$$

$$= \frac{1}{n}\left((L-z^n)\sum_{-\infty}^{\infty}(\varepsilon z)^{-r-n}L^{r/n}\right)_-.$$

This is zero as becomes clear if we open the brackets in $(L-z^n)$. The same holds for the second operator. □

1.7.5. Now we explain the name "resolvent". The operator $(L-z^n)^{-1}$ is usually called the resolvent. This is an integral operator whose kernel, a Green function, satisfies the equation $(L-z^n)\varphi = 0$ with respect to the first argument and the adjoint equation with respect to the second argument. Besides, there are some conditions of discontinuity on the kernel and its $n-1$ derivatives on the diagonal. The operators which we call "resolvents" relate to kernels which satisfy the same equations but not the discontinuity conditions, $(L-z^n)T$ and $T(L-z^n)$ being no more unity but purely differential operators. The sum of all the basic resolvents is just $(L-z^n)^{-1}$.

1.7.6. Definition. The following mapping $R_-/R_{(-\infty,-n-1)} \to R_{(0,n-1)} X \in R_-/R_{(-\infty,-n-1)} \mapsto H^{(z)}(X) = (\hat{L}X)_+\hat{L} - \hat{L}(X\hat{L})_+ = -(\hat{L}X)_-\hat{L} + \hat{L}(X\hat{L})_- \in R_{(0,n-1)}; \hat{L} = L - z^n$, where z is a fixed real or complex number, is called the Adler mapping. In fact, we have a family of mappings depending on the parameters.

1.7.7. Exercise. Why is $H^{(z)}$ well defined on $R_-/R_{(-\infty,-n-1)}$ and is a mapping to $R_{(0,n-1)}$?

$H^{(z)}$ depends on z linearly. We write

$$H^{(z)}(X) = H^{(0)}(X) + z^n H^{(\infty)}(X).$$

The superscript (z) is usually dropped.

The series $\sum_{k_0}^{\infty} X_k z^{-k}$, where $X_k \in R_+^{(n)}$, form the space $R_+^{(n)}((z^{-1}))$. Similarly we define $R_{(-\infty,-n)}((z^{-1})), R_-/R_{(-\infty,-n-1)}((z^{-1}))$ etc. The Adler mapping can be defined with the same formula 1.7.6. as a mapping $R_-/R_{(-\infty,-n-1)}((z^{-1})) \to R_{(0,n-1)}((z^{-1}))$, but here z will be not a fixed number but a parameter entering the series $\sum X_k z^{-k}$; thus we do not have a family of mappings but a single mapping.

1.7.8. Proposition. The resolvents considered as elements of $R_-/R_{(-\infty,-n-1)}((z^{-1}))$ belong to the kernel of the Adler mapping H.

Proof: $H(T_\varepsilon) = -(\hat{L}T_\varepsilon)_-\hat{L} + \hat{L}(T_\varepsilon \hat{L})_- = 0$ according to 1.7.4. □

1.7.9. Proposition. The kernel of the Adler mapping consists of linear combinations $\sum_{(\varepsilon)} c_\varepsilon T_\varepsilon$, where c_ε are formal series $c_\varepsilon = \sum_{k_0}^{\infty} c_{\varepsilon k} z^{-k}$ with constant coefficients. We begin the proof with two lemmas.

1.7.10. Lemma. If $X = \sum_1^n a_i \partial^{-i} \in R_{(-n,-1)}, [L, X]_+ = 0$ and the differential polynomials a_i belong to $\mathcal{A}_0$ (i.e. do not contain constants), then $X = a_n \partial^{-n}$, i.e. $a_1 = \ldots = a_{n-1} = 0$.

Proof: Let k be the least of the numbers $i = 1, \ldots, n-1$ such that $a_i \neq 0$. In the expression $[L, X]_+$ the coefficient in ∂^{n-k-1} is na'_k; hence $a'_k = 0$ and $a_k = 0$ since $a_k \in \mathcal{A}_0$. This contradicts the assumption. Thus $a_1 = \ldots = a_{n-1} = 0$. □

1.7.11. Lemma. If $X = \sum_{j=j_0}^{\infty} X_j z^{-j} = \sum_{j=j_0}^{\infty} \sum_{i=1}^{n} X_{ji} z^{-j} \partial^{-i} \in R_{(-n,-1)}((z^{-1})), X_{ji} \in \mathcal{A}_0$ and $H(X) = 0$, then $X = 0$.

Proof: The equation $H(X) = 0$ can be written as a recurrence one:

$$(1.7.12) \qquad H^{(0}(X_j) + H^{(\infty)}(X_{j+n}) = 0.$$

This relation connects the numbers j which differ by n. Thus there are n chains of the operators X_j connected with each other. For the first term of each chain we have $H^{(\infty)}(X_{j_1}) = 0$, i.e. $[L, X_{j_1}]_+ = 0$. According to 1.7.10 this yields $X_{j_1} = X_{j_1,n} \partial^{-n}$. The next equation in the chain is $H^{(0)}(X_{j_1,n} \partial^{-n}) + H^{(\infty)}(X_{j_1+n}) = 0$. The coefficient in the highest derivative ∂^{n-1} is $nX'_{j_1,n}$, i.e. $X_{j_1,n} = 0$, since $X_{j_1,n} \in \mathcal{A}_0$. Then $X_{j_1} = 0$, which contradicts the assumption. □

Now we can prove the proposition 1.7.8. We have seen that an element of the kernel of the Adler mapping is determined by the constants in its coefficients. Thus, we have to prove that the linear combinations $\sum c_\varepsilon T_\varepsilon$ can have any set of constants in their coefficients. The PDO $L^{r/n}$ is the sum of ∂^r and an operator whose coefficients do not contain constants. Hence

T_ε (considered as an element of $R_-/R_{(-\infty,-n-1)}((z^{-1})))$ is

$$T_\varepsilon = \frac{1}{n}\sum_{r=-n}^{-1}(\varepsilon z)^{-r-n}\partial^r + T_\varepsilon^*$$

where T_ε^* is an operator without constants in its coefficients. It is convenient to pass from $\{T_\varepsilon\}$ to another basis. Let ε_0 be a primitive root of 1. All the roots are $\varepsilon = \varepsilon_0^k, k = 0,\ldots, n-1$. Let $S_l = \sum_{k=0}^{n-1}\varepsilon_0^{lk}T_{\varepsilon_0^k}, l = 0,\ldots, n-1$. All the T_ε can be expressed in terms of S_l. We have

$$S_l = z^{-l}\partial^{l-n} + S_l^*$$

S_l^* are also operators without constants in their coefficients. It is easy to see that the linear combinations $X = \sum_{l=0}^{n-1} c_l(z)S_l$, where $c_l(z) = \sum_{k_0}^{\infty} c_{lk}z^{-k}$, ensure any set of constants in the coefficients X_{ji}. □

Any linear combination $T = \sum_{(\varepsilon)} c_\varepsilon T_\varepsilon$, i.e. any element of the kernel ker H, is also called the resolvent. If $T = \sum T_j z^{-j}$, then $\{T_j\}$ satisfies Eq. (1.7.12).

1.7.13. Exercise. For $n = 2, L = \partial^2 + u$, obtain the set of equations for the coefficients of the resolvent $T = S_1\partial^{-1} + S_2\partial^{-2}$ (using $H(T) = 0$).
Answer:

(1.7.14)
$$S_1''' + 4uS_1' + 2u'S_1 - 4z^2 S_1' = 0,$$
$$S_2' = -(1/2)S_i''.$$

1.7.15. Remark to bibliography. The generalization of the KdV equation connected with the transition from the linear differential second order operator to the n-th order operator appeared first in the works by Krichever [15] and Gelfand and Dickey [16]. The last paper contained also the idea of using the fractional powers of operators for this purpose. The exposition in this paper was cumbersome. Now it can be made much more transparent after the works of Manin [108], Adler [17], Wilson [18] and others. The resolvent was suggested by Gelfand and Dickey in Ref [19] for $n = 2$ and in Ref [20] for any n. However, the simple and beautiful equation for this purpose, $H(T) = 0$, was invented by Adler (l.c.). First integrals appeared in the same papers. As to the soliton solutions for these general equations we hesitate to say who was the author of the exposed method. We have extracted it basically from a paper by Manin [21], but it seems that he had referred to Drinfeld [22] and Krichever.

CHAPTER 2. Hamiltonian Structures.

2.1. Finite-dimensional case.

2.1.1. To make the book self-contained we permit ourselves to give some well-known facts about Hamiltonian structures. (An excellent account on this the reader can find in Arnold [23]).

The usual classical mechanical definition of the Hamiltonian system is the following. There is a special, "canonical", set of independent variables $(q^1, \ldots, q^n, p^1, \ldots, p^n)$ consisting of two groups, $\{q^i\}$ the coordinates and $\{p_i\}$ the momenta. There is also a function $\mathcal{H}(q, p)$, the Hamiltonian. Canonical Hamilton equations are

$$(2.1.2) \qquad \dot{q}^i = \partial \mathcal{H}/\partial p^i \, , \dot{p}^i = -\partial \mathcal{H}/\partial q^i \, .$$

This definition depends on special variables. Not every change of variables preserves the form of these equations. The changes having this property are called canonical (it is not an exact definition but here it does not matter). We shall define Hamiltonian systems in an invariant way, independent of the coordinate systems.

We expect from the reader only knowledge of the first notions of the analysis on manifolds (see e.g. Ref. [24]): the manifold, the tangent space, vector fields as linear differential operators $\xi \in TM$ which can act on functions, i.e. ξf. Vector fields form a Lie algebra with respect to the commutator $[\xi, \eta] = \xi\eta - \eta\xi$. Further, we use the cotangent space at the point

$x \in M, T_x^*M$. The coupling between the elements $\xi \in T_xM$ and $\omega \in T_x^*M$ is denoted as $\langle \xi, \omega \rangle$. We shall also use the following: differential forms, inner product of a vector and a form (if ξ is a vector and ω a n-form, $i(\xi)\omega$ is a $(n-1)$-form defined as $\forall \xi_1, \ldots, \xi_{n-1} \in T_xM$ $(i(\xi)\omega)(\xi_1, \ldots, \xi_{n-1}) = \omega(\xi, \xi_1, \ldots, \xi_{n-1})$). We need also a coordinateless formula for the differential $d\omega$ (formula Lee, see Ref. [24]):

(2.1.3)
$$\forall \xi_1, \ldots, \xi_{n+1}(d\omega)(\xi_1, \ldots, \xi_{n+1}) = \sum_i (-1)^{i+1} \xi_i \omega(\xi_1, \ldots, \hat{\xi}_i, \ldots, \xi_{n+1})$$
$$+ \sum_{i<j} (-1)^{i+j} \omega([\xi_i, \xi_j], \xi_1, \ldots, \hat{\xi}_i, \ldots, \hat{\xi}_j, \ldots \xi_{n+1}).$$

2.1.4. Remark. At first sight it is not at all clear that this formula defines $d\omega$ as a differential form. The value of a differential form at a point must depend only on the values of the vector fields at that same point and not at other points of its vicinity, which appears to fail for (2.1.3). In actual fact (see Ref. [24]), however, everything is right here.

The Lie derivative L_ξ of a form ω along a vector field ξ is

$$L_\xi \omega = (di(\xi) + i(\xi)d)\omega.$$

In the particular case of 0-forms, i.e. functions f, we have $L_\xi f = i(\xi)df = \xi f$.

2.1.5. Proposition. Let ω be a r-form and $\xi, \xi_1, \ldots, \xi_r$ be vector fields. Then

$$\xi(\omega(\xi_1, \ldots, \xi_r))$$
$$= (L_\xi \omega)(\xi_1, \ldots, \xi_r) + \sum_{i=1}^{r} \omega(\xi_1, \ldots, [\xi, \xi_1], \ldots, \xi_r).$$

Proof: We have

$$(L_\xi \omega)(\xi_1, \ldots, \xi_r) = (i(\xi)d\omega)(\xi_1, \ldots, \xi_r) + (di(\xi)\omega)(\xi_1, \ldots, \xi_r)$$
$$= d\omega(\xi, \xi_1, \ldots, \xi_r) + (d(i(\xi)\omega))(\xi_1, \ldots, \xi_r)$$
$$= \xi\omega(\xi_1, \ldots, \xi_r) + \sum_{i=1}^{r}(-1)^i \xi_i \omega(\xi, \xi_1, \ldots, \hat{\xi}_i, \ldots, \xi_r)$$
$$+ \sum_{i=1}^{r}(-1)^i \omega([\xi, \xi_i], \xi_1, \ldots, \hat{\xi}_i, \ldots, \xi_r)$$

$$+ \sum_{i<j}(-1)^{i+j}\omega([\xi_i,\xi_j],\xi,\xi_1,\ldots,\hat{\xi}_i,\ldots,\hat{\xi}_j,\ldots,\xi_r)$$

$$+ \sum_{i=1}^{r}(-1)^{i-1}\xi_i\omega(\xi,\xi_1,\ldots,\hat{\xi}_i,\ldots,\xi_r)$$

$$+ \sum_{i<j}(-1)^{i+j}\omega(\xi,[\xi_i,\xi_j],\xi_1,\ldots,\hat{\xi}_i,\ldots,\hat{\xi}_j,\ldots,\xi_r)$$

$$= \xi\omega(\xi_1,\ldots,\xi_r) + \sum_{i=1}^{r}(-1)^{i}\omega([\xi,\xi_i],\xi_1,\ldots,\hat{\xi}_i,\ldots,\xi_r)$$

which is equivalent to the required equality. □

2.1.6. Corollary. If $\xi, \eta \in TM$ and $\omega \in T^*M$, then

$$\eta\langle\omega,\xi\rangle = \langle L_\eta\omega,\xi\rangle + \langle\omega,[\eta,\xi]\rangle, (\langle\omega,\xi\rangle = \omega(\xi)).$$

A system of ordinary differential equations

$$\dot{x}(t) = \xi(x(t))$$

corresponds to each vector field $\xi(x)$; here x is a point of the manifold, $\dot{x}(t) \in TM, (\dot{x}(t)f = \dfrac{d}{d\tau}f(x(t+\tau))|_{\tau=0})$.

Every nondegenerate 2-form permits to identify the tangent and the cotangent spaces, $T_xM \to T_x^*M$:

(2.1.7) $\qquad\qquad\qquad \xi \in T_\alpha M \mapsto -i(\xi)\omega \in T_x^*M$

(the non-essential minus sign stands here for convenience of what follows). In the coordinate form the operation $i(\xi)\omega$ is the lowering of the superscripts with the help of the tensor ω_{ij}, where $\omega = \sum \omega_{ij} dx^i \wedge dx^j$. The mapping (2.1.7) is a bijection. The inverse mapping will be denoted as $H : \alpha \in T_x^*M \mapsto \xi \in T_xM, \alpha = -i(\xi)\omega$ (in the coordinate form this is the lifting of subscripts with the help of the tensor ω^{ij}, the matrix (ω^{ij}) being inverse to (ω_{ij})).

2.1.8. Definition. A nondegenerate and closed ($d\omega = 0$) form ω is said to be symplectic.

These forms can exist only on even-dimensional manifolds. A manifold with a given symplectic form will be called a phase space.

A vector field ξ which preserves the symplectic form ω, i.e. $L_\xi \omega = 0$, is called a Hamiltonian field. For such a field

$$0 = L_\xi \omega = (di(\xi) + i(\xi)d)\omega = d(i(\xi)\omega).$$

According to the Poincaré lemma a function $\mathcal{H}(x)$ exists (at least locally) such that

(2.1.9) $$d\mathcal{H}(x) = -i(\xi)\omega.$$

The function $\mathcal{H}(x)$ is the Hamiltonian of the system.

Conversely, to every function $\mathcal{H}(x)$ we can find a Hamiltonian vector field $\xi_\mathcal{H}$ whose Hamiltonian is $\mathcal{H}$, i.e. $d\mathcal{H} = -i(\xi_\mathcal{H})\omega$, since the mapping (2.1.7) is a bijection. The field is Hamiltonian because

$$L_{\xi_\mathcal{H}}\omega = d(i(\xi_\mathcal{H})\omega) = -d(d\mathcal{H}) = 0.$$

A differential equation which corresponds to a Hamiltonian vector field

(2.1.10) $$\dot{x} = \xi_\mathcal{H}(x)$$

is called a Hamilton equation.

2.1.11. Exercise. To show that the Hamilton equation corresponding to the form $\omega = \sum dp^i \wedge dq^i$ and the Hamiltonian $\mathcal{H}(p,q)$ is Eq. (2.1.2).

2.1.12. Lemma. If h is a Hamiltonian, i.e. a function, and ξ is a vector field, then

(2.1.13) $$\xi h = \omega(\xi, \xi_h).$$

Proof:
$$\xi h = i(\xi)dh = -i(\xi)i(\xi_h)\omega = -\omega(\xi_h, \xi) = \omega(\xi, \xi_h). \square$$

As a particular case, take $\xi = \xi_g$, where g is another Hamiltonian. Then

$$\xi_g h = \omega(\xi_g, \xi_h) = -\omega(\xi_h, \xi_g) = -\xi_h g.$$

2.1.14. Definition. The Poisson bracket of two Hamiltonians is

$$\{h, g\} = \xi_h g = \omega(\xi_h, \xi_g).$$

2.1.15. Proposition.

$$\xi_{\{g,h\}} = [\xi_g, \xi_h].$$

Proof. The form ω is closed. Therefore,

$$\begin{aligned}
\forall \xi \, 0 &= (d\omega)(\xi_g, \xi_h, \xi) = \xi_g \omega(\xi_h, \xi) - \xi_h \omega(\xi_g, \xi) \\
&\quad + \xi \omega(\xi_g, \xi_h) - \omega([\xi_g, \xi_h], \xi) + \omega([\xi_g, \xi], \xi_h) \\
&\quad - \omega([\xi_h, \xi], \xi_g) = -\xi_g \xi h + \xi_h \xi g + \xi\{g, h\} \\
&\quad - \omega([\xi_g, \xi_h], \xi) + [\xi_g, \xi]h - [\xi_h, \xi]g \\
&= -\xi \xi_g h + \xi \xi_h g + \xi\{g, h\} - \omega([\xi_g, \xi_h], \xi) \\
&= -\xi\{g, h\} - (i([\xi_g, \xi_h])\omega)(\xi) = (-d\{g, h\} - i([\xi_g, \xi_h])\omega)(\xi).
\end{aligned}$$

Since ξ is arbitrary, this yields

$$d\{g, h\} = -i([\xi_g, \xi_h])\omega$$

which is equivalent to the required identity. □

2.1.16. Proposition. The Poisson bracket has the following properties:

(i) $\quad \{g, h\} = -\{h, g\}$

(ii) $\quad \{f, g \cdot h\} = g\{f, h\} + h\{f, g\}$

(iii) $\quad \{h_1, \{h_2, h_3\}\} + \text{c.p.} = 0$

where c.p. symbolizes adding of all the cyclic permutations.

Proof: Only (iii) needs verification:

$$\begin{aligned}
0 &= (d\omega)(\xi_{h_1}, \xi_{h_2}, \xi_{h_3}) = \xi_{h_1} \omega(\xi_{h_2}, \xi_{h_3}) \\
&\quad - \omega([\xi_{h_1}, \xi_{h_2}], \xi_{h_3}) + \text{c.p.} = \xi_{h_1}\{h_2, h_3\} - [\zeta_{h_1}, \zeta_{h_2}]h_3 + \text{c.p.} \\
&= \{h_1, \{h_2, h_3\}\} - \xi_{\{h_1, h_2\}} h_3 + \text{c.p.} \\
&= \{h_1, \{h_2, h_3\}\} - \{\{h_1, h_2\}, h_3\} + \text{c.p.} = 2\{h_1, \{h_2, h_3\}\} + \text{c.p.} □
\end{aligned}$$

The Poisson bracket turns the space of all functions into a Lie algebra. Proposition 2.1.15 means that the mapping $h \to \xi_h$ is a homomorphism of the Lie algebra of functions with the Poisson bracket as a commutator to the Lie algebra of vector fields.

2.1.17. Proposition. If f is a function, then, in virtue of the differential equation $\dot{x} = \xi_{\mathcal{H}}(x)$,

$$\partial_t f = \{\mathcal{H}, f\}$$

holds.

Proof: The differentiation with respect to the parameter t by virtue of the equation is equivalent to the action of the vector field $\xi_{\mathcal{H}}$:

$$\partial_t f = \xi_{\mathcal{H}} f = \{\mathcal{H}, f\}. \square$$

2.1.18. Corollary. A function f is a first integral of the equation if and only if it commutes with the Hamiltonian, i.e. $\{\mathcal{H}, f\} = 0$.

2.1.19. Definition. Two functions are in involution (i.e. are involutive) if $\{f, g\} = 0$.

Thus the involutiveness is the neccessary and sufficient condition for the corresponding vector fields to be commutative.

2.2. Hamilton mapping.

2.2.1. As we have seen, a non-degenerate form ω generates a bijection: $T_x M \to T_x^* M$. The inverse mapping we denoted as H. The form ω can be expressed in terms of H:

$$\forall \alpha, \beta \in T_x^* M, \ \omega(H\alpha, H\beta) = \langle H\alpha, \beta \rangle.$$

A question arises: what are the conditions on H equivalent to the fact that ω is symplectic? H must be skew symmetric. The condition that ω is closed is more complicated (see [26–28] for more detail).

2.2.2. Definition. The Schouten bracket of two skew symmetric mappings $H, J : T^* M \to TM$ is a trilinear mapping

$$[H, K] : T^* M \times T^* M \times T^* M \to \mathcal{F}(M)$$

($\mathcal{F}(M)$ is the ring of functions on M) defined by

$$\forall \alpha_1, \alpha_2, \alpha_3 \in T^* M [H, K](\alpha_1, \alpha_2, \alpha_3) = \langle KL_{H\alpha_1} \alpha_2, \alpha_3 \rangle \\ + \langle HL_{K\alpha_1} \alpha_2, \alpha_3 \rangle + \text{c.p.}$$

2.2.3. Proposition. The form ω is closed if and only if $[H, H] = 0$.

Proof: Let $\xi_i = H\alpha_i, i = 1, 2, 3$.

$$d\omega(\xi_1, \xi_2, \xi_3) = \xi_1 \omega(\xi_2, \xi_3) - \omega([\xi_1, \xi_2], \xi_3) + \text{c.p.}$$
$$= H\alpha_1 \langle H\alpha_2, \alpha_3 \rangle - \langle [H\alpha_1, H\alpha_2], \alpha_3 \rangle + \text{c.p.} .$$

Transform the first term using 2.1.6:

$$d\omega(\xi_1, \xi_2, \xi_3) = \langle [H\alpha_1, H\alpha_2], \alpha_3 \rangle + \langle H\alpha_2, L_{H\alpha_1}\alpha_3 \rangle$$
$$- \langle [H\alpha_1, H\alpha_2], \alpha_3 \rangle + \text{c.p.} = -\langle L_{H\alpha_1}\alpha_3, H\alpha_2 \rangle + \text{c.p.}$$
$$= \langle HL_{H\alpha_1}\alpha_3, \alpha_2 \rangle + \text{c.p.} = \frac{1}{2}[H.H](\alpha_1, \alpha_2, \alpha_3) .$$

It remains to note that α_1, α_2 and α_3 are arbitrary. □

2.2.4. *Proposition.* The correspondence $f \mapsto \xi_f$ can be expressed as

$$\xi_f = Hdf.$$

Proof: $df = -i(\xi_f)\omega$ is equivalent to $\xi_f = Hdf$. □

2.2.5. *Proposition.*

$$\{f, g\} = \langle Hdf, dg \rangle.$$

Proof:

$$\{f, g\} = \xi_f g = \langle \xi_f, dg \rangle = \langle Hdf, dg \rangle .$$

2.3. Variational principles.

2.3.1. What are usual sources of symplectic forms? For example, the natural symplectic form in the cotangent bundle is well-known. We shall not speak in detail about this form since we do not use it directly. Briefly the main point is the following. Elements of the cotangent bundle are pairs: a point of the manifold, $x \in M$, and a covector, i.e. an element $\alpha \in T_x^*M$, at the same point. The symplectic form is defined in the tangent space to the cotangent bundle at its point (x, α). The tangent vector will be given if we specify a shift of the point x, i.e. an element $\xi \in T_xM$, and a shift of the covector α. First we define the natural 1-form $\omega^1 = \langle \xi, \alpha \rangle$. Then the symplectic form ω will be $\omega = d\omega^1$.

2.3.2. Now we describe another source of symplectic forms: variational principles of mechanics. We look for an extremal of the functional of action

$S = \int_\alpha^\beta \Lambda dt$ where the Lagrangian Λ depends on the coordinates $\{x_i\}, i = 1,\ldots, n$ on the manifold as well as on their derivatives with respect to the parameter $t : \{x_i, \dot x_i, \ddot x_i, \ldots, x_i^{(N_i)}\}$. (This is in a sense an old-fashioned exposition; the modern one requires jet bundles, but we do not worry about it here.) It is well-known that this problem leads to the neccessary condition for the extremum: the Euler-Lagrange equation. Integrating by parts we obtain

$$0 = \delta S = \int_\alpha^\beta \delta\Lambda \cdot dt = \int_\alpha^\beta \sum_{i,j} \frac{\partial \Lambda}{\partial x_i^{(j)}} \delta x_i^{(j)} dt = \int_\alpha^\beta \sum_{i,j} (-\partial)^j \frac{\partial \Lambda}{\partial x_i^{(j)}} \cdot \delta x_i dt$$
$$= \int_\alpha^\beta \sum_i \frac{\delta \Lambda}{\delta x_i} \delta x_i dt \Rightarrow \left\{ \frac{\delta \Lambda}{\delta x_i} = 0 \right\}, i = 1,\ldots, n; \Leftrightarrow \frac{\delta \Lambda}{\delta x} = 0,$$
$$x = (x_1, \ldots, x_n).$$

It is less known that this procedure also yields the symplectic form and the Hamilton representation of the equation $\delta\Lambda/\delta x = 0$.

Not making this notion more exact we shall assume that the Lagrangian is non-degenerate, i.e. there is a generic case. For example, we shall assume that the highest derivatives do not enter the Lagrangian linearly and the order of the Lagrangian cannot be reduced by integration by parts.

In the nondegenerate case the variational system is of the Cauchy-Kovalevsky type, for which the highest derivatives can be expressed in terms of the lower ones. For example, when $n = 2$,

$$0 = \delta\Lambda/\delta x_1 = ax_1^{(2N_1)} + bx_2^{(N_1+N_2)} + \ldots$$
$$0 = \delta\Lambda/\delta x_2 = cx_1^{(N_1+N_2)} + dx_2^{(2N_2)} + \ldots$$

where the coefficients a, b, c and d as well as of the remaining terms contain derivatives of lower orders. Let $N_1 > N_2$ and differentiate the second equation $N_1 - N_2$ times:

$$0 = cx_1^{(N_1+N_2+1)} + dx_2^{(2N_2+1)} + \ldots$$
$$\cdots\cdots\cdots\cdots\cdots\cdots\cdots\cdots\cdots\cdots$$
$$0 = cx_1^{(2N_1)} + dx_2^{(N_1+N_2)} + \ldots.$$

If $\begin{vmatrix} a & b \\ c & d \end{vmatrix} \neq 0$ (the condition must be included into the notion of non-degeneracy) then the first and the last of all the equations written here permit to express $x_1^{(2N_1)}$ and $x_2^{(N_1+N_2)}$ in terms of the lower derivatives; the rest of the equations give expressions for $x_2^{(N_1+N_2-1)}, \ldots, x_2^{(2N_2)}$ in terms of the lower derivatives, until only $x_1, \ldots, x_1^{(2N_1-1)}, x_2, \ldots, x_2^{(2N_2-1)}$ are left. (This procedure will be described in more detail for a concrete case in Chap. 14.)

Thus, the variables

(2.3.3) $\qquad x_i^{(j)}, i = 1, \ldots, n; j = 0, 1, \ldots, 2N_i - 1$

can be accepted as the coordinates in the phase space of the equation $\delta\Lambda/\delta x = 0$. A vector field ξ corresponds to the equation $\delta\Lambda/\delta x = 0$ in the phase space and its action on the coordinates is the following: $\xi x_i^{(j)} = x_i^{(j+1)}$, i.e. this is the differentiation ∂_t. However the "extra" derivatives $x_i^{(2N_i)}$ must be excluded with the help of the equations $\delta\Lambda/\delta x = 0$. This implies that the action of ξ on an arbitrary function $f(x_i^{(j)})$ is the differentiation ∂_t with the exclusion of the "extra" derivatives.

Let us consider in the phase space differential forms $\Omega = \sum a_{ij\ldots}^{kl\ldots} dx_i^{(k)} \wedge dx_j^{(l)} \wedge \ldots$. The Lie derivative of this form along the vector field ξ is

$$L_\xi \Omega = \sum (\xi a_{ij\ldots}^{kl\ldots}) dx_i^{(k)} \wedge \ldots + \sum a_{ij\ldots}^{kl\ldots} d(\xi x_i^{(k)}) \wedge \ldots + \ldots$$
$$= \sum \dot{a}_{ij\ldots}^{kl\ldots} dx_i^{(k)} \wedge \ldots + \sum a_{ij\ldots}^{kl\ldots} dx_i^{(k+1)} \wedge \ldots + \ldots$$

with the exclusion of the extra derivatives.

2.3.4. Return to the procedure of integration by parts in the process of deducing the variational equation. This integration means that $\delta\Lambda$ is represented in the form

(2.3.5) $\qquad \delta\Lambda = \sum A_i \delta x_i + \partial_t \omega$.

The first term contains only variations of the coordinates $\{x_i\}$ and not of their derivatives; the second term is a derivative ∂_t of a form $\omega^{(1)} = \sum a_i^k \delta x_i^{(k)}$ (∂_t acts on both the coefficients a_i^j and the differentials $\partial_t \delta x_i^{(k)} = \delta x_i^{(k+1)}$). The form $\omega^{(1)}$ was insignificant when we deduced the equation $\delta\Lambda/\delta x = 0$; now it plays the decisive role.

Equation (2.3.5) is an identity. If the extra derivatives are excluded with the help of the equation $\delta\Lambda/\delta x = 0$, then ∂_t turns to the Lie derivative L_ξ

and Eq. (2.3.5) takes the form (we write here the more common symbol d instead of δ):

$$(2.3.6) \qquad d\Lambda = L_\xi \omega^{(1)}, \omega^{(1)} = \sum a_i^{(1)} dx_i^{(j)}, j < 2N_i.$$

Put $\omega = d\omega^{(1)}$. This is closed and non-degenerate in the generic case form, i.e. symplectic.

2.3.7. Proposition. The equation $\delta\Lambda/\delta x = 0$ can be written as a Hamilton equation $d\mathcal{H} = -i(\xi)\omega$ with respect to the form ω introduced above. The Hamiltonian is

$$(2.3.8) \qquad \mathcal{H} = -\Lambda + i(\xi)\omega^{(1)}$$

($i(\xi)\omega^{(1)}$ being calculated thus: $i(\xi)dx_i^{(j)} = \xi x_i^{(j)} = x_i^{(j+1)}$ with the elimination of the extra derivatives).

Proof: Let us rewrite Eq. (2.3.6) as

$$d\Lambda = (di(\xi) + i(\xi)d)\omega^{(1)} = d(i(\xi)\omega^{(1)}) + i(\xi)\omega.$$

This implies $d(-\Lambda + i(\xi)\omega^{(1)}) = -i(\xi)\omega$. $\square$

2.3.9. The equation is written in the Hamiltonian form without canonical variables. If needed they can be found readily. Denote

$$(2.3.10) \qquad p_i^{(j-1)} = \delta\Lambda/\delta x_i^{(j)} = \sum_{k=0}^{\infty}(-\partial_t)^k \partial\Lambda/\partial x_i^{(j+k)}, i = 0,\ldots,N_i.$$

We call $p_i^{(j)}, j = 0,\ldots,N_i - 1$ the momentum adjoint to the coordinate $x_i^{(j)}$ of the phase space. We have $p_i^{(-1)} = \delta\Lambda/\delta x_i$. The following recurrence formulas

$$p_i^{(j-1)} = \partial\Lambda/\partial x_i^{(j)} - \partial_t p_i^{(j)}$$

obviously hold. In $d\Lambda = \sum_{i,j}(\partial\Lambda/\partial x_i^{(j)})dx_i^{(j)}$, let us eliminate $\partial\Lambda/\partial x_i^{(j)}$ with the help of the recurrence relation:

$$d\Lambda = \sum_{i,j}(p_i^{(j-1)} + \partial_t p_i^{(j)})dx_i^{(j)}$$
$$= \sum_{i,j} p_i^{(j-1)} dx_i^{(j)} + \partial_t \sum_{i,j} p_i^{(j)} dx_i^{(j)} - \sum_{i,j} p_i^{(j)} dx_i^{(j+1)}$$
$$= \sum_i p_i^{(-1)} dx_i + \partial_t \sum_{i,j} p_i^{(j)} dx_i^{(j)} = \sum_i (\delta\Lambda/\delta x_i) dx_i + \partial_t \sum_{i,j} p_i^{(j)} dx_i^{(j)}$$

which implies $\omega^{(1)} = \sum p_i^{(j)} dx_i^{(j)}$. Taking into account that $\omega = d\omega^{(1)}$ we obtain the proposition below.

2.3.11. *Proposition.* The symplectic form ω can be written in the "coordinate-momentum" variables as

$$\omega = \sum_{i,j} dp_i^{(j)} \wedge dx_i^{(j)}.$$

2.3.12. *Corollary.* The variational equation $\delta\Lambda/\delta x = 0$ in the same variables has the form

$$\dot{x}_i^{(j)} = \partial\mathcal{H}/\partial p_i^{(j)}$$
$$\dot{p}_i^{(j)} = -\partial\mathcal{H}/\partial x_i^{(j)}, i = 1,\ldots,n; j = 0,\ldots,N_i - 1.$$

with the Hamiltonian $\mathcal{H} = -\Lambda + \sum p_i^{(j)} x_i^{(j+1)}$ where the old variables are expressed in terms of the canonical ones.

2.4. Symplectic form on an orbit of the coadjoint representation of a Lie group.

2.4.1. Another example of a symplectic manifold yields an orbit of the coadjoint representation of a Lie group (see also Ref. [23], App. 2).

Let G be a Lie group and $\mathfrak{G}$ its Lie algebra which can be identified with $T_e G$. There is a homomorphism of the group G to the group $Aut\,\mathfrak{G}$ of linear non-degenerate transformations of the linear space $\mathfrak{G}$ which is denoted as $g \in G \mapsto Ad(g) \in Aut\,\mathfrak{G}$. We have $Ad(gh) = Ad(g)Ad(h)$. The mapping $g \mapsto Ad(g)$ is the adjoint representation of a group[a].

Every representation of a group generates a representation of its algebra. Thus we obtain the adjoint representation of a Lie algebra, $\alpha \in \mathfrak{G} \mapsto ad(\alpha) \in End\,\mathfrak{G}$. We have $ad([\alpha,\beta]) = [ad(\alpha), ad(\beta)]$. The adjoint representation of a Lie algebra is expressed by a simple formula $\forall \alpha, \beta \in \mathfrak{G}, ad(\alpha)\beta = [\alpha,\beta]$.

Let $\mathfrak{G}^*$ be the dual to the linear space $\mathfrak{G}$, i.e. a coupling $\alpha \in \mathfrak{G}, m \in \mathfrak{G}^*; \langle \alpha, m \rangle \in \mathbb{R}(\mathbb{C})$ is defined. This coupling is a bilinear function of α and

---

[a]Recall the construction Ad. Let $L_g: G \to G$ be the left translation on the group, $L_g h = gh$, and R_g the right translation, $R_g h = hg$. They induce linear mappings of the tangent spaces L_{g*} and R_{g*}. Then $L_g R_h = R_h L_g$; $L_{gh*} = L_{g*} L_{h*}$, $R_{gh*} = R_{h*} R_{g*}$ and $Ad(g) = R_{g^{-1}*} \circ L_{g*}: T_e G \to T_e G$.

m. The adjoint operators $Ad^*(g) : \mathfrak{G}^* \to \mathfrak{G}^*$ and $ad^*(\alpha) : \mathfrak{G}^* \to \mathfrak{G}^*$ are defined as usual:

$$\langle m, Ad(g)\alpha \rangle = \langle Ad^*(g)m, \alpha \rangle, \langle m, ad(\beta)\alpha \rangle = \langle ad^*(\beta)m, \alpha \rangle.$$

Obviously,

(2.4.2) $\qquad Ad^*(gh) = Ad^*(h)Ad^*(g), ad^*([\alpha, \beta]) = [ad^*(\beta), ad^*(\alpha)].$

The space $\mathfrak{G}^*$ is called a Lie coalgebra, mappings $g \mapsto Ad^*(g)$, and $\alpha \mapsto ad^*(\alpha)$ are coadjoint representations of a group and its algebra, respectively.

2.4.3. Let $m_0 \in \mathfrak{G}^*$ be a fixed element. The set of elements $m = Ad^*(g)m_0$ for all $g \in G$ is an orbit of the coadjoint representation. Two orbits either coincide or have no intersection. Thus, the space $\mathfrak{G}^*$ foliates to orbits. The symplectic form will be defined on the orbits.

The group G acts on an orbit as a transitive group of right transformations (right means that (2.4.2) holds). Passing from the group to the algebra, we see that a vector field $\xi(m) = ad^*(\alpha)m$ corresponds to each $\alpha \in \mathfrak{G}^*$. This is a right representation of the Lie algebra by vector fields:

(2.4.4) $\qquad [ad^*(\alpha)m, ad^*(\beta)m] = [ad^*(\alpha), ad^*(\beta)]m = ad^*([\beta, \alpha])m.$

(The commutator has different meanings in the three parts of this equality: in the left-hand side this is the commutator of vector fields, in the central – the commutator of linear transformations, and in the right-hand side the commutator in the Lie algebra $\mathfrak{G}$.)

Thus, the tangent space to an orbit at a point m_0, i.e. $T_{m_0}M$, consists of vectors $\xi = ad^*(\alpha)m_0$. This space is in a natural way inserted into $\mathfrak{G}^*$. What is the adjoint space $T^*_{m_0}M$? The elements of the Lie algebra $\mathfrak{G}$ can be considered as linear functionals on $T_{m_0}M$: if $\alpha \in \mathfrak{G}, \xi \in T_{m_0}M$ then $\alpha(\xi) = \langle \alpha, \xi \rangle$. Thus, there is a linear mapping $\mathfrak{G} \to T^*_{m_0}M$ (it is easy to see that this is an epimorphism, mapping onto $T^*_{m_0}M$). This mapping has a kernel $N_{m_0} = \{\alpha \in \mathfrak{G} \,|\, \langle \alpha, ad^*(\beta)m_0 \rangle = 0 \,\forall \beta \in \mathfrak{G}\}$, which can be written more simply: if for every $\beta \in \mathfrak{G}$ a relation $0 = \langle \alpha, ad^*(\beta)m_0 \rangle = \langle ad(\beta)\alpha, m_0 \rangle = \langle [\beta, \alpha], m_0 \rangle = -\langle [\alpha, \beta], m_0 \rangle = -\langle \beta, ad^*(\alpha)m_0 \rangle$ holds then $ad^*(\alpha)m_0 = 0$, and vice versa. Hence $N_{m_0} = \{\alpha \in \mathfrak{G} \,|\, ad^*(\alpha)m_0 = 0\}$ and $T^*_{m_0}M = \mathfrak{G}/N_{m_0}$.

Elements of the space $\mathfrak{G}/N_{m_0}$, i.e. the cosets, will be denoted by the same letters as their representatives, $\alpha, \beta, \ldots$.

Define a mapping

$$H : T^*_{m_0}M \to T_{m_0}M, \alpha \in T^*_{m_0}M \mapsto ad^*(\alpha)m_0 \in T_{m_0}M$$

and a two-form in $T_{m_0}M$:

$$\omega(H\alpha, H\beta) = \langle H\alpha, \beta\rangle.$$

In other words

(2.4.5) $\qquad \omega(ad^*(\alpha)m, ad^*(\beta)m) = \langle ad^*(\beta)m, \beta\rangle == \langle m, [\alpha, \beta]\rangle.$

It remains to prove that the form is symplectic.

2.4.6. *Remark.* It is easy to calculate the result of action of a vector field on a function on the orbit having a special form $f(m) = \langle m, \alpha\rangle$, m being a point on the orbit and $\alpha \in \mathfrak{G}$ a fixed element. If $\xi \in T_m M$ then $\xi f(m) = \langle \xi, \alpha\rangle$.

2.4.7. *Proposition.* The form (2.4.5) is symplectic.

Proof: Non-degeneracy and skew symmetry are clear from (2.4.5). It remains to prove that the form is closed. Let $\xi_i(m_0) = ad^*(\alpha_i)m_0, i = 1, 2, 3$. In order to apply the Lie formula for the differential (2.1.3), the vector fields ξ_i must be prolonged form the point m_0 to the neighbouring points of the orbit; how it will be done makes no difference (see the remark 2.1.4). We do it in the simplest way: $\xi_i(m) = ad^*(\alpha_i)m$, where α_i are fixed, the same as for the point m_0. We find

$$(d\omega)(\xi_1, \xi_2, \xi_3) = \xi_1 \omega(\xi_2, \xi_3) - \omega([\xi_1, \xi_2], \xi_3) + \text{c.p.}$$
$$= (ad^*(\alpha_1)m)\langle m, [\alpha_2, \alpha_3]\rangle - \langle ad^*([\alpha_2, \alpha_1])m, \alpha_3\rangle + \text{c.p.}$$
$$= \langle m, ad(\alpha_1)[\alpha_2, \alpha_3]\rangle - \langle m, ad([\alpha_2, \alpha_1])\alpha_3\rangle + \text{c.p.}$$
$$= \langle m, [\alpha_1, [\alpha_2, \alpha_3]]\rangle - \langle m, [[\alpha_2, \alpha_1], \alpha_3]\rangle + \text{c.p.} = 0.$$

2.4.8. *Corollary.* In the Hamilton structure corresponding to the form ω constructed above, each function on the orbit $f(m)$ generates a vector field $\xi_f = ad^*(df)m, (df \in T^*_m M)$.

The function $f(m)$ is the Hamiltonian of this field.

2.4.9. *Corollary.* The relevant Poisson bracket is

$$\{f, g\} = \xi_f g = (ad^*(df)m)g = \langle ad^*(df)m, dg\rangle = \langle m, [df, dg]\rangle.$$

2.4.10. *Remark.* The preceding formula can define the Poisson bracket not only on the orbit but in the whole space $\mathfrak{G}^*$. However in this case the bracket will be degenerate.

The bracket 2.4.9 is called the Poisson-Lie-Berezin-Kirillov-Kostant bracket. The symplectic form ω was suggested by Kirillov and Kostant.

2.5. Purely algebraic treatment of the Hamiltonian structure.

2.5.1. When we pass to an infinite-dimensional case the geometrical objects used earlier maintain only a relative meaning. Instead of making the definitions more precise we single out a bare algebraic scheme which remains valid in the infinite-dimensional case.

We introduce the following: An arbitrary Lie algebra $\mathfrak{E}$ (in the preceding theory $\mathfrak{E}$ played the role of the Lie algebra of vector fields). Further, Ω^0, a left $\mathfrak{E}$-module, i.e. a linear space in which the elements of $\mathfrak{E}$ act as left operators; in other words, $\forall \xi \in \mathfrak{E}, f \in \Omega^0 \mapsto \xi f \in \Omega^0$ and

(2.5.2) $$(\xi_1 \xi_2 - \xi_2 \xi_1) f = [\xi_1, \xi_2] f$$

holds.

Let Ω^1 be a space of linear Ω^0-valued functionals: for all $\xi \in \mathfrak{E}$ and $\alpha \in \Omega^1$ a bilinear form $\alpha(\xi) = \langle \alpha, \xi \rangle \in \Omega^0$ is defined and the following non-degenacy condition holds. For all $\alpha \in \Omega^1$ there is $\xi \in \mathfrak{E}, \alpha(\xi) \neq 0$ and for all $\xi \in \mathfrak{E}$ there is $\alpha \in \Omega^1$ such that $\alpha(\xi) \neq 0$). We assume that Ω^1 contains all the differentials $df, f \in \Omega^0$ defined as usual: $df(\xi) = \xi f$ for all $\xi \in \mathfrak{E}$.

A skew symmetrical mapping $H : \Omega^1 \to \mathfrak{E}$ is said to be Hamiltonian if
1. $H\Omega^1 \subset \mathfrak{E}$ is a Lie subalgebra,
2. The form $\omega(H\alpha, H\beta) = \langle H\alpha, \beta \rangle = \beta(H\alpha)$ is closed with respect to the differential defined by the Lie formula

$$d\omega(H\alpha, H\beta, H\gamma) = (H\alpha)\omega(H\beta, H\gamma) - \omega([H\alpha, H\beta], H\gamma) + \text{c.p.}$$

A vector field ξ_f can be associated with every $f \in \Omega^0$:

$$\xi_f = H\, df.$$

Poisson bracket is defined as

$$f, g \in \Omega^0 : \{f, g\} = \xi_f g = dg(\xi_f) = dg(H\, df) = \langle H\, df, dg \rangle.$$

2.5.3. Proposition.
$$\xi_{\{f,g\}} = [\xi_f, \xi_g].$$

Poisson bracket satisfies the Jacobi identity $\{f_1, \{f_2, f_3\}\} + \text{c.p.} = 0$.

The proof is the same as in the finite dimensional case, 2.1.15 and 2.1.16.

These definitions are general enough to serve our needs. However we give also a more refined definition by Gelfand and Dorfman, in two variants.

Let $\Omega^1, \Omega^2, \ldots, \Omega^k, \ldots$ be spaces of k-linear Ω^c-valued functionals on $\mathfrak{E}$, i.e. $\omega^k(\xi_1, \ldots, \xi_k) \in \Omega^c$ are defined with $\omega^k \in \Omega^k$. Denote $\Omega = \Omega^0 \oplus \Omega^1 \oplus \Omega^2 \oplus \ldots$. To every $\xi \in \mathfrak{E}$ define a linear operator $i(\xi) : \Omega^k \to \Omega^{k-1}$ by

$$i(\xi)\Omega^0 = \Omega^{-1} = \{0\}, \forall \alpha \in \Omega^k, \forall \xi_1, \ldots, \xi_{k-1}, \xi \in \mathfrak{E}$$
$$(i(\xi)\alpha)(\xi_1, \ldots, \xi_{k-1}) = \alpha(\xi, \xi_1, \ldots, \xi_{k-1}).$$

We assume that the following non-degeneracy condition holds:

$$\forall \xi \in \mathfrak{E} \;\exists \alpha \in \Omega^1 \; \alpha(\xi) \neq 0 \,;\, \forall \alpha \in \Omega^1 \exists \xi \in \mathfrak{E} \quad \alpha(\xi) \neq 0.$$

Define the operator $d : \Omega^k \to \Omega^{k+1}$ by the formula

$$(d\alpha)(\xi_1, \ldots, \xi_{k+1}) = \sum_{l=1}^{k+1}(-1)^{l-1}\xi_l \alpha(\xi_1, \ldots, \hat{\xi}_l, \ldots, \xi_{k+1})$$
$$+ \sum_{\substack{l,p \\ (l<p)}} (-)^{l+p} \alpha([\xi_l, \xi_p], \xi_1, \ldots, \hat{\xi}_l, \ldots, \hat{\xi}_p, \ldots, \xi_{k+1})$$

(which is none other than the formula (2.1.3)).

2.5.4. Proposition. The operator d has the property $d^2 = 0$.

The proof is left to the reader.

Obviously $\quad \forall \xi \in \mathfrak{E}, \forall f \in \Omega^0, \xi f = (df)(\xi).$

Definition. The operator

$$L_\xi = i(\xi)d + di(\xi)$$

is called the Lie derivative in the direction ξ. For $f \in \Omega^c$ we have $L_\xi f = i(\xi)df = (df)(\xi) = \xi f$. For $\forall \alpha \in \Omega^1$ and $\forall \xi \in \mathfrak{E}, \langle \alpha, \xi \rangle$ will be used to denote $\alpha(\xi)$, the coupling of elements of $\mathfrak{E}$ and Ω'.

The space Ω is a complex with respect to the operator d. A form $\alpha \in \Omega^k$ is called a cocycle if $d\alpha = 0$, and it is a coboundary if $\exists \beta \in \Omega^{k-1} \alpha = d\beta$. Proposition 2.5.4 implies that each coboundary is a cocycle.

Let Z^k be the linear space of all k-cocycles and B^k the linear space of all k-coboundaries. The factor-space $H^k = Z^k/B^k$ is called the space of k-cohomologies.

2.5.5. *Example.* The simplest example is obtained if we put $\Omega^0 = \mathbb{R}$ (or $\mathbb{C}$) and all $\xi f = 0$, where $f \in \Omega^c$. The relevant spaces of cohomologies are usually called the cohomologies of Lie algebras (a simplest case). The formula for d becomes simpler in this case:

$$(d\alpha)(\xi_1,\ldots,\xi_{k+1}) = \sum_{l<p}(-1)^{l+p}\alpha([\xi_l,\xi_p],\xi_1,\ldots,\hat{\xi}_l,\ldots,\hat{\xi}_p,\ldots,\xi_{k+1}).$$

2.5.6. *Proposition.* $\forall \alpha \in \Omega^k$

$$\xi(\alpha(\xi_1,\ldots,\xi_k)) = (L_\xi \alpha)(\xi_1,\ldots,\xi_k)$$
$$+ \sum_{l=1}^{k}\alpha(\xi_1,\ldots,\xi_{l-1},[\xi,\xi_l],\xi_{l+1},\ldots,\xi_k).$$

Proof: See the proof of 2.1.5. $\square$

2.5.7. *Proposition.* The relations

$$i(\xi)i(\eta) + i(\eta)i(\xi) = 0, [L_\xi, i(\eta)] = i([\xi,\eta]), i(\lambda\xi + \mu\eta) = \lambda i(\xi) + \mu i(\eta)$$

hold.

Proof: The first relation immediately follows from the skew symmetry of the elements $\alpha \in \Omega^k$. The third is also clear. From 2.5.6 we obtain

$$(L_\xi(i(\eta)\alpha))(\xi_1,\ldots,\xi_{k-1}) = \xi\alpha(\eta,\xi_1,\ldots,\xi_{k-1})$$
$$- \sum_{l=1}^{k-1}\alpha(\eta,\xi_1,\ldots,\xi_{l-1},[\xi,\xi_l],\xi_{l+1},\ldots,\xi_{k-1}),$$
$$(i(\eta)(L_\xi \alpha))(\xi_1,\ldots,\xi_{k-1}) = \xi\alpha(\eta,\xi_1,\ldots,\xi_{k-1})$$
$$- \sum_{l=1}^{k-1}\alpha(\eta,\xi_1,\ldots,[\xi,\xi_l],\ldots,\xi_{k-1}) - \alpha([\xi,\eta],\xi_1,\ldots,\xi_{k-1}).$$

Subtracting one equality from the other we get

$$[L_\xi, i(\eta)]\alpha(\xi_1,\ldots,\xi_{k-1}) = \alpha([\xi,\eta],\xi_1,\ldots,\xi_{k-1}).\square$$

2.5.8. Proposition.

$$[L_\xi, L_\eta] = L_{[\xi,\eta]}$$

Proof:

$$\begin{aligned}
L_{[\xi,\eta]} &= di([\xi,\eta]) + i([\xi,\eta])d = d[L_\xi, i(\eta)] + [L_\xi, i(\eta)]d \\
&= d(L_\xi i(\eta) - i(\eta)L_\xi) + (L_\xi i(\eta) - i(\eta)L_\xi)d \\
&= di(\xi)di(\eta) - di(\eta)L_\xi + L_\xi i(\eta)d - i(\eta)di(\xi)d \\
&= L_\xi di(\eta) - di(\eta)L_\xi + L_\xi i(\eta)d - i(\eta)dL_\xi \\
&= L_\xi L_\eta - L_\eta L_\xi = [L_\xi, L_\eta] . \square
\end{aligned}$$

2.5.9. *The second variant of definition.* The complex Ω can be defined in an axiomatical way. There is a complex of linear space $(\Omega, d), \Omega = \Omega^0 \oplus \Omega^1 \oplus \ldots$ with a linear mapping $d : \Omega \to \Omega$ for which $d\Omega^k \subset \Omega^{k+1}$ and $d^2 = 0$. To each $\xi \in \mathfrak{E}$, a linear mapping $i(\xi) : \Omega \to \Omega, i(\xi)\Omega^k \subset \Omega^{k-1}, (\Omega^{-1} = \{0\})$ is attached, the relations 2.5.7 being fulfilled, where $L_\xi = i(\xi)d + di(\xi)$. Let $\forall \alpha \in \Omega^k \; \alpha(\xi_1, \ldots, \xi_k) := i(\xi_k) \ldots i(\xi_1)\alpha$. Then 2.5.8 also holds, since we used only 2.5.7 in the proof. It is easy to prove that the Lie formula for the operator d holds in this case.

2.5.10. We consider linear skew symmetric mappings $H : \Omega^1 \to \mathfrak{E}, \langle H\alpha, \beta \rangle = -\langle \alpha, H\beta \rangle$. The Schouten bracket of two such mappings is the trilinear mapping $[H, K]; \Omega^1 \times \Omega^1 \times \Omega^1 \to \Omega^0$ defined by the formula

$$\forall \alpha_1, \alpha_2, \alpha_3 \in \Omega^1 \mapsto [H, K](\alpha_1, \alpha_2, \alpha_3)$$
$$= \langle KL_{H\alpha_1}\alpha_2, \alpha_3 \rangle + \langle HL_{K\alpha_1}\alpha_2, \alpha_3 \rangle + \text{c.p..}$$

The Hamiltonian mapping is the mapping H which satisfies the condition $[H, H] = 0$.

2.5.11. *Proposition.* The conditon $[H, H] = 0$ is equivalent to

$$\forall \alpha, \beta \in \Omega^1, [H\alpha, H\beta] = H(i(H\alpha)d\beta - i(H\beta)d\alpha + di(H\alpha)\beta).$$

Proof: Let $\gamma \in \Omega^1$ be an arbitrary element. Taking into account the Lie formula for the differential and 2.5.6 we find

$$\begin{aligned}
&\langle H(i(H\alpha)d\beta - i(H\beta)d\alpha + di(H\alpha)\beta) - [H\alpha, H\beta], \gamma\rangle \\
&= \langle HL_{H\alpha}\beta, \gamma\rangle + \langle i(H\beta)d\alpha, H\gamma\rangle - \langle H\alpha, H\beta], \gamma\rangle \\
&= \langle HL_{H\alpha}\beta, \gamma\rangle + (d\alpha)(H\beta, H\gamma) - \langle [H\alpha, H\beta], \gamma\rangle \\
&= \langle HL_{H\alpha}\beta, \gamma\rangle + (H\beta)\alpha(H\gamma) - (H\gamma)\alpha(H\beta) - \alpha([H\beta, H\gamma]) \\
&\quad - \langle [H\alpha, H\beta], \gamma\rangle = \langle HL_{H\alpha}\beta, \gamma\rangle - (H\beta)\langle\gamma, H\alpha\rangle - (H\gamma)\langle\alpha, H\beta\rangle \\
&\quad - \langle [H\beta, H\gamma], \alpha\rangle - \langle [H\alpha, H\beta], \gamma\rangle = \langle HL_{H\alpha}\beta, \gamma\rangle \\
&\quad - \langle L_{H\beta}\gamma, H\alpha\rangle - \langle \gamma, [H\beta, H\alpha]\rangle - \langle L_{H\gamma}\alpha, H\beta\rangle \\
&\quad - \langle \alpha, [H\gamma, H\beta]\rangle - \langle \alpha, [H\beta, H\gamma]\rangle - \langle \gamma, [H\alpha, H\beta]\rangle \\
&= \langle HL_{H\alpha}\beta, \gamma\rangle + \text{c.p.} = \frac{1}{2}[H, H](\alpha, \beta, \gamma).
\end{aligned}$$

The rest is clear. □

2.5.12. Corollary. If H is a Hamiltonian mapping then $l = \operatorname{Im} H \subset \mathfrak{E}$ is a Lie subalgebra.

2.5.13. Definition. A form ω can be defined on l by the formula

$$\omega(H\alpha, H\beta) = \langle H\alpha, \beta\rangle = -\langle\alpha, H\beta\rangle = -\omega(H\beta, H\alpha).$$

This form is well-defined, i.e. it depends on $H\alpha$ and $H\beta$ but not on α and β. If elements of $\ker H$ are added to α or β the value of the form will not change. The mapping H can be treated as an isomorphism $H : \Omega^1/\ker H \to \operatorname{Im} H \subset \mathfrak{E}$.

2.5.14. Proposition. If H is a Hamiltonian mapping then the form ω is closed.

Proof: Repeating the proof of 2.2.3 we obtain

$$(2.5.15) \qquad d\omega(\xi_1, \xi_2, \xi_3) = \frac{1}{2}[H, H](\alpha_1, \alpha_2, \alpha_3), (\xi_i = H\alpha_i)$$

i.e. $[H, H] = 0$ imply $d\omega = 0$. □

In what sense is the converse proposition true? In contrast to Sec. 2.2, H is not a bijection here. Further, $l = \operatorname{Im} H$ is not, in general, a subalgebra (for Hamiltonian mappings H, this is true). Therefore the Lie formula for the differential has no sense. However the following assertion is true.

2.5.16. Proposition. If $l = \operatorname{Im} H$ is a Lie subalgebra of $\mathfrak{E}$ and the form ω is closed, then H is a Hamiltonian mapping.

Proof: If $\xi_i = H\alpha_i$ are arbitrary elements of l, then the formula (2.5.15) holds and $d\omega = 0$ implies $[H, H] = 0$. □

2.5.17. An element $\xi_f = Hdf \in l \subset \mathfrak{E}$ corresponds to each $f \in \Omega^0$. The mapping $f \mapsto \xi_f$ has, in general, a kernel.

The Poisson bracket of two elements $f, g \in \Omega^0$ is

$$\{f,g\} = \xi_f g = \langle dg, \xi_f \rangle = \langle Hdf, dg \rangle.$$

The Poisson bracket is skew symmetric.

2.5.18. *Proposition.*

$$\xi_{\{f,g\}} = [\xi_f, \xi_g].$$

The proof is the same as in 2.1.15. □

2.5.19. *Proposition.* The Poisson bracket satisfies the Jacobi identity

$$\{f_1, \{f_2, f_3\}\} + \text{c.p.} = 0.$$

Proof: See 2.1.16. □

Note that the assertion (ii) in 2.1.16 has no analogy here, since Ω^0 is not a ring. Moreover, we even have no manifolds. $\xi \in \mathfrak{E}$ is not a vector field on a manifold (though we shall sometimes call it so for convenience) but simply an element of a Lie algebra.

2.5.20. This algebraic approach to Hamiltonian structures was worked out in an application to the KdV-type hierarchies by Gelfand and Dickey [25,16] and formulated in the general form by Gelfand and Dorfman [26–28].

CHAPTER 3. Hamiltonian Structure of the KdV-Hierarchies.

3.1. Lie algebra $\mathfrak{E}$ and complex Ω.

3.1.1. Returning to Eqs. (1.4.3), we shall construct two symplectic forms (any linear combination of which is also a symplectic form) such that the equations are Hamiltonian with respect to them.

We define here the operator L somewhat more generally than in Chap. 1:

$$L = \partial^n + u_{n-1}\partial^{n-1} + u_{n-2}\partial^{n-2} + \ldots + u_0,$$

i.e. having an extra term $u_{n-1}\partial^{n-1}$. We shall discuss the reduction to the submanifold $u_{n-1} = 0$ later.

For the Lie algebra $\mathfrak{E}$ we take the Lie algebra of "vector fields" $\xi = \partial_a$ (Sec. 1.1) with the commutator

$$[\partial_a, \partial_b] = \partial_{(\partial_a b - \partial_b a)}.$$

The left module Ω^0 is here the space $\tilde{\mathcal{A}}$ of the functionals (Sec. 1.2). The action of vector fields on the functionals is given by lemma 1.2.2 and proposition 1.2.3. This determines the whole complex Ω.

3.1.2. This time the Lie algebra $\mathfrak{E}$ and the dual space Ω^1 are provided with additional algebraic structures which enables us to construct the Hamiltonian mapping H.

Vector fields are determined by sets $a = (a_0, \ldots, a_{n-1}), a_i \in \mathfrak{a}$. The dual space Ω^1 is determined by sets $X = (X_0, \ldots, X_{n-1}), X_i \in \mathfrak{a}$. The coupling is

$$\langle X, \partial_a \rangle = X(\partial_a) = \int \sum a_i X_i dx \in \tilde{\mathcal{A}}.$$

To each vector field ∂_a we can attach a differential operator

$$\partial_a \leftrightarrow a = a_{n-1}\partial^{n-1} + \ldots + a_0 \in R_{(0,n-1)}.$$

We often identify ∂_a and a. To an element $X \in \Omega^1$ we attach an operator

$$X = \partial^{-1}X_0 + \partial^{-2}X_1 + \ldots + \partial^{-n}X_{n-1} \quad \in R_-/R_{(-\infty,-n-1)}.$$

The coupling can be written as

$$\langle X, \partial_a \rangle = \langle X, a \rangle = \int \operatorname{res} aX \, dx.$$

Thus, the elements of $\mathfrak{E}$ and of Ω^1 can be treated as PDO. We emphasize that the commutator of differential operators a and b has nothing in common with the commutator of vector fields (differentiations) ∂_a and ∂_b. On account of Lemma 1.3.10, $\partial_a L = a$.

3.1.3. Proposition. If $\tilde{f} = \int f dx$ is an element of $\tilde{\mathcal{A}}$, then $df \in \Omega^1$ is given by the equation

$$d\tilde{f} = \delta f/\delta L, \quad \delta f/\delta L = \sum_0^{n-1} \partial^{-i-1} \delta f/\delta u_i \in R_-/R_{(-\infty,-n-1)}.$$

Proof: If $\partial_a \in \mathfrak{E}$ is arbitrary then $\langle \partial_a, d\tilde{f} \rangle = d\tilde{f}(\partial_a) = \partial_a \tilde{f} = \int \sum a_i \delta f/\delta u_i dx$. On the other hand, $\langle \partial_a, \delta f/\delta L \rangle = \int \operatorname{res} a \delta f/\delta L dx = \int \sum a_i \delta f/\delta u_i dx$ which proves the assertion. □

3.1.4. To define a Hamiltonian mapping $H : \Omega^1 \to \mathfrak{E}$, i.e. $H : R_-/R_{(-\infty,-n-1)} \to R_{(0,n-1)}$, we use the Adler mapping (1.7.6):

$$X \in R_-/R_{(-\infty,-n-1)} \mapsto H(X) = (\hat{L}X)_+\hat{L} - \hat{L}(X\hat{L})_+$$
$$= -(\hat{L}X)_-\hat{L} + \hat{L}(X\hat{L})_- \in R_{(0,n-1)}.$$

Particular cases when $z = 0$ and $z = \infty$ are

$$H^{(0)} = (LX)_+L - L(XL)_+, \quad H^{(\infty)} = [X, L]_+.$$

3.2. Proof of the Hamiltonian property of the Adler mapping.

3.2.1. Proposition. The mapping H is skew symmetric.

Proof: For any $A, B \in R$, the relations $\operatorname{res} A_+ B = \operatorname{res} A_+ B_- = \operatorname{res} AB_-$ hold. This implies

$$\langle H(X), Y \rangle = \int \operatorname{res} H(X) Y \, dx = \int \operatorname{res}((\hat{L}X)_+ \hat{L} - \hat{L}(X\hat{L})_+) Y \, dx$$

$$= \int \operatorname{res}(\hat{L}X(\hat{L}Y)_- - X\hat{L}(L)_-) \, dx$$

$$= \int \operatorname{res} X((\hat{L}Y)_- \hat{L} - \hat{L}(Y\hat{L})_-) \, dx = -\int \operatorname{res} X H(Y) \, dx = -\langle X, H(Y) \rangle . \quad \Box$$

3.2.2. Proposition. Let $l = \operatorname{Im} H$. The linear space l is a Lie subalgebra of the algebra $\mathfrak{E}$. The commutator of two elements of l is

$$[\partial_{H(X)}, \partial_{H(Y)}] = \partial_{H([X,Y]_L + \partial_{H(X)} Y - \partial_{H(Y)} X)},$$

where

$$[X, Y]_L = (-X(\hat{L}Y)_+ + (X\hat{L})_- Y)_- - (X \leftrightarrow Y)$$

with $(X \leftrightarrow Y)$ symbolizing the same expression with transposed X and Y.

Proof: According to the Sec. 3.1

$$[\partial_{H(X)}, \partial_{H(Y)}] = \partial_{(\partial_{H(X)} Y - \partial_{H(Y)} X)}.$$

We have

$$a := \partial_{H(X)} H(Y) - \partial_{H(Y)} H(X) = \partial_{H(X)}(-(\hat{L}Y)_- \hat{L} + \hat{L}(Y\hat{L})_-)$$
$$- (X \leftrightarrow Y) = -(H(X)Y)_- \hat{L} - (\hat{L}Y)_- H(X) + H(X)(Y\hat{L})_-$$
$$+ \hat{L}(YH(X))_- + H(\partial_{H(X)} Y) - (X \leftrightarrow Y) = ((\hat{L}X)_- \hat{L} - \hat{L}(X\hat{L})_-) Y)_- \hat{L}$$
$$+ (\hat{L}Y)_-((\hat{L}X)_- \hat{L} - \underline{\hat{L}(X\hat{L})_-}) - (\underline{(\hat{L}X)_- \hat{L}} - \hat{L}(X\hat{L})_-)(Y\hat{L})_-$$
$$- \hat{L}(Y((\hat{L}X)_- \hat{L} - \hat{L}(X\hat{L})_-))_- + H(\partial_{H(X)} Y) - (X \leftrightarrow Y).$$

The underlined terms cancel out if terms hidden under the symbol $(X \leftrightarrow Y)$ are taken into account. Transform two of the terms separately:

$$(((\hat{L}X)_- \hat{L}Y)_- + (\hat{L}Y)_- (\hat{L}X)_-)\hat{L} - (X \leftrightarrow Y)$$
$$= ((\hat{L}X)_- \hat{L}Y - (\hat{L}X)_-(\hat{L}Y)_-)\hat{L} - (X \leftrightarrow Y)$$
$$= ((\hat{L}X)_-(\hat{L}Y)_+)_- \hat{L} - (X \leftrightarrow Y) = (\hat{L}X(\hat{L}Y)_+)_- \hat{L} - (X \leftrightarrow Y).$$

Similarly
$$\hat{L}((X\hat{L})_-(Y\hat{L})_- + ((Y\hat{L}(X\hat{L})_-)_-) - (X \leftrightarrow Y)$$
$$= -\hat{L}((X\hat{L})_+ Y\hat{L})_- - (X \leftrightarrow Y).$$

Substitute these expressions:
$$a = (\hat{L}(X(\hat{L}Y)_+ - (X\hat{L})_- Y))_- \hat{L} - \hat{L}((X\hat{L})_+ Y - X(\hat{L}Y)_- \hat{L})_-$$
$$+ H(\partial_{H(X)}Y) - (X \leftrightarrow Y) = (\hat{L}(X(\hat{L}Y)_+ - (X\hat{L})_- Y)_- \hat{L}$$
$$- \hat{L}((X(\hat{L}Y)_+ - (X\hat{L})_- Y)_- \hat{L})_- + H(\partial_{H(X)}Y) - (X \leftrightarrow Y)$$
$$= -(\hat{L}[X,Y]_L)_- \hat{L} + \hat{L}([X,Y]_L \hat{L})_- + H(\partial_{H(X)}Y - \partial_{H(Y)}X)$$
$$= H([X,Y]_L + \partial_{H(X)}Y - \partial_{H(Y)}X). \square$$

3.2.3. Proposition. The form ω defined with the help of H:
$$\omega(\partial_{H(X)}, \partial_{H(Y)}) = \langle H(X), Y \rangle = \int \mathrm{res}\, H(X) Y \, dx$$
is closed.

Proof: Let $X, Y, Z \in \Omega^1$ be any elements. Then
$$d\omega(\partial_{H(X)}, \partial_{H(Y)}, \partial_{H(Z)})$$
$$= \partial_{H(X)}\omega(\partial_{H(Y)}, \partial_{H(Z)}) - \omega([\partial_{H(X)}, \partial_{H(Y)}], \partial_{H(Z)}) + \text{c.p.}.$$

We calculate
$$\partial_{H(X)}\omega(\partial_{H(Y)}, \partial_{H(Z)}) = \partial_{H(X)} \int \mathrm{res}\, H(Y) Z \, dx$$
$$= \partial_{H(X)} \int \mathrm{res}\, ((\hat{L}Y)_+ \hat{L} - \hat{L}(Y\hat{L})_+) Z \, dx$$
$$= \int \mathrm{res}\, ((H(X)Y)_+ \hat{L} - \hat{L}(Y H(X))_+) Z \, dx$$
$$+ \int \mathrm{res}\, ((\hat{L}Y)_+ H(X) - H(X)(Y\hat{L})_+) Z \, dx$$
$$+ \int \mathrm{res}\, (H(\partial_{H(X)}Y)Z + H(Y)\partial_{H(X)}Z) \, dx$$
$$= \int \mathrm{res}\, H(X)(Y(\hat{L}Z)_- - (Z\hat{L})_- Y + Z(\hat{L}Y)_+ - (Y\hat{L})_+ Z) \, dx$$
$$+ \int \mathrm{res}\, (-H(Z)\partial_{H(X)}Y + H(Y)\partial_{H(X)}Z) \, dx$$
$$= \int \mathrm{res}\, H(X)(-Y(\hat{L}Z)_+ - (Z\hat{L})_- Y + Z(\hat{L}Y)_+ + (Y\hat{L})_- Z) \, dx$$

$$+ \int res\,(-H(Z)\partial_{H(X)}Y + H(Y)\partial_{H(X)}Z)dx$$
$$= \int res\,H(X)[Y,Z]_L dx + \int res\,(-H(Z)\partial_{H(X)}Y + H(Y)\partial_{H(X)}Z)dx.$$

Further,

$$-\omega([\partial_{H(X)},\partial_{H(Y)}],\partial_{H(Z)})$$
$$= \omega(\partial_{H(Z)},\partial_{H([X,Y]_L+\partial_{H(X)}Y-\partial_{H(Y)}X)})$$
$$= \int res\,H(Z)([X,Y]_L + \partial_{H(X)}Y - \partial_{H(Y)}X)dx.$$

Hence

$$d\omega(\partial_{H(X)},\partial_{H(Y)},\partial_{H(Z)}) = 2\int res\,H(X)[Y,Z]_L dx + \text{c.p.}.$$

We go on calculating

$$\int res\,H(X)[Y,Z]_L dx + \text{c.p.} = \int res\,H(X)(Y(\hat{L}Z)_+ - (Y\hat{L})_- Z)dx + \text{p.}$$
$$= \int res\,(-(\hat{L}X)_-\hat{L} + \underline{\hat{L}(X\hat{L})_-})Y(\hat{L}Z)_+ dx$$
$$+ \int res\,(-\underline{(\hat{L}X)_+\hat{L}} + \hat{L}(X\hat{L})_+)(Y\hat{L})_- Z dx + \text{p.}$$

(p. symbolizes adding all the permutations; if a permutation is odd the sign changes). The underlined terms are:

$$\int res\,(\hat{L}(X\hat{L})_- Y(\hat{L}Z)_+ - (\hat{L}X)_+(\hat{L}(Y\hat{L})_- Z)dx + \text{p.}$$
$$= \int res\,((\hat{L}Z)_+\hat{L}(X\hat{L})_- Y - (\hat{L}X)_+\hat{L}(Y\hat{L})_- Z)dx + \text{p.} = 0.$$

The rest of the terms are

$$\int res\,(-(\hat{L}X)_-\hat{L}Y(\hat{L}Z)_+)dx + \int res\,Z\hat{L}(X\hat{L})_+(Y\hat{L})_- dx + \text{p.}.$$

This expression vanishes if the following lemma is taken into account.

3.2.4. *Lemma.* For any $A, B, C \in R$

$$\int \operatorname{res} AB_+C_- \, dx + \text{c.p.} = \int \operatorname{res} ABC \, dx.$$

Proof:
$$\int \operatorname{res} AB_+C_- \, dx + \text{c.p.} = \int \operatorname{res} (A_+B_+C_- + A_-B_+C_-) dx + \text{c.p.}$$
$$= \int \operatorname{res} (A_+B_+C + A_-BC_-) dx + \text{c.p.}$$
$$= \frac{1}{3} \int \operatorname{res} (2AB_+C_- + A_+B_+C + A_-BC_-) dx + \text{c.p.}$$
$$= \frac{1}{3} \int \operatorname{res} (2AB_+C_- + AB_+C_+ + AB_-C_-) dx + \text{c.p.}$$
$$= \frac{1}{3} \int \operatorname{res} (AB_+C + ABC_-) dx + \text{c.p.}$$
$$= \frac{1}{3} \int \operatorname{res} (AB_+C + AB_-C) dx + \text{c.p.}$$
$$= \frac{1}{3} \int \operatorname{res} ABC \, dx + \text{c.p.} = \int \operatorname{res} ABC \, dx. \ \Box$$

This completes the proof of the proposition. $\Box$

According to 2.5.16 this shows that the mapping H is Hamiltonian.

In particular the forms

(3.2.5)
$$\omega^{(0)}(\partial_{H^{(0)}(X)}, \partial_{H^{(0)}(Y)}) = \int \operatorname{res}((LX)_+L - L(XL)_+)Y \, dx,$$
$$\omega^{(\infty)}(\partial_{H^{(0)}(X)}, \partial_{H^{(\infty)}(Y)}) = -\int \operatorname{res}[L, X]Y \, dx$$

are closed.

Thus, a pair of mappings is constructed having the property that each linear combinations of them, $H^{(0)} + z^n H^{(\infty)}$, is a Hamiltonian mapping. Such pairs are called Hamiltonian.

3.2.6. The space R_- is a Lie algebra with respect to the commutator $[X, Y]$. $R_{(-\infty, -n-1)}$ is an ideal of this algebra, i.e. $R_-/R_{(-\infty, -n-1)}$ is a Lie algebra. Another commutator can be introduced into this algebra. We turn reader's attention to proposition 3.2.2. The expression

$$[X, Y]^*_L = [X, Y]_L + \partial_{H(X)}Y - \partial_{H(Y)}X$$

is a commutator in $R_-/R_{(-\infty,-n-1)}$, turning it into a Lie algebra. In fact, it is easy to see that this expression can be considered as a mapping $\Omega^1 \times \Omega^1 \to \Omega^1 (\Omega^1 = R_-/R_{(-\infty,-n-1)})$. Proposition 3.2.2 implies

$$\partial_{H([[X,Y]_L^*,Z]_L^* + \text{c.p.})} = [[\partial_{H(X)}, \partial_{H(Y)}], \partial_{H(Z)}] + \text{c.p.} = 0.$$

Hence $[[X,Y]_L^*, Z]_L^* + \text{c.p.}$ belongs to the kernel of the mapping H, i.e. this is a resolvent. It suffices to prove that this expression does not contain constants in the coefficients of expansion in ∂^{-1} and z^{-1}, then this resolvent vanishes, according to 1.7.11.

The constants in $[X,Y]_L^*$ can arise only from the terms

$$(-X^c(\partial^n Y^c)_+ + (X^c \partial^n)_- Y^c)_- + \partial_{(\partial^n X^c)_+ \partial^n - \partial^n (X^c \partial^n)_+} Y - (X \leftrightarrow Y).$$

Here X^c and Y^c are X and Y where only constants in the coefficients are left. The operators with constant coefficients commute, therefore $(\partial^n X^c)_+ \partial^n - \partial^n (X^c \partial^n)_+ = (\partial^n X^c)_+ \partial^n - \partial^n (\partial^n X^c)_+ = 0$. Further, $(-X^c(\partial^n Y^c)_+ + (X^c \partial^n)_- Y^c)_- + (Y^c(\partial^n X^c)_+ - (Y^c \partial^n)_- X^c)_- = (-X^c(\partial^n Y^c) + Y^c(\partial^n X^c))_- = 0$.

Thus, $\Omega^1 = R_-/R_{(-\infty,-n-1)}$ is a double Lie algebra with respect to the commutators $[X,Y]$ and $[X,Y]_L^*$.

3.3. Poisson bracket.

3.3.1. Proposition. The vector field corresponding to a Hamiltonian $\tilde{f} = \int f dx$ is $\xi_f = \partial_{H(\delta f/\delta L)}$.

The Poisson bracket of two Hamiltonians is

$$\{\tilde{f}, \tilde{g}\} = \xi_f \tilde{g} = \langle H(\delta f/\delta L), \delta g/\delta L \rangle = \int \text{res}\, H(\delta f/\delta L), \delta g/\delta L\, dx.$$

Proof: This immediately follows from the general theory of Chap. 2. □

For the Poisson bracket propositions 2.5.18 and 2.5.19 hold.

3.3.2. For $H^{(\infty)}$ it is not difficult to write down an explicit expression in terms of the coefficients of the operators $X = \sum_{i=0}^{n-1} \partial^{-i-1} X_i$ and $L = \sum_{i=0}^{n} u_i \partial^i (u_n = 1)$. Namely,

(3.3.3) $$H^{(\infty)}(X) = \sum_{0 \leq \alpha+\beta \leq n-1} (l_{\beta\alpha} X_\alpha) \partial^\beta,$$

where $l_{\beta\alpha} = -\tilde{l}_{\beta\alpha} + \tilde{l}^*_{\alpha\beta}$ and

$$\tilde{l}_{\beta\alpha} = \sum_{\gamma=0}^{n-1-\alpha-\beta} \binom{\gamma+\beta}{\beta} u_{\alpha+\beta+\gamma+1}\partial^\gamma, \tilde{l}^*_{\alpha\beta} = \sum_{\gamma=0}^{n-1-\alpha-\beta} \binom{\gamma+\alpha}{\alpha}$$
$$\times (-\partial)^\gamma u_{\alpha+\beta+\gamma+1}$$

(the brackets in $(l_{\beta\alpha}X_\alpha)$ have the meaning that the differential operator $l_{\alpha\beta}$ acts only on the coefficients X_α but not on any expression which can stand to the right; this is not the product of two operators, $l_{\alpha\beta}$ and the zero-order operator X_α).

3.3.4. Exercise. Deduce (3.3.3) by direct calculating. (A similar formula can be obtained for $H^{(0)}$ but it is cumbersome (see [29]).)

The corresponding Poisson bracket is

$$(3.3.5) \qquad \{\tilde{f}, \tilde{g}\}^{(\infty)} = \int \sum_{\alpha+\beta \leq n-2} (l_{\beta\alpha}\delta f/\delta u_\alpha)\delta g/\delta u_\beta dx.$$

In the particular case $n = 2$ we have $l_\infty = 2\partial, l_{01} = l_{10} = l_{11} = 0$ and the corresponding Poisson bracket is

$$(3.3.6) \qquad \{\tilde{f}, \tilde{g}\} = \int (\delta f/\delta u_0)'\delta g/\delta u_0 dx.$$

This bracket was found by Gardner [14] and Zakharov and Faddeev [9].

3.3.7. Very often (see e.g. Takhtadjan, Faddeev [4]) another expression of Poisson bracket is used:

$$\{\tilde{f}(u), \tilde{g}(u)\} = \sum_{\alpha,\beta} \int\int (\delta f/\delta u_\alpha(x)) \cdot (\delta g/\delta u_\beta(y))\{u_\alpha(x), u_\beta(y)\}dxdy$$

where $\{u_\alpha(x), u_\beta(y)\}$ is the bracket for the following local functionals, $u_\alpha(x)$ is the value of the function u_α at a fixed point x and $u_\beta(y)$ is the value of u_β at y. The bracket (3.3.6) arises if $\{u_\alpha(x), u_\beta(y)\} = \delta'(x-y)$ and the bracket (3.3.5) if $\{u_\alpha(x), u_\beta(y)\} = l_{\beta\alpha}\delta(x-y)$, where the differential operator $l_{\beta\alpha}$ acts with respect to the variable x.

3.3.8. Exercise. Prove the Jacobi identity for the bracket (3.3.6) directly using as far as possible the definition of the variational derivative only.

3.3.9. The particular cases of the Hamiltonian structure, when $z = \infty$ and when $z = 0$ are referred to as, respectively, the first and the second. The first admits a group theory interpretation in the spirit of Sec. 2.4.

The Lie algebra $\mathfrak{E}$ is here $R_- = \{\sum_0^\infty \partial^{-i-1} X_i, X_i \in a\}$ with respect to the usual commutator $[X,Y]$. The corresponding Lie group comprises the same operators and a unity. The dual space is $R_+ = \{\sum_0^\infty a_i \partial^i, a_i \in a\}$ (with finite number of terms). The coupling: $\forall a = \sum_0^\infty a_i \partial^i \in \mathfrak{E}^*, \forall X \in \mathfrak{E}, \langle a, X\rangle = \int \operatorname{res} a X dx$. We have $\forall X, Y \in \mathfrak{E}$ $\operatorname{ad}(X) Y = [X, Y]$. The coadjoint representation is defined by the formula $\forall a \in \mathfrak{E}^* \langle \operatorname{ad}^*(X) a, Y\rangle = \langle a, \operatorname{ad}(X) Y\rangle$, i.e. $\int \operatorname{res} \operatorname{ad}^*(X) a \cdot Y dx = \int \operatorname{res} a[X,Y] dx = \int \operatorname{res} Y[a,X] dx$. This implies $\operatorname{ad}^*(X) a = [a, x]_+$. In particular if $a = L$ we obtain $\operatorname{ad}^*(X) L = [L, X]_+$. Let $\tilde{f} = \int f dx$ be a Hamiltonian. Then $d\tilde{f}(L) = \delta f/\delta L$ and the vector field corresponding to the Hamiltonian $\tilde{f}$ at the "point" L is $\xi_f = [L, \delta f/\delta L]_+$. The differential equation is

$$\dot{L} = [L, \delta f/\delta L]_+$$

with the Poisson bracket

$$\{\tilde{f}, \tilde{g}\} = \int \operatorname{res} [L, \delta f/\delta L] \delta g/\delta L dx.$$

This is the first Hamiltonian structure.

In comparison to Sec. 2.4, the present situation is a little bit paradoxical. It is sufficient to find a vector field only at one "point" L of the orbit in the dual space. This determines the evolution of the generators of the differential algebra and therefore of all its elements.

3.4. *Reduction to the submanifold $u_{n-1} = 0$.*

3.4.1. At the beginning of this chapter we extended the operator L by adding a term $u_{n-1} \partial^{n-1}$, and we have constructed our structures assuming that the differential algebra a consists of differential polynomials of generators $u_0, \ldots, u_{n-1}$. What will happen if we restrict ourselves by requiring that $u_{n-1} = 0$? This requirement must be compatible with the equations, i.e. with vector fields. This means that $\partial_{H(X)} u_{n-1} = 0$. In other words, if $u_{n-1} = 0$ at the beginning of the motion this must remain so all the time. We can express this also as the requirement that $H(X)$ is a $(n-2)$-order differential operator.

3.4.2. Proposition. For the first structure the operator $H^{(\infty)}(X)$ is an operator of an order no more than $n-2$. If $X = \sum_{0}^{n-1} \partial^{-i-1} X_i$ then $H^{(\infty)}(X)$ is independent of X_{n-1}.

Proof: $H^{(\infty)}(X) = [L, X]_+$. The commutator of the operator L of order n with the operator X of order -1 has an order not greater than $n-1-1 = n-2$. The commutator of L with the term $\partial^{-n} X_{n-1}$ has an order not greater than $n - n - 1 = -1$, i.e. this gives no contribution to $[L, X]_+$. □

3.4.3. Corollary. The vector field corresponding to a Hamiltonian $\int f(u_0, \ldots, u_{n-2}) dx$ is $\partial_{H(\delta f/\delta L)}$, $\delta f/\delta L = \sum_{i=0}^{n-2} \partial^{-i-1} \delta f/\delta u_i$. Thus, the first Hamiltonian structure reduces to $u_{n-1} = 0$ automatically.

The situation with the second structure is more complicated.

3.4.4. Proposition. The vector field $\partial_{H^{(\circ)}(X)}$ is tangent to the submanifold $u_{n-1} = 0$ (i.e. $\partial_{H^{(\infty)}} u_{n-1} = 0$) if and only if

$$res[L, X] = 0.$$

Proof: This can be verified by a direct calculation:

$$H^{(0)}(X) = -(LX)_- L + L(XL)_- = -res[L, X]\partial^{n-1} + O(\partial^{n-2}).$$

Therefore, the order of $H^{(0)}(X)$ is $\leq n-2$ if $res[L, X] = 0$. □

3.4.5. Proposition. For $X = \sum_{0}^{n-1} \partial^{-i-1} X_i$, the condition 3.4.4 is equivalent to

$$\sum_{k=1}^{n} \sum_{\alpha=1}^{n} \binom{k}{\alpha} (-\partial)^{\alpha-1} u_k X_{k-\alpha} = C$$

from which X_{n-1} can be expressed as a differential polynomial in $X_i, i = 0, \ldots, n-2$. The constant C can be taken as zero.
The proof is left to the reader. □

3.4.6. Proposition. The vector field corresponding to a Hamiltonian $\tilde{f} = \int f(u_0, \ldots, u_{n-2}) dx$ in virtue of the second structure is $\partial_{H^{(0)}(\delta f/\delta L)}$, where $\delta f/\delta L = \sum_{0}^{n-1} \partial^{-i-1} \delta f/\delta u_i$; $\delta f/\delta u_i, i < n-1$ as usual and $\delta f/\delta u_{n-1} = X_{n-1}$ is by definition expressed in terms of $X_i = \delta f/\delta u_i, i < n-1$ according to the formula 3.4.5.
We leave the proof to the reader. □

3.4.7. Exercise. In the particular case $n = 2$, $u_i = 0$ write down the Poisson bracket $\{\tilde{f}, \tilde{g}\}^{(0)}$.

Answer: $\{\tilde{f}, \tilde{g}\}^{(0)} = \int [(\delta f/\delta u)'''/2 + u'\delta f/\delta u + 2u(\delta f/\delta u)'] \cdot \delta g/\delta u \cdot dx$.

3.5. Variation of the resolvent.

3.5.1. We have constructed Hamiltonian structures. However, we do not know yet whether they have any connection with the equations of Chap. 1. We must point out the Hamiltonians generating these equations.

3.5.2. Proposition.

$$\frac{\delta}{\delta L} \int \operatorname{res} L^{r/n} dx = \frac{r}{n}(L^{(r-n)/n})_- \in R_-/R_{(-\infty,-n-1)}.$$

Proof: We have

$$\delta L^{r/n} = \delta (L^{1/n})^r = \sum_{i=0}^{r-1}(L^{1/n})^i \delta(L^{1/n})(L^{1/n})^{r-i-1},$$

in particular

$$\delta L = \sum_{i=0}^{n-1}(L^{1/n})^i \delta(L^{1/n})(L^{1/n})^{n-i-1}.$$

Now

$$\delta \int \operatorname{res} L^{r/n} dx = \int \operatorname{res} \sum_{i=0}^{r-1}(L^{1/n})^i \delta(L^{1/n})(L^{1/n})^{r-i-1} dx$$

$$= r \int \operatorname{res} L^{(r-1)/n} \delta(L^{1/n}) dx = (r/n) \int \operatorname{res} L^{(r-n)/n}$$

$$\times \sum_{i=0}^{n-1}(L^{1/n})^i \delta(L^{1/n})(L^{1/n})^{n-i-1} dx = (r/n) \int \operatorname{res} L^{(r-n)/n} \delta L dx$$

which yields the required assertion. □

3.5.3. Proposition. If $T = T_\varepsilon$ is a resolvent (1.7.2) then

$$\frac{\delta}{\delta L} \int \operatorname{res} T dx = -\frac{\partial}{\partial (z^n)} T \in R_-/R_{(-\infty,-n-1)}((z^{-1})).$$

Proof:

$$\frac{\delta}{\delta L} \int \operatorname{res} T dx = -\frac{1}{n} \sum_{n=-\infty}^{\infty} (\varepsilon z)^{-r-n} \cdot \frac{r}{n} (L^{(r-n)/n})_-$$

$$= -\frac{1}{n} \sum_{-\infty}^{\infty} (\varepsilon z)^{-r-2n} \cdot \frac{r+n}{n} (L^{r/n})_-$$

$$= \frac{\partial}{\partial(z^n)} \left(\frac{1}{n} \sum_{-\infty}^{\infty} \varepsilon^{-r} (z^n)^{-(r+n)/n} L^{r/n} \right)_- = -\frac{\partial}{\partial(z^n)} T . \square$$

3.6. Hamiltonians of the KdV-hierarchies.

3.6.1. Eqs. (1.4.3) are Hamiltonian with respect to the first symplectic form; the Hamiltonians are

$$\tilde{h}_m = -(n|(m+n)) \int \operatorname{res} L^{(m+n)/n} dx .$$

Proof: We have $\delta h_m/\delta L = -L_-^{\frac{m}{n}}$ which implies $H^{(\infty)}(\delta h_m/\delta L) = -[L_-^{m/n}, L]_+$ and the equation is

$$\dot{L} = -[L_-^{m/n}, L]_+ = [L_+^{m/n}, L],$$

i.e. Eq. (1.4.3). $\square$

3.6.2. *Proposition*. Equations (1.4.3) are Hamiltonian with respect to the second structure with Hamiltonians

$$\tilde{g}_m = \int \operatorname{res} L^{m/n} dx \cdot n/m .$$

Proof: We have $\delta g_m/\delta L = L_-^{(m-n)/n}$. Thus, $\delta h_m/\delta L$ and $-\delta g_m/\delta L$ are proportional to the coefficients T_{m+n} and T_m of the expansion of the resolvent

$$T_\varepsilon = \sum_{-\infty}^{\infty} (\varepsilon z)^{-r-n} L_-^{r/n} = \sum_0^{\infty} T_r z^{-r} \in R_-/R_{(-\infty,-n-1)}$$

in Z^{-1}. Now $H^{(0)}(T_m) + H^{(\infty)}(T_{m+n}) = 0$ (see (1.7.12)) and Eq. (1.4.3) can be written as $\dot{L} = H^{(0)}(\delta \tilde{g}_m/\delta L)$. $\square$

3.6.3. Proposition. The system of elements $\{\tilde{T}_r = \int res\, T_r dx\}$ is in involution with respect to both structures (i.e. any two elements are in involution).

Proof:

$$a_{r,s} := \{\tilde{T}_r, \tilde{T}_s\}^{(0)} = \text{const.} \int res\, H^{(0)}(T_{r-n}) T_{s-n} dx$$

$$= \text{const.} \int res\, H^{(\infty)}(T_r) \cdot T_{s-n} dx = \text{const.} \int res\, T_r \cdot H^{(\infty)}(T_{s-n}) dx$$

$$= \text{const.} \int res\, T_r \cdot H^{(0)}(T_{s-2n}) dx = \text{const.} \int res\, H^{(0)}(T_r) T_{s-2n} dx$$

$$= \text{const.}\, a_{r+n, s-n},$$

all the constants being non-zero. Thus, a recurrence equation for Poisson brackets is obtained. This process can be repeated as many times as needed, the first subscript increasing and the second decreasing until the second subscript becomes negative and the Poisson bracket evidently vanishes. □

3.7. Virasoro algebra.

3.7.1. We shall generalize previous considerations. Now we shall have functionals $\tilde{f} = \int f dx$, where f belongs to the differential algebra $\mathcal{B}$ instead of $\mathcal{A}$ (see Sec. 1.1), i.e. the coefficients of differential polynomials may depend on x. The difference becomes apparent in the calculation of variational derivatives: the coefficients also must be differentiated. Now we consider $u, u', \ldots$ not as formal symbols but as smooth functions on the circle $|x| = 1$ in the complex plane. Functionals are integrals over the same circle. It can be checked that all the facts about Hamiltonian structures remain valid.

Let us find $\{\tilde{f}, \tilde{g}\}^{(0)}$ for the functionals of a special form

$$(3.7.2) \qquad \tilde{L}_k = \int u x^{k+1} dx.$$

We have $\delta L_k/\delta u = x^{k+1}$. Then $\{\tilde{L}_k, \tilde{L}_l\}^{(0)} = \int ((k+1)k(k-1)x^{k-2}/2 + u'x^{k+1} + 2u(k+1)x^k)x^{l+1} dx = \pi i(k^3 - k)\delta_{k,-l} + (k-l)\tilde{L}_{k+l}$.

3.7.3. Let us digress a little into another matter. We consider a Lie algebra with generators L_k and commutation rules $[L_k, L_l] = (k-l)L_{k+l}$ (the Jacobi relations can be easily verified). We have a cocycle (see 2.5.5) $\alpha(L_k, L_l) = c(k^3 - k)\delta_{k,-l}$ (prove that this is indeed a cocycle). A central extension of the algebra with the help of this cocycle can be constructed. Its elements are pairs (a, X), where $\{X\}$ belongs to the algebra, and $\{a\}$ are numbers. The commutator is defined by

$$[(a, X), (b, Y)] = (\alpha(X, Y), [X, Y]).$$

The Jacobi identity for the pairs is equivalent to the fact that α is a cocycle.

The extended algebra is called the Virasoro algebra with a central charge c. Thus, the Virasoro algebra is the algebra of pairs $(a, X), X = \sum c_k L_k$; the commutation rule are

(3.7.4) $\qquad [(a, L_k), (b, L_l)] = (c(k^3 - k)\delta_{k,-l}, (k - l)L_{k+l}).$

3.7.5. Now we give a realization of the Virasoro algebra by functionals. To each pair (a, X) we attach a functional: $(a, L_k) \mapsto a + \tilde{L}_k$ (we have taken the central charge $c = \pi i$). It is easy to see that this mapping is an isomorphism. The central charge may be made arbitrary by scaling $L_k \mapsto c' L_k, \{\}^{(0)} \mapsto \frac{1}{c'}\{\}^{(0)}$.

3.7.6. *Remark to bibliography.* Very soon after the discovery of the outstanding role of the KdV equation by Gardner, Green, Kruskal and Miura [8] it was established that this equation can be represented as a Hamiltonian system with respect to the Poisson bracket (3.3.6). This was done by Gardner [14] and Zakharov and Faddeev [9]. Gelfand and Dickey [16], [29] generalized the KdV-hierarchy for the case when L is an operator of an arbitraray order. In the same papers they constructed the symplectic form and proved that the equations are Hamiltonian. Their proof was very complicated. Much more transparent proof became possible after the works of Lebedev and Manin [30] and especially of Adler [17]. These authors suggested the group theory interpretation of the first structure. As to the second structure, Magri [31] found it for the KdV equation (see 3.4.7). It became clear that the second structure must exist also in the general case and play an important role: with its help a recurrence equation for involutive Hamiltonians can be written, Adler [17] has guessed what form the second structure could have: he had everything he needed for it: the mapping which we named after him. It is not clear to us why he did not prove his conjecture; this was done only in the paper by Gelfand and Dickey [29] and was given above. The idea to represent the Virasoro algebra by the algebra of functionals with the Poisson bracket as the commutator was given by Gervais and Neveu [110]. They considered the case $n = 2$. Khovanova [32] generalized this by considering the higher Poisson brackets ($n > 2$).

The recurrence equation (1.7.12) can be written in another form: $X_{j+n} + \phi(X_j) = 0$, where $\phi = (H^{(\infty)})^{-1} H^{(0)}$. The operator ϕ is called the recursion operator, and is sometimes said to have a hereditary property: to convert a Hamiltonian T_j to another T_{j+n} of the hierarchy (see the work of Focas and Fuchssteiner [33]).

CHAPTER 4. The Kupershmidt-Wilson Theorem.

4.1. A generalization of the Gardner-Zakharov-Faddeev bracket.
4.1.1. In this chapter the isomorphism of the second structure and the generalized Gardner-Zakharov-Faddeev structure will be shown. Incidentally this gives another proof of the Jacobi identity for the second Poisson bracket.

Let $\mathcal{A}$ be the differential algebra with independent differential generators $\{v_i\}, i = 1, \ldots, n$.

4.1.2. *Proposition.* The bilinear functional

$$\{\tilde{f}, \tilde{g}\} = \sum_{i=1}^{n} \int (\delta f/\delta v_i)' \delta g/\delta v_i dx$$

has all the properties of a Poisson bracket.

Proof: Let us construct all the elements of a Hamiltonian structure, according to Sec. 2.5:

1) The Lie algebra $\mathfrak{E}$ of all vector fields; every field is characterized by a set $a = (a_1, \ldots, a_n), a_i \in a$ and has the form $\partial_a = \sum_{i,j} a_i^{(j)} \partial/\partial v_i^{(j)}$.

2) The left $\mathfrak{E}$-module $\Omega^0 = \tilde{\mathcal{A}} = \mathcal{A}/\partial\mathcal{A}$. Its elements are integrals $\tilde{f} = \int f dx$.

3) The dual space Ω^1 consisting of sets $X = (X_1, \ldots, X_n), X_i \in \mathcal{A}$, with the coupling between $\mathfrak{E}$ and Ω^1 given by $\langle \partial_a, X \rangle =$

$\sum \int a_i X_i \, dx = \int a X \, dx$.

4) The mapping $H : \Omega^1 \to \mathfrak{E} : X \mapsto \partial_{H(X)} = \partial_{X'}$.
5) The form $\omega(\partial_{X'}, \partial_{Y'}) = \int X' Y \, dx$.

4.1.3. Lemma. $l = \operatorname{Im} H$ is a Lie subalgebra of $\mathfrak{E}$.

Proof:
$$[\partial_{X'}, \partial_{Y'}] = \partial_{\partial_{X'} Y' - \partial_{Y'} X'} = \partial_{(\partial_{X'} Y - \partial_{Y'} X)'} . \square$$

4.1.4. Lemma. The form ω is closed.

Proof: $\forall X, Y, Z \in \Omega^1$

$$d\omega(\partial_{X'}, \partial_{Y'}, \partial_{Z'}) = \partial_{X'} \omega(\partial_{Y'}, \partial_{Z'}) - \omega([\partial_{X'}, \partial_{Y'}], \partial_{Z'}) + \text{c.p.}$$

$$= \partial_{X'} \int Y' \cdot Z \, dx - \int (\partial_{X'} Y' - \partial_{Y'} X') Z \, dx + \text{c.p.}$$

$$= \int (\partial_{X'} Y' \cdot Z - \partial_{X'} Z' \cdot Y - \partial_{X'} Y' \cdot Z - \partial_{Y'} X' \cdot Z) dx + \text{c.p.} = 0 . \square$$

4.1.5. Lemma. H is a Hamiltonian mapping.

Proof: This immediately follows from 4.1.3, 4.1.4 and 2.5.16. $\square$

6) The operator $d : \Omega^0 \to \Omega^1$:

$$d\tilde{f} = \delta f / \delta v = (\delta f / \delta v_1, \ldots, \delta f / \delta v_n) .$$

7) The vector field $\xi_{\tilde{f}}$ corresponding to a Hamiltonian $\tilde{f}$:

$$\xi_{\tilde{f}} = H(d\tilde{f}) = \partial_{(\delta f / \delta v)'} .$$

8) The Poisson bracket

$$\{\tilde{f}, \tilde{g}\} = \partial_{(\delta f / \delta v)'} \tilde{g} = \int (\delta f / \delta v)' (\delta g / \delta v) dx$$

$$= \sum \int (\delta f / \delta v_i)' (\delta g / \delta v_i) dx .$$

The general theory implies that this bracket has all the needed properties, including the Jacobi identity and

$$\xi_{\{\tilde{f}, \tilde{g}\}} = [\xi_{\tilde{f}}, \xi_{\tilde{g}}] .$$

4.2. Miura transformation. The Kupershmidt-Wilson theorem.

4.2.1. Let L be, as usual, the operator $L = \sum_0^n u_i \partial^i, u_n = 1$. Now we write it in a multiplicative form:

$$(4.2.2) \qquad \sum_0^n u_i \partial^i = (\partial - v_n)(\partial - v_{n-1})\ldots(\partial - v_1).$$

This yields an expression for each u_i as a differential polynomial in $\{v_j\}$ ($\{v_i\}$ cannot be expressed as differential polynomials in terms of $\{u_i\}$):

$$(4.2.3) \qquad u_i = Q_i(v_1,\ldots,v_n),\ i = 0,\ldots,n-1.$$

The substitution (4.2.3) given by identity (4.2.2) is called the Miura transformation.

4.2.4. Theorem of Kupershmidt and Wilson. The second Poisson bracket

$$\{\tilde{f}, \tilde{g}\} = \int \operatorname{res} \left\{ L\left(\frac{\delta f}{\delta L}L\right)_+ - \left(L\frac{\delta f}{\partial L}\right)_+ L \right\} \frac{\delta g}{\delta L} dx$$

is equal to the generalized Gardner-Zakharov-Faddeev bracket

$$\int (\delta f/\delta v)'(\delta g/\delta v) dx,$$

where $\{u_i\}$ and $\{v_i\}$ are connected by the Miura transformation.

Proof: We express $\{\delta f/\delta v_i\}$ in terms of $\{\delta f/\delta u_i\}$.

$$\delta \tilde{f} = \int \operatorname{res}(\delta f/\delta L)\delta L dx = \int \sum (\delta f/\delta v_i)\delta v_i dx.$$

Denoting $\partial_i = \partial - v_i$ we obtain

$$\int \operatorname{res}(\delta f/\delta L)\delta L dx = -\int \operatorname{res}(\delta f/\delta L) \sum_{i=1}^n \partial_n \ldots \partial_{i+1} \delta v_i \partial_{i-1} \ldots \partial_1 dx$$

$$= -\int \sum_{i=1}^n \operatorname{res} \partial_{i-1}\ldots\partial_1(\delta f/\delta L)\partial_n\ldots\partial_{i+1}\delta v_i dx$$

which implies

$$(4.2.5) \qquad \delta f/\delta v_i = -\operatorname{res}\partial_{i-1}\ldots\partial_1(\delta f/\delta L)\partial_n\ldots\partial_{i+1}.$$

Now

$$\int \sum (\delta f/\delta v_i)(\delta g/\delta v_i)' dx = \int \sum res(\partial_{i-1}\ldots\partial_1(\delta f/\delta L)\partial_n\ldots\partial_{i+1})$$
$$(res(\partial_{i-1}\ldots\partial_1(\delta g/\delta L)\partial_n\ldots\partial_{i+1}))' dx = \int res\Big\{\sum \partial_{i-1}$$
$$\ldots\partial_1(\delta f/\delta L)\partial_n\ldots\partial_{i+1}\Big[\partial, res(\partial_{i-1}\ldots\partial_1(\delta g/\delta L)\partial_n\ldots\partial_{i+1})\Big]\Big\} dx.$$

Nothing will change if we substitute ∂_i for ∂. Further, we note that $res\, B = (\partial B_-)_+ = (B_-\partial)_+ = (\partial_i B_-)_+ = (B_-\partial_i)_+$. We have

$$\int \sum (\delta f/\delta v_i)(\delta g/\delta v_i)' dx = \int res\Big\{\sum \partial_{i-1}\ldots\partial_1(\delta f/\delta L)\partial_n\ldots\partial_{i+1}\partial_i$$
$$((\partial_{i-1}\ldots\partial_1(\delta g/\delta L)\partial_n\ldots\partial_{i+1})_-\partial_i)_+$$
$$-\sum \partial_i\partial_{i-1}\ldots\partial_1(\delta f/\delta L)\partial_n\ldots\partial_{i+1}$$
$$(\partial_i(\partial_{i-1}\ldots\partial_1(\delta g/\delta L)\partial_n\ldots\partial_{i+1})_-)_+\Big\} dx.$$

The subscripts "−" can be dropped in this expression since the similar expression with subscripts "+" instead of "−" vanishes (in that case the external subscripts "+" are unnecessary and can be dropped; after that both the terms cancel out). Thus,

$$\int \sum (\delta f/\delta v_i)(\delta g/\delta v_i)' dx = \int res\Big\{\sum \partial_{i-1}\ldots\partial_1(\delta f/\delta L)\partial_n\ldots\partial_i$$
$$(\partial_{i-1}\ldots\partial_1(\delta g/\delta L)\partial_n\ldots\partial_i)_+ - \sum \partial_i\ldots\partial_1(\delta f/\delta L)\partial_n\ldots\partial_{i+1}$$
$$(\partial_i\ldots\partial_1(\delta g/\delta L)\partial_n\ldots\partial_{i+1})_+\Big\} dx = \int res\Big\{(\delta f/\delta L)\partial_n\ldots\partial_1((\delta g/\delta L)$$
$$\partial_n\ldots\partial_1)_+ - \partial_n\ldots\partial_1(\delta g/\delta L)(\partial_n\ldots\partial_1(\delta g/\delta L))_+\Big\} dx$$
$$= \int res\Big\{(\delta f/\delta L)L((\delta g/\delta L)L)_+ - L(\delta f/\delta L)(L(\delta g/\delta L))_+\Big\} dx$$
$$= \int res\,(\delta f/\delta L)\Big\{L((\delta g/\delta L)\cdot L)_+ - (L\delta g/\delta L)_+ L\Big\} dx = \{g, f\}.\square$$

4.2.6. Now we turn to the case where $u_{n-1} = 0$. This time $\{v_i\}$ cannot be independent, namely $\sum v_i = 0$. The definitions of $\delta f/\delta L$ and $\delta f/\delta v$ must be changed. As to $\delta f/\delta L$, we already know (Sec. 3.4) that in $\delta f/\delta L = \sum_{0}^{n-1} \partial^{-i-1} X_i$ only $X_i, i = n-2$ are variational derivatives; and X_{n-1} is expressed in terms of previous ones with the help of $res[L, \delta f/\delta L] = 0$, i.e. 3.4.5. Similarly, $\{\delta f/\delta v_i\}$ are no more independent. If we wish to retain the equality of rights of all the $\{v_i\}$ we must define $\delta f/\delta v_i$ as quantities a_i such that $\delta f = \sum \int a_i \delta v_i dx$ and $\sum a_i = 0$.

4.2.7. Proposition. Equation (4.2.5) remains valid in the case $u_{n-1} = 0$.

Proof: We must verify that the sum of all the expressions (4.2.5) vanishes in virtue of the equation $res[L, \delta f/\delta L] = 0$. It is sufficient to prove that the derivative of this sum vanishes. We carry out a calculation similar to that in the previous proposition:

$$[\partial, \sum res\, \partial_{i-1} \ldots \partial_1 (\delta f/\delta L) \partial_n \ldots \partial_{i+1}]$$
$$= \sum \partial_i ((\partial_{i-1} \ldots \partial_1 (\delta f/\delta L) \partial_n \ldots \partial_{i+1})_- \partial_i) +$$
$$- \sum (\partial_i (\partial_{i-1} \ldots \partial_1 (\delta f/\delta L) \partial_n \ldots \partial_{i+1})_-) + \partial_i$$
$$= \sum \partial_i (\partial_{i-1} \ldots \partial_1 (\delta f/\delta L) \partial_n \ldots \partial_i) +$$
$$- \sum (\partial_i \ldots \partial_+ (\delta f/\delta L) \partial_n \ldots \partial_{i+1}) + \partial_i$$
$$= - \sum (\partial_i (\partial_{i-1} \ldots \partial_1 (\delta f/\delta L) \partial_n \ldots \partial_i)_-) +$$
$$+ \sum ((\partial_i \ldots \partial_1 (\delta f/\delta L) \partial_n \ldots \partial_{i+1})_- \partial_i) +$$
$$= - \sum res\, (\partial_{i-1} \ldots \partial_1 (\delta f/\delta L) \partial_n \ldots \partial_i)$$
$$+ \sum res\, (\partial_i \ldots \partial_1 (\delta f/\delta L) \partial_n \ldots \partial_{i+1})$$
$$= -res((\delta f/\delta L) \partial_n \ldots \partial_1) + res(\partial_n \ldots \partial_1 \delta f/\delta L)$$
$$= res[L, \delta f/\delta L] = 0. \square$$

The Kupershmidt-Wilson theorem also remains valid in this case since we have used only (4.2.5) in the proof.

4.3. Modified KdV equation. Lie-Bäcklund transformations.

4.3.1. Exercise. Write the Miura transformation explicitly if $n = 2, L = \partial^2 + u = (\partial - v)(\partial + v)$. If u satisfies the KdV equation (1.4.6) what is the equation for v?

Answer: $u = v' - v^2$, $4\dot v = v''' - 6v^2 v'$. This is the modified KdV equation (mKdV).

4.3.2. If we know a solution u of the KdV equation and solve an auxiliary equation $(v' - v^2 = u)$ with respect to v, we shall find a solution of the mKdV equation. Such transformations of one equation to another are called Lie-Bäcklund transformations, or Bäcklund transformations.

A Bäcklund transformation of the KdV equation to itself can also be found. The mKdV equation is invariant with respect to the substitution $v \mapsto -v$. This leads to the following.

4.3.3. Proposition. Let u be a solution of the KdV equation. If v is a solution of the Riccati equation $v' - v^2 = u$ and $u_1 = -v' - v^2$, then u_1 is another solution of the KdV equation.

4.3.4. The notion of the modified KdV equation can be generalized. Let $\dot{L} = [P, L]$, where $P = L_+^{m/n}$ is any equation (1.4.3). The Miura transformation (4.2.2) expresses $\{u_i\}$ in terms of $\{v_i\}$. Then $\{v_i\}$ satisfies some differential equation which is called the modified KdV equation.

4.3.5. Proposition. If the Hamiltonian $\tilde{f}$ of the KdV equation $\dot{L} = [P, L] = H^{(0)}(\delta f/\delta L)$ with respect to the second structure is expressed in terms of $\{v_i\}$ by the Miura transformation, then the corresponding modified equation will be

$$\dot{v}_i = (\delta f/\delta v_i)'.$$

Proof: This is a corollary of the Kupershmidt-Wilson theorem. □

4.3.6. Now we shall learn another form of the mKdV (Adler [34]). Let operators L and P be expressed in terms of $\{v_i\}$. Denote by L_Ω and P_Ω the operators obtained by a cyclic permutation of the variables; $\Omega : v_1 \mapsto v_2, v_2 \mapsto v_3, \ldots, v_n \mapsto v_1$. Thus, $L_\Omega = (\partial - v_1)(\partial - v_n)\ldots(\partial - v_2)$. Similarly, L_{Ω^j} and P_{Ω^j} are operators obtained by the permutation Ω^j. The operators P_{Ω^j} and L_{Ω^j} form a P, L-pair for all j. Evidently, $L_{\Omega^n} = L$ and $P_{\Omega^n} = P$.

4.3.7. Proposition (Adler). Operators $B_j = P_{\Omega^j}\partial_j - \partial_j P_{\Omega^{j-1}}$ are of order zero. The system of equations $\dot{v}_j = -B_j$ is equivalent to $\dot{L} = [P, L]$, i.e. the mKdV equation.

Proof: Let the order of the operators B_j be $s > 0$. We write these operators as $B_j = b_j \partial^s + c_j \partial^{s-1} + \ldots$. Let $M = \sum_1^n \partial_n \ldots \partial_{j+1} B_j \partial_{j-1} \ldots \partial_1$. It is easily seen that $M = [P, L]$, and the order of M is $\leq n - 2$ since P and L form a P, L-pair. On the other hand, each term in the definition of M is the operator of order $n - 1 + s \geq n$. Hence at least the two highest of the terms must cancel out. This implies

$$\sum_j b_j = 0, \sum_j (n-j)b'_j + \sum_j c_j - \sum_{\substack{m,j \\ (m \neq j)}} b_j v_m = 0.$$

More general equations can be obtained by cyclic permutations:

$$\sum_j b_{j+k} = 0, \quad \sum_j (n-j)b'_{j+k} + \sum_j c_{j+k} - \sum_{m,j} b_{j+k} v_{m+k} = 0$$

(where $b_{j+n} = b_j, c_{j+n} = c_j$). All the terms, besides $\sum jb'_{j+k}$, do not depend on k. Hence $\sum_1^n jb'_j = \sum_1^n jb'_{j+1}$ and $\sum_1^n jb_j = \sum_2^n (j-1)b_j + nb_1 + \text{const}$. This implies $\sum_2^n b_j = (n-1)b_1 + \text{const}$ and $b_1 = \text{const}$. Then $b_j = \text{const}$, the constant being independent of j. From $\sum b_j = 0$ we obtain $b_j = 0$, which contradicts our assumption. Thus B_j are of zero order.

We can construct equations $\dot{v}_j = -B_j$. In virtue of these equations

$$\dot{L} = -\sum \partial_n \ldots \partial_{j+1} \dot{v}_j \partial_{j-1} \ldots \partial_1 = \sum \partial_n \ldots \partial_{j+1} B_j \partial_{j-1} \ldots \partial_1 = [P, L]$$

which coincides with (1.4.3). □

4.3.8. *Proposition.* The mKdV equation is invariant under the cyclic permutation $\Omega : v_1 \mapsto v_2, \ldots, v_n \mapsto v_1$.
Proof: This is clear from the form of these equations:

$$\dot{v}_j = -B_j. \ \square$$

4.3.9. *Corollary.* The cyclic permutation $\Omega : v_1 \mapsto v_2, \ldots, v_n \mapsto v_1$ generates the Bäcklund transformation of the KdV equation $\dot{L} = [P, L]$, the n-th power of which is an identical transformation.

4.3.10. The theorem 4.2.4 was proved by Kupershmidt and Wilson in Ref. [35]. We give here a much shorter proof from our paper [36]. We shall return to the Miura transformation and the Kupershmidt-Wilson theorem in Chap. 10.

CHAPTER 5. The KP-Hierarchy.

5.1. Introductory notes.

In Chap. 1 for every integer n a hierarchy of equations was constructed. A problem arises as to what is the relation between these hierarchies and whether it is possible to "stabilize" the theory embedding the space of operators of a given order into the space of operators of a greater order. In this direct form the idea cannot be correct, e.g. for the reason that the highest coefficient of the operator L is 1. The more effective idea of mathematicians of the Kyoto school (Sato, and others, see the survey [37]) was the following. Not operators of distinct orders must be identified but roots of them: $L_1 = L^{1/n}$ is in each case a PDO of order 1. We consider now the coefficients of the PDO L_1 as the generators of the differential algebra $\mathcal{A}$. Every one of the previous hierarchies can be obtained by a reduction of the new, vast hierarchy (which is called the KP-hierarchy) to a submanifold determined by the requirement that the n-th power of the operator L_1 is a purely differential operator. The next simple and beautiful idea is that various equations of the hierarchy can be integrated together since their vector fields commute. Its own parameter, "time", can be chosen for every vector field, and solutions can be found which depend on all these parameters (or, better to say, on a finite number of "times" not fixed beforehand). The third idea (also occurred to Japanese mathematicians, beginning with a much earlier work by Hirota [38]) is to construct some universal τ-function in terms of which the solutions can be expressed.

Besides the Japanese work we also use the different version of τ-function given by Segal and Wilson [39].

5.2. Definition of the KP-hierarchy.

5.2.1. Let

$$L = \partial + u_0 \partial^{-1} + u_1 \partial^{-2} + \ldots$$

be a PDO (see Sec. 1.2) whose coefficients $u_0, u_1, \ldots$ we shall consider as the generators of a differential algebra $\mathcal{A}$.

Similarly to that in Sec. 1.3 we construct a set of differential equations, or flows

(5.2.2) $$\partial_m L = [B_m, L], \partial_m = \partial/\partial x_m$$

where x_m are parameters ("times"), $x_1 = x$, and

$$B_m = L_+^m.$$

5.2.3. Proposition. Equations (5.2.2) imply

(5.2.4) $$\partial_m B_n - \partial_n B_m = [B_m, B_n]$$

(equations having such a form are called "zero curvature equations").
Proof:

$$\partial_m B_n - \partial_n B_m - [B_m, B_n] = ((\partial_m L) \cdot L^{n-1} + L(\partial_m L)L^{n-2} + \ldots$$
$$+ L^{n-1}(\partial_m L) - (\partial_n L)L^{m-1} - \ldots - L^{m-1}(\partial_n L) - [B_m, B_n])_+$$
$$= \left(\sum_{i=0}^{n-1} L^i[B_m, L]L^{n-i-1} - \sum_{i=0}^{m-1} L^i[B_n, L]L^{m-i-1} - [B_m, B_n]\right)_+$$
$$= ([B_m, L^n] - [B_n, L^m] - [B_m, B_n])_+$$
$$= [B_n - L^n, B_m - L^m]_+ = [L_-^m, L_-^n]_+ = 0. \square$$

5.2.5. Remark. Every equation (5.2.2) is equivalent to an infinite system of differential equations with an infinite number of unknown quantities $u_0, u_1, \ldots$ and two independent variables (the first equation, $n = 1$, is trivial). Everyone of equations (5.2.4) is equivalent to a finite system of equations, the number of equations being equal to the number of unknown

quantities, i.e. the system is closed. There are three independent variables, x, x_m and x_n. From the point of view of the analysis, the system (5.2.2) cannot be considered as a definite one; as to the system (5.2.4), this is a classical system in partial derivatives. In connection with this remark a problem of terminology arises. It is customary to consider as a principal one the system (5.2.2) which is called the KP-hierarchy. It would be better, perhaps, to call "the KP-hierarchy" the set of equations (5.2.4), especially since one of these equation is, in fact, the KP-equation (see 5.2.7 below). In any case, every time we speak about the KP-hierarchy we shall indicate coherent system we have in mind, (5.2.2) or (5.2.4).

5.2.6. *Exercise.* Let $n > m$. Show that Eqs. (5.2.4) form a system of $n-1$ equations with the same quantity of unknown variables $u_0, \ldots, u_{n-2}$.

5.2.7. *Exercise.* Let $n=3, m=2$. Write the system of equations for u_0 and u_1 obtained from (5.2.4).

Answer: Denote $2u_0 = u, x_2 = y, x_1 = t$. Then

$$u_y = u'' + 4u_1'$$
$$u_y' + 2(u_1)_y - \frac{2}{3}u_t = \frac{1}{3}u''' + 2(u_1)'' - uu'$$

(the primes denote derivatives with respect to x). Eliminating u_1 we obtain

(5.2.8) $$3u_{yy} = (4u_t - u''' - 6uu')'.$$

This equation is called the Kadomtsev-Petviashvili (KP) equation which name is also given to the whole hierarchy.

5.2.9. *Proposition.* Vector fields $\{\partial_m\}$ defined by (5.2.2) commute.

Proof:

$$\partial_m(\partial_n L) - \partial_n(\partial_m L) = \partial_m[B_n, L] - \partial_n[B_m, L]$$
$$= [\partial_m B_n - \partial_n B_m, L] + [B_n, [B_m, L]] - [B_m, [B_n, L]]$$
$$= [[B_m, B_n], L] + [B_n, [B_m, L]] - [B_m, [B_n, L]] = 0$$

according to the Jacobi identity. □

5.2.10. The fact that the vector fields ∂_n and ∂_m are commutative means that each of Eqs. (5.2.2) generates a symmetry of every other equations, i.e. translation of a solution of one of these equations along another vector field transforms this solution into another solution of the same equation. In other words, if L is a solution of the n-th equation (5.2.2) then $L + \varepsilon\partial_m L$ (up to $O(\varepsilon^2)$) is also a solution, for any m. The equation $\partial_k L = [B_k, L]$ generates also a symmetry of any Eq. (5.2.4) in the same sense; this is easy to check.

5.3. Reduction of the KP-hierarchy to the KdV.

5.3.1. Let us explain what is a reduction. Put on some constraints

(5.3.2) $$Q : \{Q_k(u) = 0\}, k = 1, 2, \ldots \; Q_k(u) \in a.$$

The $\{u_k\}$ ceases to be independent generators of the differential algebra. Let J_Q be a differential ideal in the differential algebra a generated by $\{u_k\}$ consisting of $\left\{\sum a_{k,i} Q_k^{(i)}\right\}$, $a_{k,i} \in \mathcal{A}$, $Q_k^{(i)} = \partial^i Q_k$. By virtue of Eqs. (5.3.2) we must identify the whole ideal J_Q with zero; the set of all these equations will be all the corollaries of Eqs. (5.3.2). In other words we must consider the quotient-algebra $\mathcal{A}_Q = \mathcal{A}/J_Q$. The differentiation ∂ can be transferred to this quotient since $\partial J_Q \subset J_Q$. Thus, the elements from $\mathcal{A}_Q$ are the differential polynomials to within the equivalence generated by the relations (5.3.2).

We say that a system of equations, i.e. a vector field ∂_a, admits reduction to a manifold defined by Eqs. (5.3.2), if $\partial_a J_Q \subset J_Q$. This enables us to transfer the vector field to $\mathcal{A}_Q$; we shall also say that the vector field is tangent to $\mathcal{A}_Q$. An equality of two elements f and g modulo the ideal J_Q will be denoted as $f = g \bmod J_Q$ or as $f \stackrel{Q}{=} g$.

5.3.3. Now we specify the constraints (5.3.2). Let

(5.3.4) $$Q_n : L_-^n = 0, n = 2, 3, \ldots.$$

We consider these equations not together but to each of n there will be its own reduction.

5.3.5. *Proposition.* Vector fields (5.2.2) admit reduction to each of manifolds (5.3.4).

Proof:
$$\partial_m(L_-^n) = [B_m, L^n]_- = [B_m, L_-^n]_-.$$

The right-hand side belongs to the ideal J_{Q_n}, determined by the coefficients of the operator L_-^n. Thus, the vector field ∂_m sends the generators of the ideal (the coefficients of the operator L_-^n) into the ideal, hence $\partial_m J_{Q_n} \subset J_{Q_n}$. □

5.3.6. Proposition. By reduction on the manifold Q_n (5.3.4) the relation $\partial_{np} L = 0 \mod J_{Q_n}$ holds. This means that L does not depend on "times" t_{np}.

Proof: We have $L^n \in R_+$, hence

$$\partial_{np} L = [L_+^{np}, L] = [(L^n)_+^p, L] \stackrel{Q}{=} [L^{np}, L] = 0$$

(all the equalities are in $\mathcal{A}_Q$, i.e. modulo J_Q, in other words, in virtue of Eq. (5.3.4)). □

Some more detail of the structure of the ideal J_{Q_n} will follow.

5.3.7. Lemma. Denoting $L^n = \partial^n + v_0 \partial^{n-2} + v_1 \partial^{n-3} + \ldots$ we obtain

$$u_i = \frac{1}{n} v_i + f_i(v_{i-1}, v_{i-2}, \ldots), i = 0, 1, 2, \ldots,$$

where f_i are differential polynomials of the arguments. A similar form has the inverse expression, $v_i = n u_i + g_i(u_{i-1}, u_{i-2}, \ldots)$.
Proof: This becomes evident if we write $L^n = (\partial + u_0 \partial^{-1} + \ldots)^n$ and consider the term with ∂^{n-i-2}, i.e. $v_i \partial^{n-i-2}$. □

5.3.8. Corollary. In the differential algebra $\mathcal{A}$ elements $\{v_i\}, i = 0, 1, \ldots$ can be taken as generators. The ideal J_{Q_n} comprises the differential polynomials in $\{v_i\}$, each term of which contains at least one of the multiplicators $v_i^{(j)}, i > n-2$. Thus, reduction to the manifold $L_-^n = 0$ consists in identifying all the $v_i, i > n-2$ with zero.

5.3.9. Corollary. The equations $\partial_m L^n = [L_+^m, L^n]$, which are equivalent to $\partial_m L = [L_+^m, L]$, can be written as a system of equations in $v_0, \ldots, v_{n-2}$. This is nothing but the n-th KdV-hierarchy of Chap. 1. In particular, the KdV equation is a reduction of the equation $\partial_3 L^2 = [L_+^3, L^2]$ to the manifold $L_-^2 = 0$.

5.4. First integrals.

5.4.1. Proposition. The quantities

$$J_k = \int \operatorname{res} L^k dx$$

are first integrals of Eqs. (5.2.2).

Proof:
$$\partial_m J_k = \int res\,[B_m, L^k]dx = 0\,.\square$$

It is clear that the same quantities are first integrals of the equations reduced to the submanifolds $L_-^n = 0$. They can be expressed in terms of the variables $\{v_i\}$.

5.5. Soliton solutions.

5.5.1. N-soliton solution for the KP-hierarchy (5.2.2) is constructed in almost the same way as in Sec. 1.6 for the equations of the KdV hierarchy. The solutions will depend on any number of times $\{x_m\}$ and will satisfy a corresponding equation with respect to each time. The possibility of the existence of such solutions follows from the fact that the vector fields $\{\partial_m\}$ are commutative.

Here and further we use the notation

$$\xi(x,\alpha) = x_1\alpha + x_2\alpha^2 + x_3\alpha^3 + \dots.$$

Let $\alpha_k, \beta_k, a_k, k = 1, \dots, N$ being distinct complex constants. Put[a]

$$y_k(x) = \exp\xi(x,\alpha_k) + a_k \exp\xi(x,\beta_k), k = 1,\dots,N\,.$$

They have the property

(5.5.2) $$\partial_m y_k = \partial^m y_k\,.$$

Let

$$\Phi = \frac{1}{\Delta}\begin{vmatrix} y_1, & \dots, & y_N, & 1 \\ y_1', & \dots, & y_N', & \partial \\ & \dots & & \\ y_1^{(N)}, & \dots, & y_N^{(N)}, & \partial^N \end{vmatrix},$$

where Δ is the Wronskian of the functions $y_1, \dots, y_N$. The equation

(5.5.3) $$\phi y_k = 0, \quad k = 1,\dots,N$$

---

[a]Since we speak here about genuine functions and not about formal series, and on the other hand we do not require the convergence of the series $\xi(x,\alpha)$ we shall consider it as a finite sum with no fixed number of terms. The operator L will satisfy as many equations as what $\{x_k\}$ we have left.

holds. Let
$$L = \phi \partial \phi^{-1}.$$
If $K = \phi \partial^{-N}$, this equality can be written as $L = K \partial K^{-1}$.

5.5.4. Proposition. The operator L is a solution of all the equations (5.2.2). The proof is similar to that of 1.6.5. For any m we have
$$L_+^m \phi - \phi \partial^m = -L_-^m \phi.$$
In the left-hand side is a differential operator and in the right-hand side an operator of the order less than N. Now,
$$0 = \partial_m(\phi y_k) = \partial_m \phi \cdot y_k + \phi \cdot \partial_m y_k = \partial_m \phi \cdot y_k + \phi \cdot \partial^m y_k$$
$$= \partial_m \phi \cdot y_k + L_+^m \phi y_k + L_-^m \phi y_k = (\partial_m \phi + L_-^m \phi) y_k.$$
The differential operator $\partial_m \phi + L_-^m \phi$ of the order less than N annuls N linearly independent functions y_k, hence $\partial_m \phi = -L_-^m \phi$. Further,
$$\partial_m L = (\partial_m \phi) \partial \phi^{-1} - \phi \partial \phi^{-1} (\partial_m \phi) \phi^{-1} = -L_-^m \phi \partial \phi^{-1} + \phi \partial \phi^{-1} L_-^m$$
$$= [\phi \partial \phi^{-1}, L_-^m] = [L, L_-^m] = [L_+^m, L]. \square$$

5.5.5. Proposition. If $\{\alpha_k\}$ and $\{\beta_k\}$ are connected by $\beta_k = \varepsilon \alpha_k$, where $\varepsilon^n = 1$, then $L_-^n = 0$ will hold and L^n satisfies all the equations of the n-th KdV-hierarchy.

Proof: This is exactly the assertion proved in 1.6.5 if $L^n = \phi \partial^n \phi^{-1}$ is taken into account. $\square$

5.5.6. Remark. Proving the proposition we were not able to conclude from $\phi y_k = 0$ that $L_-^m \phi y_k = 0$ since the application of the formal PDO to functions is defined only for differential operators. Otherwise "proofs" of the type $y_k = \phi^{-1} \phi y_k = 0$ were possible.

5.5.7. Exercise. Find the 1-soliton ($N = 1$) solution of the KP-hierarchy. Write a formula at least for u_0.
Answer:
$$(5.5.8) \quad u_0 = \partial^2 \ln y = \frac{(\beta - \alpha)^2}{2} \cdot \frac{1}{\cosh^2((\xi(x,\beta) - \xi(x,\alpha))/2 + \ln a)}$$
(cf. (1.6.8)). The general formula for $\{u_k\}$ can be written, e.g. in the following form. let $\chi = -\partial \ln y$. The Faà di Bruno polynomials are differential polynomials $P_k(\chi)$ of a variable χ, satisfying the recurrence equations:
$$(5.5.9) \quad P_0(\chi) = 1, P_{k+1}(\chi) = P_k'(\chi) + \chi P_k(\chi),$$

i.e. $P_k(\chi) = (\partial + \chi)^k 1$. For example, $P_0 = 1, P_1 = \chi, P_2 = \chi' + \chi^2$. Then $L = \partial + \sum_0^\infty (-1)^k \chi' P_k(\chi) \partial^{-k-1}$, i.e.

(5.5.10) $$u_k = \chi' P_k(\chi) \cdot (-1)^k .$$

If we leave here only the times $x_2 = y, x_3 = t$, the formula (5.5.8) will give a soliton for the KP-equation (5.2.8). If $\beta = -\alpha$ the solution is independent of y and becomes a solution of the KdV-equation.

CHAPTER 6. Hamiltonian Structure of the KP-Hierarchy.

6.1. Hamiltonian structure.

6.1.1. We shall discuss the Hamiltonian structure of Eqs. (5.2.2). The first structure was suggested by Watanabe [40] by analogy with that for the KdV-hierarchies (Chap. 3). In our note [41] the second structure was added. The full solution of this problem was obtained by Radul [42] who noticed that this pair of Hamiltonian structures is nothing more than the beginning of the infinite series of the Hamiltonian pairs. More exactly, the Hamiltonian pair of every KdV-hierarchy induces a pair in the phase space of the KP-hierarchy. Reduced to the submanifold $L_-^n = 0$, this pair transforms into the initial Hamiltonian pair.

As to the hierarchy of Eqs. (5.2.4) involving three independent variables, there are no grounds to distinguish one of them as an evolutionary time; it is more natural to consider these equations in a field theory. In Chaps. 17, 18 such a field theoretical Lagrange-Hamiltonian formalism will be built.

6.1.2. Let

$$L = \partial + u_{-1} + u_0 \partial^{-1} + u_1 \partial^{-2} + \ldots$$

(in comparison with 5.2.1 here is an additional term u_{-1}).
Let

$$L^n = \partial^n + v_{n-1} \partial^{n-1} + v_{n-2} \partial^{n-2} + \ldots.$$

As in 5.3.7 we can find expressions for $\{v_i\}$ in terms of $\{u_i\}$:

$$v_i = nu_{n-2-i} + Q_i(\{u_\alpha\}), \alpha < n-2-i, i = n-1, n-2, \ldots.$$

Now we shall construct all the elements of a Hamiltonian structure.

As in the Lie algebra $\mathfrak{E}$ we take differentiations $\partial_a = \sum_{i=-\infty}^{n-1} \sum_{j=0}^{\infty} a_i^{(j)} \partial /\partial v_i^{(j)}$, being in one-to-one corrrespondence with PDO

$$a = a_{n-1}\partial^{n-1} + a_{n-2}\partial^{n-2} + \ldots = \sum_{-\infty}^{n-1} a_i \partial^i \in R_{(-\infty, n-1)}.$$

The dual space Ω^1 consists of operators

$$X = \partial^{-n} X_{n-1} + \partial^{-n+1} X_{n-2} + \ldots = \sum_{-\infty}^{n-1} \partial^{-i-1} X_i \in R/R_{(-\infty, -n-1)}$$

(the sums are finite). The coupling is

$$\langle \partial_a, X \rangle = \int \operatorname{res} aX \, dx = \int \sum a_i X_i \, dx.$$

The module Ω^0 is $\tilde{a} = \{\tilde{f} = \int f(v) dx\}$ as usual. The mapping $d : \Omega^0 \to \Omega^1$ is, as can easily be seen,

$$d\tilde{f} = \delta f/\delta L^n = \sum_{-\infty}^{n-1} \partial^{-i-1} \delta f/\delta v_i.$$

The Hamiltonian mapping H^n is defined by

$$X \mapsto H^n(X) = (\widehat{L^n}X)_+ \widehat{L^n} - \widehat{L^n}(X\widehat{L^n})_+ \in R_{(-\infty, n-1)}, \widehat{L^n} = L^n - z^n.$$

We have
$$H^n(X) = H^{n(0)}(X) + z^n H^{n(\infty)},$$

where $H^{n(0)} = (L^n X)_+ L^n - L^n(XL^n)_+$ and $H^{n(\infty)}(X) = -[X_+, L^n] - [L^n, X]_+ = -[X_+, L^n_+] - [X_+, L^n_-] - [L^n_+, X]_+ - [L^n_-, X]_+ = [L^n_-, X_+]_- - [L^n_+, X_-]_+$.

Thus,

(6.1.3)
$$H^{n(0)}(X) = (L^n X)_+ L^n - L^n(XL^n),$$
$$H^{n(\infty)}(X) = [L^n_-, X_+]_- - [L^n_+, X_-]_+.$$

The form ω is

$$\omega(\partial_{H^n(X)}, \partial_{H^n(Y)}) = \langle H^n(X), Y \rangle.$$

6.1.4. Proposition. A relation

$$[\partial_{H^n(X)}, \partial_{H^n(Y)}] = \partial_{H^n([X,Y]_L + \partial_{H^n(X)} Y - \partial_{H^n(Y)} X)},$$

holds where

$$[X, Y]_L = (-X(\widehat{L^n}Y)_+ + (X\widehat{L^n})_- Y)_- - (X \leftrightarrow Y).$$

The proof is similar to that in 3.2.2. We have not used the fact that the operator L is differential. □

6.1.5. Proposition. The form ω is closed.
Proof: See 3.2.3. □

The correspondence between Hamiltonians and vector fields is, as usual, $\tilde{f} \mapsto \partial_{H^n(\delta f/\delta L^n)} = \partial_{\tilde{f}}$ and the Poisson bracket is

$$\{\tilde{f}, \tilde{g}\} = \partial_{\tilde{f}} \tilde{g} = \int \operatorname{res} H^n(\delta f/\delta L^n) \cdot \delta g/\delta L^n dx.$$

6.1.6. The Poisson bracket can be written in terms of $\{u_i\}$ instead of $\{v_i\}$. The relevant formulas are implicit, nevertheless we shall say a few words about them.

First we find the connection between variational derivatives $\delta f/\delta L^n$ and $\delta f/\delta L$.

$$\int \operatorname{res}(\delta f/\delta L^n) \delta L^n dx = \int \operatorname{res}(\delta f/\delta L^n) \sum_{i=0}^{n-1} L^{n-i-1}(\delta f/\delta L^n) L^i dx$$

which implies

$$\delta f/\delta L = \sum_{i=0}^{n-1} L^{n-i-1}(\delta f/\delta L^n) L^i \in R/R_{(-\infty,-2)}.$$

The mapping $A \mapsto \sum_{i=0}^{n-1} L^{n-i-1}AL^i = B$ is uniquely invertible. Indeed, if $B = \sum_{-1}^{N} b_i \partial^i$ then for the coefficients of the operator $A = \sum_{-\infty}^{N-n+1} a_i \partial^i$ recurrence equations

$$b_N = n \cdot a_{N-n+1}$$
$$b_{N-1} = n(n-1)/2 \cdot a'_{N-n+1} + nu_{-1}a_{N-n+1} + na_{N-n}$$
$$\ldots$$

arise. The coefficients $a_{N-n+1}, a_{N-n}, \ldots$ can be found in succession. We use the fact that the series in ∂^i break to the positive side; but since they can break at different places there is no analytical formula. In any case there is an inverse operator $A = \mathcal{F}^n(B)$, the coefficients of A being elements of the algebra $\mathcal{A}_u$. Thus, $\delta f / \delta L^n = \mathcal{F}^n(\delta f / \delta L)$. The Poisson bracket is

$$\{\tilde{f}, \tilde{g}\} = \int \operatorname{res} H^n(\mathcal{F}^n(\delta f/\delta L)) \cdot \mathcal{F}^n(\delta g/\delta L) dx \,.$$

It is more convenient to use the variables $\{v_i\}$.

In reduction to the submanifold $L_-^n = 0$ the Poisson bracket turns into that for the corresponding KdV-hierarchy.

6.2. Resolvent.

6.2.1. Extending the mapping H^n to a series in z^{-1} we shall find its kernel.

The formal series

$$T^n(z) = -\frac{1}{n} \sum_{-\infty}^{\infty} z^{-i-n} L^i \in R/R_{(-\infty, -n-1)}((z^{-1}))$$

is called the resolvent.

6.2.2. Lemma. $(\widehat{L^n T^n})_+ = 0$.
Proof:

$$R/R_- \ni \widehat{L^n T^n} = -\frac{1}{n} \sum_{-\infty}^{\infty} z^{-i-n} L^{i+n} + \frac{1}{n} \sum_{-\infty}^{\infty} z^{-i} L^i = 0 \Rightarrow (\widehat{L^n T^n})_+ = 0. \square$$

6.2.3. Proposition.
$$H^n(T^n(\varepsilon z)) = 0,$$

where ε is a root of 1; $\varepsilon^n = 1$.
Proof:
$$H^n(T^n(z)) = (\widehat{L^n}T^n)_+ \widehat{L^n} - \widehat{L^n}(T^n \widehat{L^n})_+ = 0.$$

If εz is substituted for z the operator $\widehat{L^n}$ remains the same. This is why $H^n(T^n(\varepsilon z)) = 0$. □

6.2.4. Proposition.
$$H^n(\partial^{-n}) = 0.$$

Proof: $H^n(\partial^{-n}) = (\widehat{L^n}\partial^{-n})_+ \widehat{L^n} - \widehat{L^n}(\partial^{-n}\widehat{L^n})_+ = \widehat{L^n} - \widehat{L^n} = 0$. □

6.2.5. Proposition. The kernel of the mapping
$$H^n : R/R_{(-\infty,-n-1)}((z^{-1})) \to R_{(-\infty,n-1)}((z^{-1}))$$
consists of linear combinations
$$\sum_{(\varepsilon)} c_\varepsilon T^n(\varepsilon z) + b\partial^{-n},$$

where c_ε and b are series in z^{-1}, $\sum\limits_{k_0}^{\infty} c_{\varepsilon,k} z^{-k}$ and $\sum\limits_{k_0}^{\infty} b_k z^{-k}$ with constant coefficients.

Proof: We start with lemmas:

6.2.6. Lemma. If $Y = \sum\limits_0^N y_k \partial^k$, $y_k \in \mathcal{A}$ and $[L^n, Y]_- = 0$ then $Y = y_0 = $ const.

Proof: In the expansion of the commutator $[L^n, Y]$ the coefficient of ∂^{N-i-1} contains the variables u_j with the smallest index $j = -i$; it is involved in the term $+iy'_N u_{-i} + Nu'_{-i}y_N$. This expression must vanish if $i \geq N$. This yields $y_N = 0$ unless $N = 0$ and $y'_N = 0$. □

6.2.7. Lemma. If
$$X(z) = \sum_{i_0}^{\infty} X_i z^{-i-1} = \sum_{i_0}^{\infty} \sum_{-\infty}^{\infty} X_{i\alpha} z^{-i-1}$$
$$\times \partial^\alpha \in R/R_{(-\infty,-n-1)}; H^n(X(z)) = 0,$$

and the constants in all $X_{i,-n}, X_{i,-n+1}, \ldots, X_{i,0}$ are zero then $X(z) = 0$.
Proof: The equation $H^n(X(z)) = 0$ can be written as a recurrence equation:

(6.2.8) $\qquad H^{n(0)}(X_i) + H^{n(\infty)}(X_{i+n}) = 0.$

Let X_{i_1} be the first nonvanishing term in a recurrence chain. Then $H^{n(\infty)}(X_{i_1}) = 0$, i.e. $[L^n, (X_{i_1})_+]_- - [L^n, (X_{i_1})_-]_+ = 0$, i.e. $[L^n, (X_{i_1})_+]_- = 0$ and $[L^n, (X_{i_1})_-]_+ = 0$. The first of these equalities implies $(X_{i_1})_+ = $ const, according to the previous lemma. The assumptions of the present lemma imply that the constant is zero, $(X_{i_1})_+ = 0$. The second of the equalities, together with 1.7.11, yields $(X_{i_1})_- = a\partial^{-n}$. the recurrence equation (6.2.8) gives, further, $H^{n(0)}(X_{i_1}) + H^{n(\infty)}(X_{i_1+n}) = 0$, i.e.

$$(L^n a\partial^{-n})_+ L^n - L^n (a\partial^{-n} L^n)_+ = -[L^n, (X_{i_1+n})_+]_- + [L^n, (X_{i_1+n})_-]_+.$$

Taking the projection on R_+ we obtain

$$[a, L^n]_+ = [L^n, (X_{i_1+n})_-]_+.$$

In the right-hand side stands an operator of the order $\leq n-2$, in the left-hand side there is a term $na'\partial^{n-1}$, hence $a' = 0$, and, according to the assumption, $a = 0$, i.e. $X_{i_1} = 0$. $\square$

Now we shall prove the proposition 6.2.5. We must prove that the linear combinations we speak of can have any set of constants in the coefficients of $\partial^{-n}, \ldots, \partial^0$; according to the second lemma these coefficients uniquely determine a solution. All the constants in $T^n(\varepsilon z)$ form the operator $-\frac{1}{n}\sum_{-n}^{\infty}(\varepsilon z)^{-i-n}\partial^i$. Let ε be a primitive root of 1. All the roots are $1, \varepsilon, \ldots, \varepsilon^{n-1}$. Denote $t_j = -\frac{1}{n}\sum_{-n}^{0}(\varepsilon^j z)^{-i-n}\partial^i, j = 0, \ldots, n-1$ and let $X(z)$ be an element of the kernel of the mapping H^n. Single out its constant part, $\sum_{-n}^{0} a_i(z)\partial^i$. We need to prove that the equation

$$\sum_{-n}^{0} c_j(z)t_j + b(z)\partial^{-n} = \sum_{-n}^{0} a_i(z)\partial^i$$

is always solvable with respect to the series c_j and b with constant coefficients. This yields a system of equations with the determinant

$$\begin{vmatrix} z^n, & z^{n-1}, & \ldots, & 1 \\ (\varepsilon z)^n, & (\varepsilon z)^{n-1}, & \ldots, & 1 \\ & \ldots & & \\ (\varepsilon^{n-1} z)^n, & (\varepsilon^{n-1} z)^{n-1}, & \ldots, & 1 \\ 1, & 0, & \ldots, & 0 \end{vmatrix} \neq 0.$$

The system is solvable. □

6.3. Hamiltonians of the KP-hierarchies.

6.3.1. We take as Hamiltonians the coefficients of the expansion in powers of z^{-1} of the expression

$$\int \text{res}\, T^n(z)dx = -\frac{1}{n}\sum_0^\infty z^{-i-n}\int \text{res}\, L^i dx.$$

6.3.2. *Proposition.* The relation

$$\frac{\delta}{\delta L^n}\int \text{res}\, L^r dx = \frac{r}{n}L^{r-n} \in R/R_{(-\infty,-n-1)}$$

holds. This can be written also as

$$\frac{\delta}{\delta L^n}\int \text{res}\, T^{(n)}(z)dx = -\frac{\partial}{\partial (z^n)}T^{(n)}(z).$$

The proof is the same as that for the propositions 3.5.2 and 3.5.3. □

6.3.3. *Proposition.* The Hamiltonians $\hat{f}_i = \int \text{res}\, L^i dx$ generate, in virtue of the structures $H^{n(0)}$ and $H^{n(\infty)}$ the vector fields

$$\frac{i}{n}\partial_{[L^n, L_+^{i-n}]}, -\frac{i}{n}\partial_{[L^n, L_+^{i-2n}]},$$

i.e. the equations

$$\dot{L}^n = -\frac{i}{n}[L_+^{i-n}, L^n], \dot{L}^n = \frac{i}{n}[L_+^{i-2n}, L^n]$$

which are equivalent to

$$\dot{L} = -\frac{i}{n}[L_+^{i-n}, L], \dot{L} = \frac{i}{n}[L_+^{i-2n}, L].$$

These equations differ from Eqs. (5.2.2) by unimportant contants.
Proof: For $H^{n(\infty)}$ this follows from

$$H^{n(\infty)}(d\tilde{f}_i) = ([L^n, L_+^{i-n}]_- - [L^n, L_-^{i-n}]_+)\cdot\frac{i}{n}$$

$$= ([L^n, L_+^{i-n}]_- + [L^n, L_+^{i-n}]_+)\cdot\frac{i}{n} = [L^n, L_+^{i-n}]\cdot\frac{i}{n}.$$

For $H^{n(0)}$ this follows from recurrence equations (6.2.8) for the resolvent. □

6.3.4. Proposition. The Hamiltonians $\tilde{f}_i = \int \operatorname{res} L^i\, dx$ are in involution with respect to all Poisson brackets.

Proof: This follows from the recurrence relations (6.2.8) in the same manner as in 3.6.3. □

Now we come to the reduction to the submanifold $u_{-1} = 0$. In $\delta f/\delta L^n = \sum_{-\infty}^{n-1} \partial^{-i-1} X_i$ the coefficient X_{n-1} becomes indefinite. The situation is similar to that in Chap. 3. Namely, this coefficient is not involved in $H^{n(\infty)}(\delta f/\delta L)$; besides, the last expression is an operator of the order not greater than $n-2$; thus, this Hamiltonian mapping can be restricted to the submanifold $u_{-1} = 0$ (i.e. $v_{n-1} = 0$). As to $H^{n(0)}$ the requirement that $H^{n(0)}(\delta f/\delta L)$ should be of an order not greater than $n-2$ yields the expression of X_{n-1} in terms of the other coefficients

$$(6.3.5) \qquad \operatorname{res}[L^n, \delta f/\delta L^n] = 0$$

as it can be easily seen. The expression (6.3.5) is given in detail:

$$\operatorname{res}\left[\sum_{-\infty}^{n} v_i \partial^i, \sum_{-\infty}^{n-1} \partial^{-j-1} X_j\right] = \sum_{-\infty}^{n-1} v_i X_i - \operatorname{res} \sum_{j=-\infty}^{n-1} \sum_{i=-\infty}^{n} \partial^{-j-1} X_j v_i \partial^i$$

$$= \sum_{-\infty}^{n-1} v_i X_i - \operatorname{res} \sum_{i=-\infty}^{-1} \sum_{j=-\infty}^{i} \sum_{\alpha=0}^{-j-1} \binom{-j-1}{\alpha} (X_j v_i)^{(\alpha)} \partial^{i-j-1-\alpha}$$

$$- \operatorname{res} \sum_{i=0}^{n} \sum_{j=0}^{i} \sum_{\alpha=0}^{\infty} \binom{-j-1}{\alpha} (X_j v_i)^{(\alpha)} \partial^{i-j-1-\alpha}$$

$$= \sum_{-\infty}^{n-1} v_i X_i - \sum_{i=-\infty}^{-1} \sum_{j=-\infty}^{i} \binom{-j-1}{i-j} (X_j v_i)^{(i-j)}$$

$$- \sum_{i=0}^{n} \sum_{j=0}^{i} (-1)^{i-j} \binom{i}{j} (X_j v_i)^{(i-j)} = -\sum_{m=1}^{m} \sum_{p=1}^{m} \binom{m}{p}$$

$$\times (X_{-m-1} v_{-m-1+p})^{(p)} - \sum_{m=0}^{n} \sum_{p=1}^{m} (-1)^p \binom{m}{p} (X_{m-p} v_m)^{(p)} = 0,$$

where $v_n = 1$ and $X_n = 0$. By integrating we obtain

$$\sum_{m=1}^{\infty} \sum_{p=1}^{m} \binom{m}{p} (X_{-m-1} v_{-m-1+p})^{(p-1)} + \sum_{m=0}^{n} \sum_{p=1}^{m} (-1)^p$$

$$\times \binom{m}{p}(X_{m-p}v_m)^{(p-1)} = \text{const}.$$

This gives the required expression for X_{n-1}. The constant can be taken as zero.

CHAPTER 7. Baker Function, τ-Function.

7.1. Dressing.

7.1.1. Constructing soliton solutions in Secs. 1.6 and 5.5, we represented the operator L of the n-th and 1st order as $L = \phi \partial^n \phi^{-1}$ and $L = \phi \partial \phi^{-1}$ respectively. This representation is called "dressing" of the operators ∂^n and ∂. Let $L = \partial + u_0 \partial^{-1} + u_1 \partial^{-2} + \ldots$ be written as

$$(7.1.2) \qquad L = \phi \partial \phi^{-1}, \phi = 1 + \sum_0^\infty w_i \partial^{-i-1}.$$

The series ϕ is formal. Eq. (7.1.2) implies expressions of all the $\{u_i\}, i = 0, 1, \ldots$ in terms of differential polynomials in $\{w_i\}$. For example $u_0 = -w_0', u_1 = -w_1' + w_0 w_0' + w_0^3$.

The series ϕ is determined up to the right multipliers $1 + \sum_0^\infty a_i \partial^{-i-1}, a = $ const.

7.1.3. Lemma. The expressions of $\{u_i\}$ in terms of $\{w_i\}$ have the form

$$u_i = -w_i' + Q_i(w_0, \ldots, w_{i-1})$$

where Q_i are differential polynomials. Proof: We have

$$\phi^{-1} = \sum_{\alpha=0}^\infty (-1)^\alpha (w_0 \partial^{-1} + w_1 \partial^{-2} + \ldots)^\alpha = 1 + \sum_0^\infty a_i \partial^{-i-1}.$$

The greatest number j which has $\{w_j\}$ involved in a_i is $j = i$, namely, $a_i = -w_i + R_i(w_0, \ldots, w_{i-1})$, R_i being differential polynomials. Now

$$L = \phi \partial \phi^{-1} = \left(1 + \sum_0^\infty w_i \partial^{-i-1}\right) \partial \left(1 + \sum_0^\infty (-w_j + R_j)\partial^{-j-1}\right)$$

$$= \partial + \sum_0^\infty (-w_i' + Q_i(\{w_j\}))\partial^{-i-1}$$

where Q_i contain only w_j with $j < i$. □

7.1.4. Lemma. Any differential polynomial in $\{w_i\}$ can be written as a differential polynomial in $\{u_i, w_i\}$, where $\{w_i\}$ are involved not in a differential form, i.e. $\{w_i\}$ are present but not their derivatives.

Proof: This is a simple corollary of the previous lemma. □

7.1.5. We can construct a new differential algebra $\mathcal{A}_w$ with generators $\{w_i\}$; Lemma 7.1.3 gives the embedding of differential algebras $\mathcal{A}_u \xhookrightarrow{i} \mathcal{A}_w$. If the generators $\{u_i, w_i\}$ are used (lemma 7.1.4) then by this embedding the image of $\mathcal{A}_u$ will consist of the differential polynomials without $\{w_i\}$.

7.1.6. Proposition. The vector fields ∂_m can be extended from the algebra $\mathcal{A}_u$ to the whole algebra $\mathcal{A}_w$ by

$$(7.1.7) \qquad \partial_m \phi = -L_-^m \phi.$$

They remain commutative.

Proof: Let ∂_m be defined on $\mathcal{A}_w$ by Eq. (7.1.7). Let us show that, being restricted to $\mathcal{A}_u$, they coincide with the former ∂_m;

$$\partial_m L = \partial_m(\phi \partial \phi^{-1}) = (\partial_m \phi) \cdot \partial \phi^{-1} - \phi \partial \phi^{-1} \cdot (\partial_m \phi) \cdot \phi^{-1}$$
$$= -L_-^m \phi \partial \phi^{-1} + \phi \partial \phi^{-1} L_-^m = [L, L_-^m] = [L_+^m, L].$$

Now we prove that they commute

$$\partial_m(\partial_n \phi_-) = -(\partial_m L_-^n)\phi - L_-^n(\partial_m \phi) = -[L_+^m, L^n]_- \cdot \phi + L_-^n L_-^m \phi$$
$$= [L_-^m, L^n]_- \phi + L_-^n \cdot L_-^m \phi = [L_-^m, L_-^n]_- + [L_-^m, L_+^n]_- \phi$$
$$+ L_-^n L_-^m \phi = L_-^m \cdot L_-^n \phi - [L_+^n, L^m]_- \phi = \partial_n(\partial_m \phi) \cdot \Box$$

7.1.8. Remark about the notion of the integral. We have an embedding of differential algebras $\mathcal{A}_u \xhookrightarrow{i} \mathcal{A}_w$ which commutes with the operator ∂, i.e.

$i\partial = \partial i$. This embedding induces a mapping of the linear spaces $\tilde{\mathcal{A}}_u \xrightarrow{\tilde{i}} \tilde{\mathcal{A}}_w$ making the diagram

$$\begin{array}{ccc} \mathcal{A}_u & \xrightarrow{i} & \mathcal{A}_w \\ \int\downarrow & & \downarrow\int \\ \tilde{\mathcal{A}}_u & \xrightarrow{\tilde{i}} & \tilde{\mathcal{A}}_w \end{array}$$

commutative. However $\tilde{i}$ is not an embedding, it can have a kernel. For example, the algebra $\mathcal{A}_u$ generated by u can be embedded into $\mathcal{A}_\varphi$ generated by φ according to the formula $u = \varphi'$. Then $\int u\,dx \xrightarrow{\tilde{i}} \int \varphi'\,dx = 0$. What are the consequences of this fact? When we apply these theories, we consider, instead of algebras of formal symbols $u_i^{(j)}$, classes of functions $\{u_i(x)\}$. The property $\int f'\,dx = 0$ is required. Now the question is: which of the differential polynomials in $\{u_i\}$ or in $\{v_i\}$, is to satisfy this property. For example, we consider either $\{u_i(x)\}$ or $\{v_i(x)\}$ to be rapidly decreasing. If v_i are such ones then so are u_i, but not vice versa. This is why it is important always to know in what sense the integral is understood.

7.2. Baker function.

7.2.1. Recall the designation $\xi(x,z) = \sum_1^\infty x_i z^i$. Till now the action of PDO on functions was not defined (except of the differential operators). Now we define the action of

$$\partial^k, k \in \mathbb{Z} \text{ and } \partial_m \text{ on } \exp\{\pm\xi(x,z)\}:$$
$$\partial^k \exp\{\pm\xi(x,z)\} = (\pm z)^k \exp\{\pm\xi(x,z)\}, \partial_m \exp\{\pm\xi(x,z)\}$$
$$= \pm z^m \exp\{\pm\xi(x,z)\}.$$

This implies $\sum X_i \partial^i e^{\pm\xi(x,z)} = \sum X_i(\pm z)^i e^{\pm\xi(x,z)}$. The series is, as usual, formal; $\partial_m(ae^{\pm\xi(x,z)}) = (\partial_m a \pm z^m a)e^{\pm\xi(x,z)}$.

7.2.2. Definition. The Baker function (also the Baker-Akhiezer function or wave function) is

$$w = \Phi e^{\xi(x,z)} = (1 + w_0 z^{-1} + w_1 z^{-2} + \ldots)e^{\xi(x,z)} = \hat{w}(z)e^{\xi(x,z)},$$

where Φ is the dressing operator from (7.1.2). w is determined up to a multiplier $1 + \sum_0^\infty a_i z^{-i-1}, a_i = \text{const}$.

7.2.3. Proposition. The Baker function satisfies the equations

$$Lw = zw,$$
$$\partial_m w = B_m w, (B_m = L_+^m).$$

Proof:

$$Lw = L\phi e^{\xi(x,z)} = \phi \partial e^{\xi(x,z)} = z\phi e^{\xi(x,z)} = zw.$$
$$\partial_m w = \partial_m(\phi e^{\xi(x,z)}) = (\partial_m \phi)e^{\xi(x,z)} + \phi \partial_m e^{\xi(x,z)}$$
$$= -L_-^m \phi e^{\xi(x,z)} + \phi \partial^m e^{\xi(x,z)} = (-L_-^m + L^m)\phi e^{\xi(x,z)} = L_+^m w. \square$$

7.2.4. Definition. If ϕ^* is the formal adjoint to the ϕ operator then

$$w^* = (\phi^*)^{-1} e^{-\xi(x,z)} = (1 + w_0^* z^{-1} + w_1^* z^{-2} + \ldots)e^{-\xi(x,z)} = \hat{w}^*(z)e^{-\xi(x,z)}$$

is called the adjoint Baker function.

7.2.5. Proposition. The adjoint Baker function satisfies the equations

$$L^* w^* = zw^*$$
$$\partial_m w^* = -B_m^* w^*$$

(the asterisk always denotes adjoint operator).
Proof: We have

$$L\phi = \phi \partial \Rightarrow \phi^* L^* = -\partial \phi^* \Rightarrow L^*(\phi^*)^{-1} = -(\phi^*)^{-1} \partial.$$

Proceed as in the previous proof. $\square$

7.3. Bilinear identity.

7.3.1. Denote

$$\operatorname{res}_z \sum a_i z^i = a_{-1}, \ \operatorname{res}_\partial \sum a_i \partial^i = a_{-1}.$$

7.3.2. Lemma. Let $P = \sum p_i \partial^i, Q = \sum q_i \partial^i$. Then

$$\operatorname{res}_z[(Pe^{xz}) \cdot (Qe^{-xz})] = \operatorname{res}_\partial PQ^*.$$

Proof: Take into account that for $a, b \in \mathcal{A}$ relations $\operatorname{res}_\partial a\partial^{-1}b = ab$, $\operatorname{res}_\partial a\partial^r b = 0, (r \neq -1)$ hold. We have

$$\operatorname{res}_z[(Pe^{xz}) \cdot (Qe^{-xz})] = \operatorname{res}_z(\sum p_i z^i \sum q_j(-z)^j) = \sum_{i+j=-1}(-1)^j p_i q_j,$$

$$\operatorname{res}_\partial(P \cdot Q^*) = \operatorname{res}_\partial \sum_{i,j} p_i \partial^i (-\partial)^j q_j = \sum_{i+j=-1}(-1)^j p_i q_j. \;\square$$

7.3.3. Proposition. The following "bilinear identity"

$$\operatorname{res}_z(\partial_1^{i_1} \ldots \partial_m^{i_m} w) \cdot w^* = 0$$

holds for any $(i_1, \ldots, i_m) = (i), i_j \geq 0$.

Proof: Since $\partial_m w = B_m w$, it is sufficient to prove this equality when $(i) = (i, 0, \ldots, 0), i \geq 0$.

$$\operatorname{res}_z(\partial^i w) \cdot w^* = \operatorname{res}_z(\partial^i \phi e^{\xi(x,z)})(\phi^*)^{-1} e^{-\xi(x,z)}$$
$$= \operatorname{res}_z(\partial^i \phi e^{xz})((\phi^*)^{-1} e^{-xz}) = \operatorname{res}_\partial \partial^i \phi \cdot \phi^{-1} = \operatorname{res}_\partial \partial^i = 0. \;\square$$

7.3.4. Remark. The identity 7.3.3 can be written in the symbolic form:

$$\operatorname{res}_z w(x', z) w^*(x, z) = 0$$

where $f(x')$ should be understood as an expansion in $x' - x$:

$$f(x') = \sum_{(i)} (x_1' - x_1)^{i_1} \ldots (x_m' - x_m)^{i_m} \partial^{(i)} f(x) / i_1! \ldots i_m!$$

(a formal series).

The bilinear identity can be formulated and proven in the dual form $\operatorname{res}_z w \cdot \partial^{(i)} w^* = 0$.

Now we formulate the converse, in a sense, theorem. Here $\{w_i\}$ are the functions of $\{x_i\}$ (not formal symbols).

7.3.5. Proposition. Let

$$w = (1 + \sum_0^\infty w_i z^{-i-1}) e^{\xi(x,z)}, w^* = (1 + \sum_0^\infty w_i^* z^{-i-1}) e^{-\xi(x,z)}$$

be formal series, $\{w_i, w_i^*\}$ be functions of some variables $\{x_j\}$. Let $\mathrm{res}_z(\partial^{(i)}w)w^* = 0$ for any multiindex $(i) = (i_1, \ldots, i_m), i_j \geq 0$. Then an operator $L = \partial + u_0\partial^{-1} + u_1^{-2} + \ldots$ can be constructed with coefficients depending on $\{x_j\}$ satisfying Eq. (5.2.2), w and w^* being its Baker and adjoint Baker functions.

Proof: Let

$$\phi = 1 + \sum_0^\infty w_i \partial^{-i-1}, \psi = 1 + \sum_0^\infty w_i^*(-\partial)^{-i-1}$$

Evidently, $w = \phi e^{\xi(x,z)}, w^* = \psi e^{-\xi(x,z)}$. Then

$$\mathrm{res}_\partial \partial^i \phi \psi^* = \mathrm{res}_z (\partial^i \phi e^{\xi(x,z)})(\psi e^{-\xi(x,z)}) = \mathrm{res}_z \partial^i w \cdot w^* = 0,$$

according to the assumptions. This is true for all $i \geq 0$. But $\phi\psi^* = 1 + X$ (where $X \in R_-$) implies $\mathrm{res}_\partial \partial^i X = 0$, from which $X = 0$ and $\phi\psi^* = 1$, i.e. $\psi = (\phi^*)^{-1}$.

Further, let $L = \phi\partial\phi^{-1}, B_m = L_+^m$. Let us show that $\partial_m L = [B_m, L]$. It is sufficient to show that $\partial_m \phi = -L_-^m \phi$ (see 7.1.6). We have

$$((\partial_m \phi) + L_-^m \phi)e^{\xi(x,z)} = (\partial_m \cdot \phi - \phi \cdot \partial_m + L_-^m \phi)e^{\xi(x,z)}$$
$$= (\partial_m \cdot \phi - \phi \partial^m + L_-^m \phi)e^{\xi(x,z)} = (\partial_m \cdot \phi - L^m \phi + L_-^m \phi)e^{\xi(x,z)}$$
$$= (\partial_m - L_+^m)\phi e^{\xi(x,z)}$$

Then $\mathrm{res}_z \partial^i (\partial_m \phi - L_+^m \phi)e^{\xi(x,z)} \cdot \psi e^{-\xi(x,z)} = 0$ (which is true according to the assumption) implies

$$\mathrm{res}_z \partial^i ((\partial_m \phi) + L_-^m \phi)e^{\xi(x,z)} \cdot \psi e^{-\xi(x,z)} = 0$$

and

$$\mathrm{res}_\partial \partial^i ((\partial_m \phi) + L_-^m \phi)\psi^* = 0, \mathrm{res}_\partial \partial^i ((\partial_m \phi) + L_-^m \phi)\phi^{-1} = 0.$$

This yields $((\partial_m \phi) + L_-^m \phi)\phi^{-1} = 0$ and $\partial_m \phi + L_-^m \phi = 0.\square$

A proposition can be formulated that is converse to 7.3.3 in a more direct sense, i.e. in the language of identities between differential polynomials.

7.3.6. *Proposition.* Let $\mathcal{A}_w$ be a differential algebra with generators $\{w_i\}, i = 0, 1, \ldots$ and some commuting differentiations $\{\partial_m\}, m =$

$1, 2, \ldots, \partial_1 = \partial$. Let $\phi = 1 + \sum_0^\infty w_i \partial^{-i-1}$ and $(\phi^*)^{-1} = 1 + \sum_0^\infty w_i^* \partial^{-i-1}$. The coefficients $\{w_i^*\}$ can be expressed as differential polynomials in $\{w_i\}$; thus, they belong to $\mathcal{A}_w$. Finally, let $L = \phi \partial \phi^{-1}$. Then the differential ideal generated by all the coefficients of the operators $\partial_m L - [L_+^m, L]$ coincides with the ideal generated by all the expressions $\operatorname{res}_z(\partial^{(i)} w) w^*$, where $w = \phi e^{\xi(x,z)}, w^* = (\phi^*)^{-1} e^{-\xi(x,z)}$ and (i) is any non-negative multiindex.

We give no proof; it can be obtained by a suitable re-formulation of the proofs of 7.3.3 and 7.3.5.

7.4. Riccati equation.

7.4.1. In this section, w, which depend only on x, will attract our attention. Let
$$\chi = w'/w = z + \hat\chi = z + \sum_{-1}^\infty \chi_i z^{-i-1}$$
where w is the Baker function. We shall write an equation for χ. We consider two cases: one of the KP-hierarchy, $L = \partial + \sum_{-1}^\infty u_i \partial^{-i-1}$ and one of a KdV-hierarchy, $L_n = \sum_0^n u_i \partial^i, u_n = 1$.

We start with the latter. Recall the definition of the Faà di Bruno polynomials (5.5.9): $P_0(\chi) = 1, P_{i+1}(\chi) = (\partial + \chi) P_i(\chi)$

7.4.2. **Prosposition.** If w is the Baker function of the operator L_n, i.e. $L_n w = z^n w$, then χ satisfies the equation
$$\sum_0^n u_k P_k(\chi) = z^n$$
which is called the Riccati equation.

Proof: Let us check by induction the formula

(7.4.3) $\qquad\qquad w^{(k)} = P_k(\chi) w.$

It is trivial for $k = 0$. Let it be true for k. Then

$$w^{(k+1)} = \partial(P_k(\chi) w) = P_k'(\chi) w + P_k(\chi) \chi w = ((\partial + \chi) P_k(\chi)) w = P_{k+1}(\chi) w.$$

This proves the formula. Now substitute this expression into $L_n w = z^n w$ and we obtain the required equation.

Note that if $\chi(z)$ is a solution of the Riccati equation than so is $\chi(\varepsilon z)$, where $\varepsilon^n = 1$. There are n roots ε, the set of all $\chi(\varepsilon z)$ exhausts all the solutions.

7.4.4. Corollary. The coefficients $\{\chi_i\}$ of the expression $\chi = w'/w$ are differential polynomials in $\{u_k\}$.

(The meaning of this corollary becomes clear if we remember that the coefficients of w are not differential polynomials in $\{u_k\}$).

Proof: Substituting the series $\chi = z + \sum_{-1}^{\infty} \chi_i z^{-i-1}$ into the Riccati equation we obtain a recurrence relation for $\{\chi_i\}$ of the form $\chi_i = Q_i(\{u_k\}, \chi_0, \ldots, \chi_{i-1})$, where Q_i are differential polynomials.

In order to pass to the KP-hierarchy we must extend the definition of $P_k(\chi)$ to the negative k:

(7.4.5) $$P_{-k}(\chi) = (\partial + \chi)^{-k} 1, \, k = 1, 2, \ldots$$

where $(\partial + \chi)^{-k}$ is understood as a formal series in χ^{-1}, e.g.

$$P_{-1}(\chi) = (\partial + \chi)^{-1} 1 = (\chi^{-1} - \chi^{-1}\partial \chi^{-1} + \chi^{-1}\partial \chi^{-1}\partial \chi^{-1} - \ldots)1$$
$$= \chi^{-1} + \chi^{-2}\chi' + \ldots .$$

If a series in z^{-1} is substituted for χ this will yield a series in z^{-1} every coefficient of which requires a finite number of operations for its computation.

7.4.6. Lemma. The recurrence relation

$$(\partial + \chi)P_{-k-1}(\chi) = P_{-k}(\chi), \, k \in \mathbb{Z}$$

holds.

The proof is obvious. □

Recall that while, in general, the action of ∂^{-k-1} on functions is not defined, the action of ∂^{-k-1} on the Baker function is defined by the fact that $w = \hat{w}e^{\xi(x,z)}, \partial^{-k-1}e^{\xi(x,z)} = z^{-k-1}e^{\xi(x,z)}$, and there is a commutation relation of ∂^{-k-1} with a function.

7.4.7. Lemma.

$$\partial^{-k-1}w = P_{-k-1}(\chi)w, \, k = 0, 1, 2, \ldots$$

which generalizes the formula (7.4.3).

Proof: Let $\partial^{-k}w = P_{-k}(\chi)w$ be already proved. Then

$$(P_{-k-1}(\chi)w)' = (P_{-k-1}(\chi))'w + P_{-k-1}(\chi) \cdot w'$$
$$= ((\partial + \chi)P_{-k-1}(\chi))w = -P_{-k}(\chi)w = \partial^{-k}w.$$

Together with the fact that constants in both sides of the formula which we try to prove are equal ($= z^{-k-1}$) this completes the proof. □

7.4.8. Proposition. If w is the Baker function of the operator $L = \partial + \sum_{-1}^{\infty} u_k \partial^{-k-1}$, i.e. $Lw = zw$, then $\chi = w'/w$ satisfies the Riccati equation

$$\chi + \sum_{-1}^{\infty} u_k P_{-k-1}(\chi) = z.$$

Proof: This immediately follows from the previous lemma and the equation for w. □

7.4.9. Corollary. The coefficients $\{\chi_k\}$ can be expressed as differential polynomials in $\{u_k\}$.

7.4.10. The Riccati equation appears also in another context. Represent the operator $L_n - z^n$ in a multiplicative form

$$L_n - z^n = (\partial - \varphi_n(z))\ldots(\partial - \varphi_1(z))$$

where $\varphi_i(z) = \sum_{\alpha=-1}^{\infty} \varphi_{i\alpha} z^{-\alpha}, \varphi_{i,-1} = 1$.

7.4.11. Proposition. The function $\varphi_1(z)$ satisfies the Riccati equation

$$\sum_{0}^{n} u_k P_k(\varphi_1) = z^n.$$

For simplicity of writing substitute u_0 for $u_0 - z^n$. We then apply induction. Let this already be proved for the operator $L_n = (\partial - \varphi_n)\ldots(\partial - \varphi_1)$ of the n-th order, i.e. $\sum_{0}^{n} u_k P_k(\varphi_1) = 0$ (for $n = 1$ this is evident). Let $L_{n+1} = (\partial - \varphi_{n+1})L_n = \sum_{0}^{n+1} \tilde{u}_k \partial^k$, then

$$\tilde{u}_{k+1} = u_k - \varphi_{n+1}u_{k+1} + u'_{k+1}, k = -1, 0, 1, \ldots, n$$

where $u_{-1} = u_{n+1} = 0$. We have

$$\sum_0^{n+1} \tilde{u}_k P_k(\varphi_1) = \sum_0^{n+1} (u_{k-1} - \varphi_{n+1} u_k + u'_k) P_k(\varphi_1)$$

$$= \sum_0^{n+1} u_{k-1} P_k(\varphi_1) + \sum_0^{n+1} u'_k P_k(\varphi_1) = \sum_0^{n+1} u_{k-1} P_k(\varphi_1) - \sum_0^{n+1} u_k P'_k(\varphi_1)$$

$$= \sum_0^{n+1} u_{k-1} P_k(\varphi_1) - \sum_0^{n+1} u_k P_{k+1}(\varphi_1) + \sum_0^n u_k \varphi_1 P_k(\varphi_1) = 0. \square$$

Note that φ_1 can be any solution of the Riccati equation since the substitution $z \mapsto \varepsilon z$ conserves the operator $L_n - z^n$ and sends a solution $\varphi_1(z)$ of the Riccati equation to another solution. The operator $L_n - z^n$ can be represented in the form $\mathcal{D}_{n-1}(\partial - \chi)$, where $\mathcal{D}_{n-1}$ is a $(n-1)$-th order operator and χ any solution of the Riccati equation.

7.4.12. Besides $\chi = w^{-1}/w$ there is another combination of the Baker functions which can be expressed in terms of differential polynomials in $\{u_k\}$. This is

$$S = ww^*.$$

Let us write a differential equation for this function too. We consider now the case of $L_n = \sum_0^n u_k \partial^k$. We have $L_n w = z^n w$ and $L_n^* w^* = z^n w^*$. Denote $L_n^* = \sum_0^n (-\partial)^k u_k = \sum_0^n v_k \partial^k$ whence

(7.4.13) $$v_k = \sum_{l=k}^n (-1)^l \binom{l}{k} u_l^{(l-k)}.$$

7.4.14. *Proposition.* The quantity $S = ww^*$ satisfies the equation

$$\sum_0^n v_k (\partial - \chi)^k S = z^n S.$$

Proof: First we prove the formula

(7.4.15) $$w^{*(k)} = w^{-1}(\partial - \chi)^k S.$$

For $k = 0$ this is evident. Let this be true for k. Then

$$w^{*(k+1)} = \partial(w^{-1}(\partial - \chi)^k S) = -w' w^{-2}(\partial - \chi)^k S$$
$$+ w^{-1}(\partial - \chi)^{k+1} S + w^{-1} \chi (\partial - \chi)^k S = w^{-1}(\partial - \chi)^{k+1} S$$

as required. Substituting this into $L^*w^* = z^n w^*$ we obtain the statement. □

It can be verified that this equation does not give the needed expansion of S into a series yet: calculation of coefficients requires integration. Now we shall do this integration in a general form.

7.4.16. Lemma. A relation

$$P_k(\chi) = \sum_{l=0}^{k} (-1)^l \binom{k}{l} \partial^{k-l} (\partial - \chi)^l$$

holds (in the right-hand side there is a differential operator; the lemma, in particular, states that this differential operator is of zero order, i.e. all the terms cancle except the zero order term).

Proof: It is easy to apply induction, using $P_{k+1} = P_k' + \chi P_k = \partial \circ P_k - P_k(\partial - \chi)$.□

7.4.17. Proposition. The quantity S satisfies the equation:

$$\sum_{\substack{\alpha,\beta,\gamma \geq 0 \\ \alpha+\beta+\gamma \leq n-1}} (-1)^{\alpha+\beta} \frac{(\alpha+\beta+\gamma+1)!}{\alpha!\beta!(\gamma+1)!} \partial^\gamma \circ u_{\alpha+\beta+\gamma+1}^{(\alpha)} (\partial - \chi)^\beta S = c(z).$$

(The operator standing to the left of S acts on S).

Proof: We multiply the Riccati equation 7.4.2 by S and subtract Eq. (7.4.14):

$$\sum_{0}^{n} u_k P_k(\chi) S - \sum_{k=0}^{n} v_k (\partial - \chi)^k S = 0.$$

Use 7.4.16 and 7.4.13:

$$0 = \sum_{k=0}^{0} \sum_{l=0}^{k} u_k (-1)^l \binom{k}{l} \partial^{k-1}(\partial-\chi)^l S - \sum_{k=0}^{n} \sum_{l=k}^{n} (-1)^l \binom{l}{k}$$

$$u_l^{(l-k)}(\partial-\chi)^k S = \sum_{k=0}^{n} \sum_{l=0}^{k} u_k (-1)^l \binom{k}{l} \partial^{k-l}(\partial-\chi)^l S -$$

$$\sum_{k=0}^{n} \sum_{l=0}^{k} (-1)^k \binom{k}{l} u_k^{(k-l)} (\partial-\chi)^l S.$$

In the first term, transfer ∂^{k-l} to the left using

$$f\partial^{k-l} = \sum_{\alpha=0}^{k-l}(-1)^\alpha \binom{k-l}{\alpha} \partial^{k-l-\alpha} \circ f^{(\alpha)}.$$

One of the terms ($\alpha = k-l$) cancels with the second term; we then have

$$\sum_{k=1}^{n}\sum_{l=0}^{k-1}\sum_{\alpha=0}^{k-l-1} (-1)^{\alpha+l} \frac{k!}{l!\alpha!(k-l-\alpha)!} \partial^{k-l-\alpha-1} \circ u_k^{(\alpha)}(\partial - \chi)^l S = 0.$$

This can be integrated

$$\sum_{k=1}^{n}\sum_{l=0}^{k-1}\sum_{\alpha=0}^{k-l-1} (-1)^{\alpha+l} \frac{k!}{l!\alpha!(k-l-\alpha)!} \partial^{k-l-\alpha-1} \circ u_k^{(\alpha)}(\partial - \chi)^l S = c(z),$$

$c(z)$ is a constant series. This is equivalent to the equation we set out to prove. □

7.4.18. Corollary. S can be expanded into a series with coefficients being differential polynomials in $\{u_k\}$.

Proof: The series $S = 1 + \sum_{0}^{\infty} S_i z^{-i}$ must be substituted into the equation and the recurrency equations will appear from which everything can be obtained. □

7.4.19. Exercise. Obtain an equation similar to 7.4.17 for $S = ww^*$ where w and w^* are Baker functions for the KP-hierarchy.

The author does not know the answer. The equation similar to 7.4.14 can be obtained, namely

$$-(\partial - \chi)S + \sum_{-1}^{\infty} v_k (\partial - \chi)^{-k-1} S = 0$$

where $(\partial - \chi)^{-k} = \sum_{\alpha=0}^{\infty} \binom{k+\alpha-1}{\alpha}(\partial - \hat\chi)^\alpha z^{-\alpha-k}$, $v_k = \sum_{l=0}^{k}(-1)^{k-1} \binom{k}{k-l} u_l^{(k-l)}$. The last step is hindered by the fact that it is difficult to find a formula similar to 7.4.16 for negative k.

Nevertheless, the fact that $S = ww^*$ can be represented as $1 + \sum_{0}^{\infty} S_k z^{-k-1}$, where S_k are differential polynomials in $\{u_k\}$, we can prove easily.

7.4.20. Proposition.
$$S = ww^* = \operatorname{res}_\partial \sum_{-\infty}^{\infty} L^i z^{-i-1}.$$

This means that S is the residue of a resolvent (6.2.1).

Proof: If $S = 1 + \sum_0^\infty S_k z^{-k-1}$ then
$$S_k = \operatorname{res}_z z^k ww^* = \operatorname{res}_\partial \phi \partial^k \phi^{-1} = \operatorname{res}_\partial L^k$$
and $S = 1 + \operatorname{res}_\partial \sum_0^\infty L^k z^{-k-1} = \operatorname{res}_\partial \sum_{-\infty}^\infty L^k z^{-k-1}$. □

7.5. List of useful formulas for the Faà di Bruno polynomials.

We produce a few formulas of the type of 7.4.16 which can be easily proved (see [59]).

(7.5.1) $$(\partial + \chi)^k = \sum_{\alpha=0}^k \binom{k}{\alpha} P_{k-\alpha}(\chi) \partial^\alpha.$$

(7.5.2) $$(\partial - \chi)^k = \sum_{\alpha=0}^k (-1)^{k-\alpha} \binom{k}{\alpha} \partial^\alpha \circ P_{k-\alpha}(\chi).$$

(7.5.3) $$P_k(\chi) = \sum_{\alpha=0}^k (-1)^{k-\alpha} \binom{k}{\alpha} \partial^\alpha (\partial - \chi)^{k-\alpha}.$$

(7.5.4) $$P_k(\chi) = \sum_{\alpha=0}^k (-1)^\alpha \binom{k}{\alpha} (\partial + \chi)^{k-\alpha} \partial^\alpha.$$

(7.5.5) $$\partial_k = \sum_{\alpha=0}^k \binom{k}{\alpha} P_\alpha(\chi)(\partial - \chi)^{k-\alpha}$$

(7.5.6) $$\partial_k = \sum_{\alpha=0}^k (-1)^\alpha \binom{k}{\alpha} (\partial + \chi)^{k-\alpha} \circ P_\alpha(\chi)$$

(7.5.7) $$P_k(\chi) = \sum_{i_1 + 2i_2 + \ldots + l i_l = k} \frac{k!}{i_1! \ldots i_l!} \left(\frac{\chi}{1!}\right)^{i_1} \left(\frac{\chi'}{2!}\right)^{i_2} \cdots \left(\frac{\chi^{(l-1)}}{l!}\right)^{i_l}, [43]$$

7.6. Resolvent and Baker function.

7.6.1. In 7.4.20 we already have found a connection between the resolvent and the Baker function for the KP-hierarchy. Now we discuss KdV-hierarchies. The connection is much more profound: not only the residue of the resolvent but even the resolvent itself can be expressed in terms of the Baker function.

7.6.2. Proposition. The resolvent (1.7.2) can be written as

$$T_\varepsilon = \frac{(\varepsilon z)^{-n+1}}{n} w(\varepsilon z) \partial^{-1} w^*(\varepsilon z).$$

Proof: It suffices to prove this formula for $\varepsilon = 1$ since T_ε arises from T_1 by the substitution $z_1 \mapsto \varepsilon z$. First we note that the coefficients of the double expansion of the written expression in ∂^{-1} and in z^{-1} belong to $\mathcal{A}_u$, i.e. are differential polynomials in $\{u_i\}$. Indeed,

(7.6.3)
$$w\partial^{-1}w^* = \sum_{\alpha=0}^{\infty} \partial^{-1-\alpha} w^{(\alpha)} w^* = \sum_{\alpha=0}^{\infty} \partial^{-1-\alpha} P_\alpha(\chi) w w^*$$
$$= \sum_{\alpha=0}^{\infty} \partial^{-1-\alpha} P_\alpha(\chi) S$$

(see (7.4.3)), and χ and S can be expanded in z^{-1} with the coefficients in $\mathcal{A}_u$ (7.4.4 and 7.4.18).

Now we verify the fact that the written expression is a resolvent, i.e. it belongs to the kernel of the Adler mapping. It is sufficient to show that the products of this expression by $\hat{L} = L - z^n$ to the left and to the right are purely differential operators.

7.6.4. Lemma. If D is any differential operator and f a function (element of $\mathcal{A}$ or $\mathcal{A}((z^{-1}))$) then $D \circ f = (Df) + D_1 \partial$ and $f \circ D = (D^*f) + \partial D_2$, where D_1 and D_2 are differential operators.

Proof of the lemma: The first immediately follows from the rule of commutation of the operator ∂ and the operator of multiplication on functions. Further, $D^* \circ f = (D^*f) + D_3 \partial$. Taking the conjugate equality we get the second relation. □

Now continue our proof of the proposition.

$$\hat{L} \circ w \partial^{-1} w^* = (\hat{L}w) \cdot \partial^{-1} w^* + D_1 \partial \cdot \partial^{-1} w^*.$$

We have $\hat{L}w = 0$, according to the definition of the Baker function. The second term is a differential operator. Similarly

$$w\partial^{-1}w^* \circ \hat{L} = w\partial^{-1}(\hat{L}^*w^*) + w\partial^{-1}\partial D_2 = wD_2.$$

It remains to prove that the resolvent 7.6.2 is identical with the resolvent (1.7.2). We know that a resolvent is uniquely determined by constants in the coefficients of its expansion in ∂^{-1} and z^{-1}. Taking into account (7.6.3) we can write all the constant terms. We have $P_\alpha(\chi) = \chi^\alpha + \ldots$ and $\chi = z + \ldots$ (the terms without constants are denoted by dots). Hence

$$T_\varepsilon = n^{-1}(\varepsilon z)^{-n+1} \sum_{\alpha=0}^\infty \partial^{-1-\alpha}(\varepsilon z)^\alpha + \ldots.$$

It is easy to see that the resolvent (1.7.2) has just the same constants. □

7.6.5. We give another form of the resolvent, in terms of χ and S. We use the following designations. The quantity $(\partial - \chi)^{-1}$ can be considered as a series in positive powers of ∂ and in negative powers of χ, or vice versa. Let

$$(\partial - \chi)_p^{-1} = -\chi^{-1} - \chi^{-1}\partial\chi^{-1} - \chi^{-1}\partial\chi^{-1}\partial\chi^{-1} - \ldots$$
$$(\partial - \chi)_n^{-1} = \partial^{-1} + \partial^{-1}\chi\partial^{-1} + \partial^{-1}\chi\partial^{-1}\chi\partial^{-1} + \ldots$$

The obvious identity $(\partial - \chi) \circ w = w\partial$ implies $w\partial^{-1} = (\partial - \chi)_n^{-1} \circ w$. This equation is obtained by multiplication by $(\partial - \chi)_n^{-1}$ to the left and by ∂^{-1} to the right. Multiplication by $(\partial - \chi)_n^{-1}$ has meaning in the class of series in ∂^{-1} with functional coefficients. Under the functions we understand here one-way infinite series in z^{-1}, and also $\exp(zx)$ (no more than one such multiplier). Note that from $(\partial - \chi) \circ w = w\partial$ it does not follow that $(\partial - \chi)_p^{-1} \circ w = w\partial^{-1}$, and, what is more, $(\partial - \chi)_p^{-1} \circ w$ has no meaning, since infinitely many terms must be added in the calculation of a coefficient in any ∂^{-r}. We understand $(\partial - \chi)_p^{-1}$ as a series $\sum_0^\infty P_k z^{-k-1}$, where P_k are some differential operators (not to confuse with the Faà di Bruno polynomials having the same designation). This can be multiplied by a one-way infinite series in z^{-1}; w is not such a series due to $\exp(xz)$.

7.6.6. *Proposition.* The resolvent (1.7.2) can be written as

$$T_\varepsilon = \frac{(\varepsilon z)^{1-n}}{n}(\partial - \chi)_n^{-1} S.$$

Proof: This follows from 7.6.2 and $w\partial^{-1} = (\partial - \chi)_n^{-1} w$. □

This expression for a resolvent was suggested by Cherednik [105].

7.6.7. Proposition. The series in z^{-1}:

$$K = \frac{(\varepsilon z)^{1-n}}{n}[(\partial - \chi)_n^{-1} - (\partial - \chi)_p^{-1}] \circ S = \frac{(\varepsilon z)^{1-n}}{n} \sum_{k=-\infty}^{\infty} \partial^{-1}(\partial\chi^{-1})^k \circ S$$

(two-way infinite) with coefficients which are one-way infinite series in ∂^{-1} has properties $(L - z^n)K = 0$ and $K(L - z^n) = 0$. Therefore it coincides with

(7.6.8) $$K = \frac{(\varepsilon z)^{1-n}}{n} \sum_{r=-\infty}^{\infty} (\varepsilon z)^{-r-n} L^{r/n}.$$

Its integral part coincides with the resolvent, and the differential one is a generator for the P-operators, together with L giving Lax pairs.

Proof: According to a note at the end of 7.4.11, $L - z^n$ can be written as $D_{n-1}(\partial - \chi)$ which implies $(L - z^n)K = n^{-1}(\varepsilon z)^{1-n} D_{n-1}(1 - 1)S = 0$. In order to prove the second equality we denote $\chi^* = w^{*'}/w^*$. The function w^* satisfies the equation $L^* w^* = z^n w^*$. Hence, denoting $L^* = \sum_0^n v_k \partial^k$, χ^* satisfies the Riccati equation $\sum_0^n v_k P_k(\chi^*) = z^n$. Factorizing the operator $L^* = D_{n-1}(\partial - \chi^*)$ and noting that

$$(\partial - \chi)_p^{-1} S = S(\partial + \chi^*)_p^{-1}, (\partial - \chi)_n^{-1} S = S(\partial + \chi^*)_n^{-1}$$

(the proof of this fact: $(\partial - \chi) w w^* = w w^*(\partial - \chi) + \chi w w^* + \chi^* w w^* = w w^*(\partial + \chi^*)$, both the parts can be multiplied by $(\partial - \chi)_p^{-1}$ to the left and by $(\partial + \chi^*)_p^{-1}$ to the right, these multipliers being considered as one-way infinite series in z^{-1} with operator coefficients, and by $(\partial - \chi)_n^{-1}$ and $(\partial + \chi^*)_n^{-1}$, respectively, considering them as one-way infinite series in ∂^{-1} with coefficients which are series in z^{-1}) we obtain

$$(K\hat{L})^* = D_{n-1}(\partial - \chi^*)[(-\partial + \chi^*)_n^{-1} - (-\partial + \chi^*)_p^{-1}]S = 0;$$

passing to conjugate operators we obtain the second relation.

7.7. τ-function.

7.7.1. Proposition. If

$$w = \hat{w}e^{\xi(x,z)} = \left(1 + \sum_0^\infty w_i z^{-i-1}\right) e^{\xi(x,z)}, x = x_1, x_2, \ldots.$$

is the Baker function of the KP-hierarchy, then a function $\tau(x_1, x_2, \ldots)$ exists such that

(7.7.2) $\quad \hat{w}(x,z) = \tau\left(x_1 - \frac{1}{z}, x_2 - \frac{1}{2z^2}, x_3 - \frac{1}{3z^3}, \ldots\right) / \tau(x_1, x_2, \ldots)$

(the expression must be expanded in z^{-1}).

Proof: We give a proof close to the original one [37]. We start from a lemma:

7.7.3. Lemma. If $f(z) = 1 + \sum_0^\infty a_i z^{-i-1}$ is a formal series, then

$$\operatorname{res}_z f(z) \cdot z^{-1}\left(1 - \frac{z}{t}\right)^{-1} = f(t), \ \operatorname{res}_z f(z)(1 - z/t)^{-1} = t(f(t) - 1).$$

If $f(z) = \sum_{-\infty}^\infty a_i z^{-i}$, then

$$\operatorname{res}_z (t^{-1}(1 - z/t)^{-1} + z^{-1}(1 - t/z)^{-1}) f(z) = f(t).$$

Proof of the lemma: We find

$$\operatorname{res}_z f(z) z^{-1}(1 - z/t)^{-1} = \operatorname{res}_z \left(1 + \sum_0^\infty a_i z^{-i-1}\right) \sum_0^\infty (z/t)^j z^{-1}$$

$$= 1 + \sum_0^\infty a_j t^{-j-1} = f(t),$$

$$\operatorname{res}_z f(z)(1 - z/t)^{-1} = \operatorname{res}_z \left(1 + \sum_0^\infty a_i z^{-i-1}\right) \sum_0^\infty (z/t)^j$$

$$= \sum_0^\infty a_i t^{-i} = t(f(t) - 1),$$

and

$$\operatorname{res}_z \sum_{-\infty}^\infty a_i z^{-i} \left(\sum_1^\infty z^{j-1} t^{-j} + \sum_1^\infty t^{j-1} z^{-j}\right) = \sum_1^\infty a_j t^{-j}$$

$$+ \sum_1^\infty a_{-j+1} t^{j-1} = f(t). \square$$

Let $G(t)$ be an operator acting on functions $f(x,z)$:

$$G(t)f(x,z) = f\left(x, -\frac{1}{t}, x_2 - \frac{1}{2t^2}, x_3 - \frac{1}{3t^3}, \ldots, z\right).$$

According to the bilinear identity (7.2.3) we have

$$\operatorname{res}_z w(x,z) G(t) w^*(x,z) = 0.$$

This residue is understood as a formal series in t^{-1}, i.e. all its terms vanish. Taking into account that $\exp(z/t + z^2/2t^2 + \ldots) = (1 - z/t)^{-1}$ we have

$$\operatorname{res}_z \hat{w}(x,z) G(t) \hat{w}^*(x,z) (1 - z/t)^{-1} = 0$$

which implies, according to the lemma,

$$t\{\hat{w}(x,t) G(t) \hat{w}^*(x,t) - 1\} = 0$$

and

(7.7.4) $$\hat{w}(x,t)^{-1} = G(t) \hat{w}^*(x,t).$$

Similarly

$$\operatorname{res}_z w(x,z) G(t_1) G(t_2) w^*(x,z) = 0$$

i.e.

$$\operatorname{res}_z \hat{w}(x,z) G(t_1) G(t_2) \hat{w}^*(x,z) \left(1 - \frac{z}{t_1}\right)^{-1} \left(1 - \frac{z}{t_2}\right)^{-1} = 0.$$

This implies

$$\operatorname{res}_z \hat{w} G(t_1) G(t_2) \hat{w}^* t_1^{-1} \left(1 - \frac{z}{t_1}\right)^{-1} = \operatorname{res}_z \hat{w} G(t_1) G(t_2)$$

$$\hat{w}^* t_2^{-1} \left(1 - \frac{z}{t_2}\right)^{-1}.$$

According to the lemma:

$$\hat{w}(x,t_1) G(t_1) G(t_2) \hat{w}^*(x,t_1) = \hat{w}(x,t_2) G(t_1) G(t_2) \hat{w}^*(x,t_2).$$

Eliminating $\hat{w}^*$ from (7.7.4) we obtain

$$\hat{w}(x,t_1)/(G(t_2)\hat{w}(x,t_1)) = \hat{w}(x,t_2)/(G(t_1)\hat{w}(x,t_2)).$$

We denote $\ln \hat{w}(x,t) = f(x,t)$. Then

$$(1 - G(t_2))f(x,t_1) = (1 - G(t_1))f(x,t_2).$$

It is rather clear from this that a function $\varphi(x)$ of $\{X_k\}$ must exist such that $f(x,t) = (1 - G(t))\varphi(x)$. However this cannot be proved immediately since there is no inverse operator $(1 - G(t))^{-1}$. Rewrite the last equality as

$$f(x,z) - G(t)f(x,z) = f(x,t) - G(z)f(x,t).$$

Let $N(z) = \sum_{j \geq 1} z^{-j-1}\partial_j - \partial/\partial z$. Evidently, for any $\varphi(x)$ we have $N(z)G(z)\varphi(x) = 0$. We apply the operator $N(z)$ to the last equality:

$$N(z)f(x,z) - G(t)N(z)f(x,z) = \sum_{j \geq 1} z^{-j-1}\partial_j f(x,t).$$

Multiply by z^i and take the residue res_z:

(7.7.5) $\quad a_i = \operatorname{res}_z z^i N(z) f(x,z) = G(t) \operatorname{res}_z z^i N(z) f(x,z) + \partial_i f(x,t).$

This yields

$$\partial_j a_i - \partial_i a_j = G(t)(\partial_j a_i - \partial_i a_j).$$

Hence $\partial_j a_i - \partial_i a_j = \text{const}$. This is a differential polynomial without constants, whence $\partial_j a_i - \partial_i a_j = 0$ and a function $\tau(x)$ exists for which $a_i = -\partial_i \ln \tau$, i.e.

(7.7.6) $\quad \partial_i \ln \tau = - \operatorname{res}_z z^i \left(\sum_{j \geq 1} z^{-j-1}\partial_j - \partial/\partial z \right) \ln \hat{w}.$

It is not difficult to express $\hat{w}$ in terms of τ. From (7.7.5) it follows that $\partial_i f = (1 - G(t))a_i$, i.e. $\partial_i f(x,t) = -(1 - G(t))\partial_i \ln \tau(x)$ and $f(x,t) = -(1 - G(t))\ln \tau(x)$. Finally, $\hat{w}(x,t) = G(t)\tau(x)/\tau(x)$ as required. $\square$

Besides (7.7.2), a formula of conversion (7.7.6) is obtained.

7.7.7. Considering the Baker function we were able to say that it depended on a finite, but not fixed, number of variables. As to the τ-function, this is not possible: the equality (7.7.2) is true if the dependence of τ on all the variables $\{x_i\}$ is involved. However, when we calculate each coefficient of $\hat{w}$ we take into account only a finite number of terms.

7.7.8. **Proposition.** The relation

$$\partial^2 \ln \tau / \partial x_1 \partial x_n = \operatorname{res}_\partial L^n$$

holds.

Proof: According to (7.7.6)

$$\partial_1 \ln \tau = -\operatorname{res}_z z \left(\sum_{j \geq 1} z^{-j-1} \partial_j - \partial/\partial z \right) \ln \hat{w}.$$

Further,

$$\partial_j \ln \hat{w} = \partial_j \hat{w}/\hat{w} = \partial_j (w e^{-\xi(x,z)})/\hat{w} = L_+^j w/w - z^j = 0(z^{j-2})$$

which shows that this term is not involved in the residue.
It remains

$$\partial_1 \ln \tau = \operatorname{res}_z z \partial \ln \hat{w}/\partial z$$

$$= \operatorname{res}_z z \cdot \sum_0^\infty (-i-1) w_i z^{-i-2} / \left(1 + \sum_0^\infty w_i z^{-i-1} \right) = -w_0.$$

Now $\partial_1 \partial_n \ln \tau = -\partial_n w_0$. On the other hand, from $\phi \partial = L \phi$ we find $w_0' = -u_0$. Further, $\partial_n L = [L, L_-^n]$. Taking the residue we obtain $\partial_n u_0 = \operatorname{res}_\partial (L_-^n)' = \operatorname{res}_\partial (L^n)'$, i.e. $(\partial_n W_0)' = -\operatorname{res}_\partial (L^n)'$, whence $\partial_n w_0 = -\operatorname{res}_\partial L^n$, which proves the assertion. □

7.7.9. **Proposition.** The functions

$$\partial^2 \ln \tau / \partial x_i \partial x_l$$

can be expressed in terms of differential polynomials in $\{u_k\}$

Proof: We have

$$\partial^2 \ln \tau / \partial x_i \partial x_l = -\operatorname{res}_z z^i \left(\sum_{j \geq 1} z^{-j-l} \partial_j - \partial/\partial z \right) \partial_l \ln \hat{w}$$

Further, $\partial_l \ln \hat{w} = (L_+^l w/w - z^l)$ (see the proof of the previous proposition), L_+^l is a differential operator, and $w^{(k)}/w = P_k(\chi)$ (7.4.3); the function χ has an expansion in z^{-1} with coefficients from $\mathcal{A}_u$ (7.4.4). This completes the proof. □

7.7.10. Proposition. The quantities

$$J_l = \partial^2 \ln \tau / \partial x_1 \partial x_l$$

are first integrals of the KP-hierarchy.

Proof: This assertion was proved earlier (5.4.1 and 7.7.8). Now we give another proof: $\partial_i J_l = \partial_1(\partial^2 \ln \tau / \partial x_i \partial x_l)$, where $\partial^2 \ln \tau / \partial x_i \partial x_l$ are also differential polynomials in $\{u_k\}$. Hence $\partial_i \int J_l dx = 0$. □

7.7.11. Remark. From the seemingly tautological fact $\partial_i \partial_1 (\partial_l \ln \tau) = \partial_1 \partial_i (\partial_l \ln \tau)$ we have obtained a non-trivial conclusion. The non-trivial part of this proof is hidden in a chain of hardly noticeable assertions. It nevertheless requires considerable effort: while $\partial_l \ln \tau$ is not a differential polynomial in $\{u_k\}$, $\partial_1 \partial_l \ln \tau$, and $\partial_i \partial_l \ln \tau$ are.

7.7.12. Remark. The Baker function is determined up to multiplication by a series in z^{-1} with constant coefficients. This corresponds to multiplication of τ by $\exp \xi(a, z)$, where $a = (a_1, a_2, \dots) = $ const.

We have spoken about formal Baker functions and τ-functions, i.e. all the series were formal. If there is a concrete solution $\{u_k(x)\}$ of the KP-hierarchy, it corresponds to a Baker function $w(x, z)$ which is a genuine functon of x and z and satisfies the equations $Lw = zw$ and $\partial_n w = B_n w$. It is determined up to a multiplier $\varphi(z)$. There is also a τ-function, connected with w by Eqs. (7.7.2) and (7.7.6).

As an example we take the multisoliton solution (Sec. 5.5).

7.7.13. Proposition. For the N-soliton solution (Sec. 5.5) the τ-function is

$$(7.7.14) \qquad \tau(x_1, x_2, \dots) = \Delta = \begin{vmatrix} y_1, & y_2, & \dots, & y_N \\ y_1', & y_2', & \dots, & y_N' \\ \vdots & \vdots & & \vdots \\ y_1^{(N-1)}, & y_2^{(N-1)}, & \dots, & y_N^{(N-1)} \end{vmatrix}$$

where y_k are the functions from Sec. 5.5.1.

Proof: In 5.5.1 a dressing operator ϕ was defined. It can be multiplied to the right by a series in ∂ with constant coefficients; we multiply it by

$\partial^{-N} : \tilde\phi = \phi \partial^{-N}$. Write the Baker function

$$w = \tilde\phi e^{\xi(x,z)}$$

(7.7.15)
$$= \frac{1}{\Delta} \begin{vmatrix} y_1, & \ldots, & y_N, & z^{-N} \\ y'_1, & \ldots, & y'_N, & z^{-N+1} \\ \multicolumn{4}{c}{\dotfill} \\ y_1^{(N)}, & \ldots, & y_N^{(N)}, & 1 \end{vmatrix} e^{\xi(x,z)}$$

Now we must prove that τ and w given by (7.7.14) and (7.7.15) are connected by the necessary condition. We find

$$G(z) y_k(x) = \exp\left\{\xi(x,\alpha_k) - \left(\frac{\alpha_k}{z} + \frac{\alpha_k^2}{2z^2} + \ldots\right)\right\} + a_k \exp\left\{\xi(x,\beta_k)\right.$$
$$\left. - \left(\frac{\beta_k}{z} + \frac{\beta_k^2}{2z^2} + \ldots\right)\right\} = \left(1 - \frac{\alpha_k}{z}\right)\exp\xi(x,\alpha_k) + a_k\left(1 - \frac{\beta_k}{z}\right)$$
$$\exp\xi(x,\beta_k) = y_k - \frac{1}{z} y'_k$$

whence

$$\frac{G(z)\tau}{\tau} = \frac{1}{\Delta}\begin{vmatrix} y_1 - \frac{1}{z}y'_1, & \ldots, & y_N - \frac{1}{z}y'_N \\ y'_1 - \frac{1}{z}y''_1, & \ldots, & y'_N - \frac{1}{z}y''_N \\ \multicolumn{3}{c}{\dotfill} \\ y_1^{(N-1)} - \frac{1}{z}y_1^{(N)}, & \ldots, & y_N^{(N-1)} - \frac{1}{z}y_N^{(N)} \end{vmatrix}.$$

The determinant in (7.7.15) can be reduced to this form if the second row divided by z is subtracted from the first one, the third from the second etc.

Even for the concrete and rather simple solution we are reminded that the τ-function depends on all the times $x_1, x_2, \ldots$ and is determined as a function only if the series $\xi(x,\alpha_k)$ and $\xi(x,\beta_k)$ converge. For the Baker function there are no such complications. Perhaps there is no need to consider τ as a function and it is better to stay on the formal level.

7.8. Additional symmetries.

7.8.1. Any equation (5.2.2) generates a symmetry for each other of these equations, in particular for Eq. (5.2.4). It turns out that there are also other symmetries of these equations. However, the class of equations under consideration must be extended by allowing an explicit dependence of their coefficients on the independent variables. The topic was started by an article by Chen, Lee, and Lin [44]. The coefficients of the equations depended on the variables linearly. Now we use the work of Orlov and Shulman [45] who generalized these symmetries by allowing the dependences of any degree.

7.8.2. Thus, we consider the KP-hierarchy (5.2.2). Represent L as the "dressed" operator ∂ (7.1.2). The operator ∂_n is defined, as usual, by (5.2.2) and its extension (7.1.7). Let $A_{n0} = \partial_n - \partial^n$ and

$$A_n = \phi A_{n0} \phi^{-1} = \partial_n - (\partial_n \phi) \phi^{-1} - L^n = \partial_n + L_-^n - L^n = \partial_n - L_+^n = \partial_n - B_n.$$

The operators ∂ and A_{n0} commute; hence the dressed operators L and A_n also commute:

(7.8.3) $$[A_n, L] = 0.$$

The main idea is to introduce the operator

(7.8.4) $$\Gamma = \sum_1^\infty k x_k \partial^{k-1}$$

and its dressing

(7.8.5) $$M = \phi \Gamma \phi^{-1}.$$

7.8.6. *Lemma.* Operators Γ and A_{n0} and therefore the dressed operators M and A_n commute: $[M, A_n] = 0$.
Proof:

$$[\Gamma, A_{n0}] = \left[\sum_1^\infty k x_k \partial^{k-1}, \partial_n - \partial^n \right] = -n \partial^{n-1} + n \partial^{n-1} = 0. \square$$

7.8.7. *Corollary.* For any $k, m \in \mathbb{Z}, k \geq 0$

$$[M^k L^m, A_n] = 0$$

holds.
Proof: Take into account (7.8.3). $\square$

Let $\tilde{A}_n = A_n + L^n = \partial_n + L_-^n$, then

(7.8.8) $\qquad [\tilde{A}_n, L] = 0$.

7.8.9. The order of $[(M^k L^m)_-, L]$ is not greater than -1. Hence we can write down the equations

(7.8.10) $\qquad \partial_{mk} L = -[(M^k L^m)_-, L], \partial_{mk} = \partial/\partial x_{mk}$,

where x_{mk} are additional times. We shall prove that these equations yield the additional symmetries, i.e. vector fields $\{\partial_{mk}\}$ commute with $\{\partial_m\}$.

7.8.11. *Lemma.* For the vector fields given by (7.8.10),

$$\partial_{mk} L_-^n = -[(M^k L^m)_-, \tilde{A}_n]$$

holds.

Proof:

$$\partial_{mk} L_-^n = -\sum_{\alpha=0}^{n-1} (L^\alpha [(M^k L^m)_-, L] L^{n-\alpha-1})_- = -[(M^k L^m)_-, L^n]_-$$

$$= -[(M^k L^m)_-, L_-^n]_- - [(M^k L^m)_-, L_+^n]_-$$

$$= -[(M^k L^m)_-, L_-^n] - [M^k L^m, L_+^n]_- .$$

We know that $\partial_n L = [L_+^n, L]$. Let us show that L here can be changed by M, $\partial_n M = [L_+^n, M]$. Indeed,

$$\partial_n M = \partial_n \left(\phi \sum_1^\infty k x_k \partial^{k-1} \phi^{-1} \right) = -L_-^n M + M L_-^n + \phi n \partial^{n-1} \phi^{-1}$$

$$= [M, L_-^n] + n L^{n-1}; [L^n, M] = \phi \left[\partial^n, \sum_1^\infty k x_k \partial^{k-1} \right] \phi^{-1}$$

$$= n \phi \partial^{n-1} \phi^{-1} = n L^{n-1}; \partial_n M = [M, L_-^n] - [M, L^n]$$

$$= -[M, L_+^n] = [L_+^n, M].$$

Then $[M^k L^m, L_+^n] = [M^k L^m, \partial_n]$. This implies

$$\partial_{mk} L_-^n = -[(M^k L^m)_-, L_-^n] - [M^k L^m, \partial_n]_-$$

$$= -[(M^k L^m)_-, L_-^n + \partial_n] = -[(M^k L^m)_-, \tilde{A}_n]. \qquad \square$$

7.8.12. Proposition.
$$[\partial_{mk}, \partial_n] = 0.$$

Proof: It suffices to prove that the action of $[\partial_{mk}, \partial_n]$ on the generators is zero.

$$\begin{aligned}
[\partial_{mk}, \partial_n]L &= -\partial_{mk}[L_-^n, L] + \partial_n[(M^k L^m)_-, L] \\
&= [\partial_n(M^k L^m)_- - \partial_{mk}L_-^n, L] + [L_-^n, [(M^k L^m)_-, L]] \\
&\quad - [(M^k L^m)_-, [L_-^n, L]] = [\partial_n(M^k L^m)_- - \partial_{mk}L_-^n, L] \\
&\quad + [[L_-^n, (M^k L^m)_-], L] \\
&= [\partial_n(M^k L^m)_- - \partial_{mk}L_-^n + [L_-^n, (M^k L^m)_-], L] \\
&= [[\partial_n + L_-^n, (M^k L^m)_-] - \partial_{mk}L_-^n, L] \\
&= [[\partial_n + L_-^n, (M^k L^m)_-] + [(M^k L^m)_-, \partial_n + L_-^n], L] = 0. \square
\end{aligned}$$

7.8.13. Remark. The vector fields ∂_{mk} do not commute between themselves.

If $k = 0$ we obtain the old symmetries ∂_m, if $k = 1$ we obtain linear with respect to $\{x_k\}$ symmetries.

7.8.14. Remark. It can be assumed that $\{u_k\}$ depend only on a finite number of the times $\{x_i\}$, taking for Γ finite sums.

7.8.15. Example. Let us write in detail the equation $\partial_{01}L = -[M_-, L]$. We have

$$[M, L] = \phi\left[\sum_1^\infty kx_k \partial^{k-1}, \partial\right]\phi^{-1} = -1.$$

Then

$$M = \phi \sum_2^\infty kx_k \partial^{k-1}\phi^{-1} + \phi x \phi^{-1} = \sum_2^\infty kx_k L^{k-1} + \phi x \phi^{-1}$$

$$= \sum_2^\infty kx_k L^{k-1} + x + \sum_0^\infty w_k(-k-1)\partial^{-k-2}\phi^{-1}$$

whence $M_+ = \sum_2 kx_k L_+^{k-1} + x$ and $[M_+, L] = \sum_2^\infty kx_k \partial_{k-1}L + [x, L]$. Therefore $[M_-, L] = -1 + [L, x] - \sum_1^\infty kx_k \partial_{k-1}L$. The equation has the form

$$\partial_{01}L = \sum_{k=0}^\infty u_k(k+1)\partial^{-k-2} + \left(\sum_2^\infty kx_k \partial_{k-1}\right)L,$$

$$\partial_{01}u_k = ku_{k-1} + \sum_2^\infty iu_i \partial_{i-1}u_k.$$

For example, let $\{u_k\}$ depend only on three variables $x, x_1 = y, x_2 = t$. Recall the KP-equation (5.2.7). The additional symmetry has the form

$$\partial_{01}u = 2yu' + 3tu_y, \partial_{01}u_1 = \frac{1}{2}u + 2yu'_1 + 3tu_{1y}.$$

7.8.16. Exercise. Check directly that this symmetry conserves Eq. (5.2.7).

7.8.17. The majority of the additional symmetries suffer from a grave shortcoming: they are expressed in terms of $\{u_k\}$ in an unlokal form. We try to avoid such objects in this book although they have the right to exist. However we have seen in (7.8.15) that in some cases they are local.

Another example: $k = 1, m = -1$. Three variables x, y and t will be taken into account. We have

$$\partial_{-1,1}L = xL' + 2y[\partial^2 + 2u_0, L] + 3t[\partial^3 + 3u_0\partial + 3u'_0 + 3u_1, L] - L^{-1}$$

7.9. τ-function and Fock representation.

7.9.1. In this section we present a wonderful connection that exists between the KP-hierarchy and the so-called Fock representation of a Clifford algebra which plays a central part in the works of the Kyoto school, (see Ref. [37]).

We consider an associative algebra **A** over $\mathbb{C}$ with generators 1 and $\{\psi_m, \psi_m^*\}, m \in \mathbb{Z}$ satisfying the defining relations

$$[\psi_m, \psi_n]_+ = [\psi_m^*, \psi_n^*]_+ = 0, [\psi_m, \psi_n^*]_+ = \delta_{mn}$$

where $[a, b]_+ = ab + ba$. This is a Clifford algebra.

We further consider a left **A**-module **A** $|vac\rangle$ with a cyclic vector denoted as $|vac\rangle$ consisting of vectors $a|vac\rangle, a \in \mathbf{A}$, the defining relations

(7.9.2) $\qquad \psi_n|vac\rangle = 0 (n < 0), \psi_n^*|vac\rangle = 0 (n \geq 0)$

being imposed. In the same way the right **A**-module is defined consisting of vectors $\langle vac|a, a \in \mathbf{A}$ with relations

(7.9.3) $\qquad \langle vac|\psi_n = 0 (n \geq 0), \langle vac|\psi_n^* = 0 (n < 0).$

A pairing between $\langle vac|\mathbf{A}$ and $\mathbf{A}|vac\rangle$ can be defined thus. Let $\langle vac|a$ and $b|vac\rangle$ be two elements. Then using commutation rules and relations (7.9.2) and (7.9.3) the expression $\langle vac|ab|vac\rangle$ can be transformed to the form $\lambda\langle vac|1|vac\rangle$ where $\lambda \in \mathbb{C}$. It remains to put $\langle vac|1|vac\rangle = 1$. Expressions $\langle vac|a|vac\rangle$ are called vacuum expectations and are denoted simply as $\langle a \rangle$. Let

$$\langle vac|a \bullet b|vac\rangle = \langle ab \rangle$$

This is the required pairing.

7.9.4. It is easy to see that

$$[\psi_m \psi_n^*, \psi_{m'}, \psi_{n'}^*] = \delta_{nm'}\psi_m \psi_{n'}^* - \delta_{mn'}\psi_{m'}\psi_n^*$$

where [,] is a usual commutator. This means that expressions $\sum c_{mn} \psi_m \psi_n^*$ form a Lie algebra denoted as $\mathfrak{G}(V, V^*)$. This algebra acts in the vector spaces $V = \sum_{m \in \mathbb{Z}} \mathbb{C}\psi_m$ and $V^* = \sum_{m \in \mathbb{Z}} \mathbb{C}\psi_m^*$ as

(7.9.5) $\qquad [\psi_m \psi_n^*, \psi_p] = \delta_{np}\psi_m, [\psi_m \psi_n^*, \psi_p^*] = -\delta_{mp}\psi_n^*$.

The spaces V and V^* can be considered as conjugated with respect to the pairing $(\psi_m, \psi_n^*) = \delta_{mn}$. (This pairing differs from the above one; $\psi_m \bullet \psi_n^*$ was equal to 1 only if $m = n < 0$, it vanished otherwise.)

It is easy to check that the actions of $\psi_m \psi_n^*$ in V and in V^* are anti-conjugated:

(7.9.6) $\qquad ([\psi_m \psi_n^*, \psi_p], \psi_q^*) = -(\psi p, [\psi_m \psi_n^*, \psi_q])$.

The Lie algebra $\mathfrak{G}(V, V^*)$ can be integrated in $\mathbf{A}$ up to a Lie group $G(V, V^*)$, e.g. a one-parametric group corresponding to $\psi_m \psi_n^*, m \neq n$, is $\exp(t\psi_m \psi_n^*) = 1 + t\psi_m \psi_n^*$. The action of the group $G(V, V^*)$ on V and V^* related to the action (7.9.6) of the algebra $\mathfrak{G}(V, V^*)$ is

$$v \mapsto gvg^{-1}, v \in V, v^* \mapsto g^{-1}v^*g$$

From (7.9.6) we have $(gv, g^{-1}, v_2^*) = (v_1, g^{-1}v_2^*g)$.

In biortogonal bases $\{\psi_m\}$ and $\{\psi_m^*\}$, operator $g \cdot g^{-1}$ and $g^{-1} \cdot g$ in V and V^* have conjugated matrices:

(7.9.7) $\qquad g\psi_n g^{-1} = \sum \psi_m a_{mn}, g^{-1}\psi_n^* g = \sum \psi_m^* a_{nm}$

7.9.8. Definitions.

1) If $x_1, x_2, \ldots$ is a set of parameters ("times") then the Hamiltonian is defined as

$$H(x) = \sum_{l=1}^{\infty} x_l \sum_{n=-\infty}^{\infty} \psi_n \psi^*_{n+l}$$

2) If $g \in G(V, V^*)$, then the time evolution of g is defined by

$$g(x) = e^{H(x)} g e^{-H(x)}$$

3) The τ-function related to g is

$$\tau(x, g) = \langle g(x) \rangle = \langle e^{H(x)} g \rangle$$

(note that the last equality $\langle g(x) \rangle = \langle e^{H(x)} g \rangle$ follows from $H(x)|vac\rangle = 0$, which is easy to verify).

We further prove the main proposition as follows.

7.9.9. Proposition. $\tau(x, g)$ is a τ-function of the KP-hierarchy.

The proof will be preceded by some preparations. A τ-function generates Baker functions w and w^*, namely,

$$\hat{w}(x, k) = G(k)\tau(x)/\tau(x), \hat{w}^*(x, k) = G^*(k)\tau(x)/\tau(x)$$

where

$$G(k)f(x_1, x_2, \ldots) = f(x_1 - k^{-1}, x_2 - \frac{1}{2}k^{-2}, x_3 - \frac{1}{3}k^{-3}, \ldots),$$

$$G^*(k)f(x_1, x_2, \ldots) = f(x_1 + k^{-1}, x_2 + \frac{1}{2}k^{-2}, x_3 + \frac{1}{3}k^{-3}, \ldots)$$

(see Sec. 7.7). In order to prove Prop. 7.9.9 we must verify the bilinear identity 7.3.3 or, what is the same, 7.3.4.

Let

$$\psi(k) = \sum_{-\infty}^{\infty} \psi_i k^i, \psi^*(k) = \sum_{-\infty}^{\infty} \psi^*_i k^{-i}.$$

7.9.10. Lemma. Relations

$$e^{H(x)} \psi(k) e^{-H(x)} = e^{\xi(x,k)} \psi(k), e^{H(x)} \psi^*(k) e^{-H(x)} = e^{-\xi(x,k)} \psi^*(k)$$

hold.

Proof: We have

$$[H(x), \psi(k)] = \sum_{i=-\infty}^{\infty} H(x)\psi_i k^i - \sum_{i=-\infty}^{\infty} \psi_i H(x) k^i$$

$$= \sum_{l=-\infty}^{\infty}\sum_{l=1}^{\infty} x_l \sum_{n=-\infty}^{\infty} \psi_n \psi_{n+l}^* \psi_i k^i - \sum_{i=-\infty}^{\infty} \psi_i H(x) k^i = \sum_{i=-\infty}^{\infty}\sum_{l=1}^{\infty} x_l \psi_{i-l} k^i$$

$$= \sum_{l=1}^{\infty} x_l k^l \sum_{i=-\infty}^{\infty} \psi_i k^i = \xi(x,k)\psi(k).$$

Let $u(t) = e^{tH}\psi(k)e^{-th}$. Then $u(0) = \psi(k)$, $du/dt = e^{tH}[H,\psi(k)]e^{-tH} = \xi(x,k)e^{tH}\psi(k)e^{-tH} = \xi(x,k)u(t)$ and $u = e^{t\xi(x,k)}\psi(k)$. Letting $t = 1$ we obtain the required equation. The second relation can be obtained similarly. □

7.9.11. Lemma. Relations

$$G(k)\langle e^{H(x)}g\rangle = \langle \psi_0^* e^{H(x)}\psi(k)g\rangle \cdot e^{-\xi(x,k)}$$
$$G^*(k)\langle e^{H(x)}g\rangle = k^{-1}\langle \psi_{-1}e^{H(x)}\psi^*(k)g\rangle e^{\xi(x,k)}$$

hold.

Proof: These relations can be proved for all $g \in \mathbf{A}$, not just for $g \in G(V,V^*)$. From the very beginning it can be assumed that in every term of g the number of multipliers ψ_i is equal to that of ψ_i^*, otherwise the term vanishes. Thus one must prove the relation $G(k)\langle e^{H(x)} \cdot \psi_{i_1}\ldots\psi_{i_l}\psi_{j_1}^*\ldots\psi_{j_l}^*\rangle = e^{-\xi(x,k)}\langle \psi_0^* e^{H(x)}\psi(k)\psi_{i_1}\ldots\psi_{j_l}^*\ldots\rangle$. One can pass to generators

$$G(k)\langle e^{H(x)}\psi(p_1)\ldots\psi(p_l)\psi^*(q_1)\ldots\psi^*(q_l)\rangle = e^{-\xi(x,k)}\langle \psi_0^* e^{H(x)}\psi(k)$$
$$\psi(p_1)\ldots\psi(p_l)\psi^*(q_1)\ldots\psi^*(q_l)\rangle.$$

According to Lemma 7.9.10, this is equivalent to

$$G(k)e^{\xi(x,p_1)+\ldots+\xi(x,p_l)-\xi(x,q_1)-\ldots-\xi(x,q_l)}\langle \psi(p_1)\ldots\psi(p_l)\psi^*(q_1)\ldots\psi^*(q_l)\rangle$$
$$= e^{\xi(x,p_1)+\ldots+\xi(x,p_l)-\xi(x,q_1)-\ldots-\xi(x,q_l)}$$
$$\langle \psi_0^*\psi(k)\psi(p_1)\ldots\psi(p_l)\psi^*(q_1)\ldots\psi^*(q_l)\rangle$$

Now we note that

$$G(k)e^{\langle \xi, p_i\rangle} = e^{\sum_{l=1}^{\infty}(x_l - \frac{1}{l}k^{-l})p_i^l} = e^{\sum_{l=1}^{\infty} x_l p_i^l}\left(1 - \frac{p_i}{k}\right) = e^{\xi(x,p_i)}\frac{k-p_i}{k}$$

$$G(k)e^{-\xi(x,q_i)} = e^{-\xi(x,q_i)} \cdot \frac{k}{k-q_i}.$$

Thus the relation

$$\langle \psi_0^* \psi(k) \psi(p_1) \ldots \psi(p_l) \psi^*(q_1) \ldots \psi^*(q_l) \rangle = \frac{(k-p_1) \ldots (k-p_l)}{(k-q_1) \ldots (k-q_l)}$$
$$\langle \psi(p_1) \ldots \psi(p_l) \psi^*(q_1) \ldots \psi^*(q_l) \rangle$$

is to be verified.

In the left-hand side, $\psi_0 + \psi_-(k) = \sum_{-\infty}^{0} \psi_i k^i$ can be substituted for $\psi(k)$. Then we note that $[\psi_-(k), \psi^*(q)]_+ = \sum_{-\infty}^{-1} k^i q^{-i} = \frac{q}{k-q}$. This implies that the left-hand side has the form $a + \sum_{s=1}^{l} b_s/(q_s - k)$ where a and b_s are independent of k. The term a is $a = \langle \psi_0^* \psi_0 \psi(p_1) \ldots \psi(p_l) \psi^*(q_1) \ldots \psi^*(q_l) \rangle = \langle \psi(p_1) \ldots \psi(p_l) \psi^*(q_1) \ldots \psi^*(q_l) \rangle$. In other words the left-hand side is $a(k^l + P_{l-1}(k))/(k - q_1) \ldots (k - q_l)$, where P_{l-1} is a polynomial of a degree $\leq l - 1$.

It remains to add that this expression vanishes if $k = p_1, \ldots, p_l$. Indeed, $\psi(k) \psi(p_1) \ldots \psi(p_l) = 0$ if $k = p_1, \ldots, p_l$.

The rest is clear: $k^l + P_{l-1} = (k - p_1) \ldots (k - p_l)$. The second relation can be proved in the same way. □

Now the proof of the proposition 7.9.9:

$$\operatorname{res}_k w(x,k) w^*(x',k) = \operatorname{res}_k k^{-1} \langle \psi_0^* e^{H(x)} \psi(k) g \rangle$$
$$\langle \psi_{-1} e^{H(x')} \psi^* g \rangle / \tau(x,g) \tau(x',g) = \sum_n \langle \psi_0^* e^{H(x)} \psi_n g \rangle$$
$$\langle \psi_{-1} e^{H(x')} \psi_n^* g \rangle / \tau(x,g) \tau(x',g)$$

Remembering (7.9.7) we obtain

$$\sum_n \sum_m \langle \psi_0^* e^{H(x)} \psi_n g \rangle \langle \psi_{-1} e^{H(x')} g \psi_m^* a_{nm} \rangle / \tau(x,g) \tau(x',g)$$
$$= \sum_m \langle \psi_0^* e^{H(x)} g \psi_m \rangle \langle \psi_{-1} e^{H(x')} g \psi_m^* \rangle / \tau(x,g) \tau(x',g) \,.$$

This expression vanishes since one of $\psi_m |vac\rangle$ and $\psi_m^* |vac\rangle$ vanishes for each m. □

This fact seems to be very striking: solutions of the KP-hierarchy arising "from nothing" are obtained.

We stop at this place, but there are many other remarkable facts in this direction, e.g. action of the group $GL(\infty)$ in the space of τ-functions etc.

CHAPTER 8. Grassmannian. τ-Function and Baker Function after Segal and Wilson. Algebraic-Geometrical Krichever's Solutions.

8.1. Grassmannian after Segal and Wilson.

8.1.1. According to an observation of Segal and Wilson [39] all examples of the special solutions of the KP-hierarchy (as well as of the KdV-hierarchies) admit of a general interpretation. A construction of an infinite-dimensional Grassmannian is considered, and to each generic element of this Grassmannian a τ-function and the Baker function (hence, a solution, too) are attached. We shall outline the main idea and refer to the reader to Ref. [39] for more detail.

Let H be a Hilbert space represented as a sum of two infinite-dimensional subspaces, $H = H_+ \oplus H_-$. The Grassmannian GrH is a set of all subspaces $W \subset H$ which differ not much from H_+ in the following sense: if p_+ and p_- are orthogonal projectors on H_+ and H_- then the restriction of p_+ to W, $p_+|_W : W \to H_+$, is a Fredholm opertor, i.e. it has a finite-dimensional kernel and cokernel, and p_- is a compact operator (the last condition seems to be not so important). Most interesting is the case where the index of the operator $p_+|_W$ is zero, i.e. the kernel and the cokernel have the same dimension. The subspace W is said to be transversal to H_-, or simply transversal, if $p_+|_W : W \to H_+$ is a bijection.

The space H is always taken as $\mathcal{L}^2(S^1)$, where $S^1 = \{z \in \mathbb{C}, |z| = 1\}$.

The subspaces H_+ and H_- are spanned on the bases $\{z^k\}, k \geq 0$ and $\{z^k\}, k < 0$ respectively.

8.1.2. Example of an element $W \in GrH$. Let W comprise functions having the form $f(z) = \sum_{-N}^{\infty} a_k z^k$ on S^1; $N > 0$ is fixed. The functions are assumed to satisfy additional conditions. Let $p = (p_1, \ldots, p_N)$ be a set of complex numbers $0 < |p_i| < 1$, with $p_i^n \neq p_j^n (i \neq j)$ for some n. Let $\lambda = (\lambda_1, \ldots, \lambda_n)$ be a set of complex numbers and ε a root of unity of the n-th degree, $\varepsilon^n = 1$. We require that

$$f(\varepsilon p_i) = \lambda_i f(p_i). \tag{8.1.3}$$

8.1.4. Proposition. For almost all sets p, λ the subspace W is a graph over H_+, i.e. $p_+|_W$ is a bijection. In other words, W is transversal.

Proof: It has to be proved that for almost all sets p, λ the following fact holds: for each $f_1(z) \in H_+$, unique set of numbers $a_{-N}, \ldots, a_{-1}$ exists such that $f = f_1 + \sum_{-N}^{-1} a_k z^k \in W$. This is equivalent to the system of equations

$$\sum_{k=-N}^{-1} a_k [(\varepsilon p_i)^k - \lambda_i p_i^k] = f_1(p_i) - f_1(\varepsilon p_i), i = 1, \ldots, N.$$

It is clear that for almost every p, λ the determinant of the system does not vanish and the system is uniquely solvable. $\square$

We denote the subspace W corresponding to the parameters p, λ as $W_{p,\lambda}$.

Now we consider some transformations of the grassmannian. Let $g(z)$ be a function holomorphic and non-vanishing in the circle $|z| \leq 1$, and $g(0) = 1$. Then $g(z) = \exp \sum_{1}^{\infty} x_k z^k = \exp \xi(x, z)$, where $\{x_k\}$ are some numbers; we assume that they are real. We consider the mapping

$$W \overset{g^{-1}}{\mapsto} g^{-1} W$$

which can be written in the block form

$$g^{-1} = \begin{pmatrix} a & b \\ 0 & c \end{pmatrix}$$

according to the decomposition $H = H_+ \oplus H_-, a : H_+ \to H_+, b : H_- \to H_+, c : H_- \to H_-$. We shall assume that for almost every set $\{x_k\}$ the subspace $g^{-1}W$ is transversal.

8.1.5. Example. In the example 8.1.2, we have $g^{-1}W_{p,\lambda} = W_{p,\lambda g^{-1}}$ where $(\lambda g^{-1})_i = \lambda_i g(p_i)/g(\varepsilon p_i)$, i.e. an element of the Grassmannian of the same type is obtained but with other λ. It is evident that almost all $g^{-1}W$ are transversal.

8.1.6. Definition. A function of z, $w_W(g,z)$, which depends on W and g (i.e. on $\{x_k\}$) as parameters is called a Baker function if

(i) $\quad w_W(g,z) \in W$, (ii) $\quad g^{-1}w_W(g,z) = 1 + \sum_{-\infty}^{-1} a_i z^i$.

The above condition of transversality implies that for a given W a unique Baker function exists for almost every g: if $g^{-1}W$ is transversal then there is a unique element $f \in g^{-1}W$ for which $p_+ f = 1$, i.e. $f = 1 + \sum_{-\infty}^{-1} a_i z^i$.

Now we give a definition to τ-function. Let W be transversal. Then a mapping $(p_+|_W)^{-1} : H_+ \to W$ exists. We consider a mapping

$$H_+ \xrightarrow{(p_+|_W)^{-1}} W \xrightarrow{g^{-1}} g^{-1}W \xrightarrow{p_+} H_+ \xrightarrow{g} H_+.$$

8.1.7. Definition. τ-function $\tau_W(g) = \tau_W(x_1, x_2, \ldots)$ (we consider it as a function on $x_1, x_2, \ldots$ depending on a parameter W) is the determinant of the above mapping:

$$\tau_W(g) = \det g p_+ g^{-1} (p_+|_W)^{-1}.$$

We do not study the existence of this determinant in the general form (the mapping is close to the identical one). If necessary, it will be simply calculated.

8.1.8. Proposition. Let W be transversal and $A : H_+ \to H_-$ be a mapping given by the graph W, i.e. $A = p_-(p_+|_W)^{-1}$, then

$$\tau_W(g) = \det(1 + a^{-1}bA).$$

Proof: Represent the elements of the space H as pairs (x,y), $x \in H_+$, $y \in H_-$. Then

$$gp_+ g^{-1}(p_+|_W)^{-1} x = gp_+ g^{-1}(x, Ax) = gp_+(ax + bAx, cAx)$$
$$= g(ax + bAx, 0) = a^{-1}(ax + bAx)$$
$$= (1 + a^{-1}bA)x \in H_+. \square$$

For the existence of the determinant it is sufficient that $a^{-1}bA$ has the trace.

8.1.9. Lemma. The identity
$$\tau_W(gg_1) = \tau_W(g)\tau_{g^{-1}W}(g_1)$$
holds.

Proof: To the right stands $\det g_1 p_+ g_1^{-1}(p_+|_{g^{-1}W})^{-1} \cdot \det g p_+ g^{-1}(p_+|w)^{-1} = \det(g_1 p_+ g_1^{-1}(p_+|_{g^{-1}W})^{-1} p_+ g^{-1}(p_+|w)^{-1}g)$. Further, if $x \in H_+$ then $(p_+|_{g^{-1}W})^{-1}gx \in W$ and $g^{-1}(p_+|_{g^{-1}W})^{-1}gx \in g^{-1}W$. Hence $(p_+|_{g^{-1}W})^{-1}$ and p_+ cancel out and we obtain

$$\det(g_1 p_+ g_1^{-1} g^{-1}(p_+|w)^{-1}g) = \det(gg_1 p_+ (gg_1)^{-1}(p_+|w)^{-1}) = \tau_W(gg_1).\ \square$$

8.1.10. Proposition. The following connection between the τ-function and the Baker function
$$g^{-1}w_W(g,z) = \tau_W(x_1 - 1/z, x_2 - 1/2z^2, x_3 - 1/3z^3, \ldots)/\tau_W(x_1, x_2, \ldots)$$
holds (which is (7.7.2)).

Proof: Let us apply the lemma 8.1.9 to $g_1(z) = 1 - z/\varsigma$, where ς is a fixed complex number, $|\varsigma| > 1$. We have $g_1(z) = \exp \log(1 - z/\varsigma) = \exp(-\sum z^k/k\varsigma^k)$, whence $g(z)g_1(z) = \exp\sum_{1}^{\infty}(x_k - 1/k\varsigma^k)z^k$ and

$$\tau_W(gg_1)/\tau_W(g) = \tau_W(x_1 - 1/\varsigma, x_2 - 1/2\varsigma^2, \ldots)/\tau_W(x_1, x_2, \ldots).$$

It remains to prove that $\tau_{g^{-1}W}(g_1) = g^{-1}w_W(g_1\varsigma)$. According to the definition 8.1.6 this means: (i) as a function of ς, $\tau_{g^{-1}W}(g_1)$ belongs to $g^{-1}W$; (ii) $\tau_{g^{-1}W} = 1 + f_0(\varsigma)$, where $f_0 \in H_-$.

Let us represent the action of $g_1^{-1} : H \to H$ in the above block form. A is the operator $H_+ \to H_-$ corresponding to the graph $g^{-1}W$. The action of b on the basis $\{z^{-k}\}$ of H_- is

$$bz^{-k} = (g_1^{-1}z^{-k})_+ = ((1 + z/\varsigma + z^2/\varsigma^2 + \ldots)z^{-k})_+ = \varsigma^{-k}g_1^{-1}(z).$$

Further, $a^{-1}bz^{-k} = g_1(z)\varsigma^{-k}g_1^{-1}(z) = \varsigma^{-k}$. Therefore, for any function $f_1(z) \in H_-$ we have $a^{-1}bf_1(z) = f_1(\varsigma)$. Thus, the operator $a^{-1}b$ sends all the functions of z to constants (ς is here a parameter), i.e. this is an operator of the rank 1. Then, so is the operator $a^{-1}bA$, and $\det(1 + a^{-1}bA) = 1 + \operatorname{tr} a^{-1}bA$. The trace will be calculated if we apply $a^{-1}bA$ to $f(z) \equiv 1, a^{-1}bA(1)$. We have $A(1) = f_0(z) \in H_-$, and $1 + f_0(z) \in g^{-1}W$. Now $\det(1 + a^{-1}bA) = 1 + a^{-1}bf_0(z) = 1 + f_0(\varsigma)$. This proves both (i) and (ii) at once. □

We have proved the formula we knew before connecting the τ-function with the Baker function. Now we shall show that the Baker function in this new sense is the Baker function of the KP-hierarchy already familiar to us.

8.1.11. Proposition. If $w = w_W$ is a Baker function (8.1.6), then for every $r \geq 2$ a unique operator $B_r = \partial^r + B_{r2}\partial^{r-2} + \ldots + B_{rr}$ exists satisfying the equation $\partial_m w = B_m w$, where $\partial_m = \partial/\partial x_m$, the coefficients $\{B_{ri}\}$ being differential polynomials in $\{a_i\}$.

Proof: It follows from $w = g\left(1 + \sum\limits_{-\infty}^{-1} a_i z^i\right)$ that

$$\partial_m w = g(z)(z^m + a_1 z^{m-1} + O(z^{m-2})),$$
$$\partial^q w = g(z)(z^q + a_1 z^{q-1} + O(z^{q-2})).$$

Therefore B_r can be chosen in such a way that

$$\partial_m w - B_m w = g(z) O(z^{-1}).$$

To the left stands an element $v \in W$. We have $g^{-1}v = O(z^{-1})$. This must vanish due to the fact that $g^{-1}W$ is transversal. □

8.1.12. Proposition. Let $\{a_i\}$ be coefficients of a Baker function w (8.1.6). Let $\phi = 1 + \sum\limits_{-\infty}^{-1} a_i \partial^i$ (where $\partial = \partial_1, x_1 = x$) and $L = \phi \partial \phi^{-1}$. Then L satisfies the equations of the KP-hierarchy, $\partial_m L = [L_+^m, L]$, and w is the Baker function of the hierarchy in the sense of 7.2.2.

Proof: We have $w = \phi g$. From 8.1.11 We get $B_m w - \partial_m w = (\partial_m \phi)g + \phi z^m g$, where $(\partial_m \phi) = \sum\limits_{-\infty}^{-1}(\partial_m a_i)\partial^i$, i.e. $((\partial_m \phi) + \phi \partial^m - B_m \phi)g = 0$. The differential operator acting on g contains differentiation with respect to $x_1 = x$ only. The multiplier $\exp \sum\limits_{2}^{\infty} x_i z^i$ can be cancelled. If for some operator P the equation $Pe^{xz} = 0$ holds then $P = 0$. Thus, $(\partial_m \phi) + \phi \partial_m - B_m \phi = 0$, which implies

(8.1.13) $\qquad B_m = (\partial_m \phi)\phi^{-1} + \phi \partial^m \phi^{-1}.$

We take the differential part of this equality $B_m = (\phi \partial^m \phi^{-1})_+ = L_+^m$. This yields $(\partial_m \phi) = -\phi \partial^m + L_+^m \phi = -L^m \phi + L_+^m \phi = -L_-^m \phi$. As we have already seen when proving 7.1.6, this equation gives $\partial_m L = [L_+^m, L]$. The rest is obvious. □

8.1.14. Proposition. Let an element of the Grassmannian $W \in GrH$ satisfy the condition $z^n W \subset W$ for some n (the subspace of such elements we denote as $Gr^{(n)}H$). Then $B_n w = z^n w, \partial_n a_i = \partial_{2n} a_i = \partial_{3n} a_i = \ldots = 0, B_n = L^n (L_-^n = 0)$.

Proof: We have

$$B_n w - z^n w = \partial_n w - z^n w = \sum_{-\infty}^{-1} (\partial_n a_i) z^i \cdot g.$$

As above, the fact that $g^{-1} W$ is transversal implies that $\sum (\partial_n a_i) z^i = 0$ and $B_n w - z^n w = 0$, i.e. $L_+^n w = z^n w$. On the other hand, $L^n w = L^n \phi g = \phi \partial^n g = z^n \phi g = z^n w$. Then $L_-^n = 0$ and $L^n = L_+^n$, hence $L^{nk} = L_+^{nk}, \partial_{nk} \phi = -L_-^{nk} \phi = 0$. Thus, L^n satisfies the equations of the n-th KdV-hierarchy

$$\partial_m L^n = [L_+^m, L^n] \text{, i.e. } \partial_m L^n = [(L^n)_+^{m/n}, L^n].$$

8.1.15. Example. Return to the example 8.1.2. Here $W \in Gr^{(n)}H$. Take the simplest case, $n = 2, N = 1$. The space W consists here of the functions having in $z = 0$ a pole of no more than the first order, regular in other points of the circle $|z| \leq 1$, and satisfying the condition $f(-p) = \lambda f(p)$ for some $\lambda \neq 0$ and $p, 0 < |p| < 1$. The Baker function has the form $w = \exp \xi(x,z)(1 + az^{-1})$ and satisfies the condition $\exp \xi(x,-p)(1 - ap^{-1}) = \lambda \exp \xi(x,p)(1 + ap^{-1})$. Denoting $\theta(x) = x_1 p + x_3 p^3 + x_5 p^5 + \ldots$ we obtain $a = -p\tanh(\theta(x) + \alpha)$. Now $L = \partial^2 + u$, where $u = -2a' = 2p^2/\cosh^2(\theta + \alpha)$. We have obtained the one-soliton solution of the KdV-hierarchy in the narrow sense, i.e. where $n = 2$:

$$u = 2p^2 \cosh^{-2}(px + p^3 x_3 + p^5 x_5 + \ldots + \alpha)$$

(see (5.5.8)). The τ-function can also be written. For the N-soliton solution it has the form of a determinant similar to that obtained earlier (7.7.14), (see also Ref. [39]). Thus, all the soliton solutions can be obtained.

8.2. Algebraic-geometrical solutions of Krichever.

8.2.1. According to Segal and Wilson's conception, the Krichever's algebraic-geometrical solutions [15] are connected with another construction of elements of the Grassmannian.

In this section, it is supposed that the reader wields some basic principles of the Riemann surfaces theory. Some definitions and formulations of the necessary theorems are listed in the Appendix to this chapter. We can recommend some books: Springer [47], an excellent book by Forster [48] (unfortunately there is no θ-function in this book), and also a very good and simply written book by Dubrovin [49] which exists only in Russian.

Let X be a compact Riemann surface of the genus g. A divisor is a linear combination with integer coefficients of a finite number of points of the surface,

$$D = \sum m_i p_i, m_i \in \mathbb{Z} \ .$$

We define the sheaf $\mathcal{O}_D$ (see Ref. [48], Sec. 16.4). Let $U \subset X$ be an open set. Then $\mathcal{O}_D(U)$ consists of all meromorphic functions f in U submitted to the divisor $-D$. The latter means that, at the points of the divisor D, if $m_i > 0$ the function f must either be regular or have poles not more than the m_i-th order. If $m_i < 0$ the function must have zeros not less than the $-m_i$-th order. The set of all $\mathcal{O}_D(U)$ is called the sheaf $\mathcal{O}_D$ (with respect to the operation of restriction of functions to subsets).

Recall also the definition of a cohomology group with coefficients in a sheaf (in this case in $\mathcal{O}_D$). Let $\{U_i\} = U$ be an open covering of X. A k-cochain is a function $C^{(k)}_{i_0 \ldots i_k} = C^{(k)}(U_{i_0}, U_{i_1}, \ldots, U_{i_k})$ which attaches to each set of neighbourhoods $U_{i_0}, U_{i_1}, \ldots, U_{i_k}$ having a non-empty intersection a function of $\mathcal{O}_D(U_{i_0} \cap U_{i_1} \cap \ldots \cap U_{i_k})$. $C^{(k)}_{i_0 \ldots i_k}$ is assumed to be skew-symmetrical with respect to $i_0, \ldots, i_k$. The set of all cochains forms an Abel group (even linear space) and is denoted as $C^k(U, \mathcal{O}_D)$. The boundary operator δ which increases the dimension of a cochain by 1 is defined by $(\delta C^{(k)})_{i_0, \ldots, i_{k+1}} = \sum_{l=0}^{k+1} (-1)^l C^{(k)}_{i_0, \ldots, \hat{i}_l, \ldots, i_{k+1}}$ (for example, $(\delta C^{(0)})_{i_0 i_1} = C^{(0)}_{i_1} - C^{(0)}_{i_0}, (\delta C^{(1)})_{i_0 i_1 i_2} = C^{(1)}_{i_1 i_2} - C^{(1)}_{i_0 i_2} + C^{(1)}_{i_0 i_1}$, etc.). It is easy to check that $\delta^2 = 0$. A cocycle is such a cochain that $\delta c = 0$. The group of the cocycles is denoted as Z^k; $Z^k = \ker(C^k \xrightarrow{\delta} C^{k+1})$. A cochain $C^{(k)}$ is called a coboundary if there is a $f^{(k-1)} \in C^{k-1}, \delta f^{(k-1)} = C^{(k)}$. The set of all the coboundaries is a group $B^k = \text{Im}(C^{k-1} \xrightarrow{\delta} C^k)$. Evidently $B^k \subset Z^k$. The k-th cohomology group is $H^k(U, \mathcal{O}_D) = Z^k/B^k$.

This group, in general, depends on the covering. However, for sufficiently small coverings, it does not depend on them. Then we denote the group as $H^k(X, \mathcal{O}_D)$. It turns out that in our case it is sufficient to take the covering with only two elements, e.g. a small neighbourhood of a point and the whole surface without this point.

The Riemann-Roch theorem (Ref. [48], Sec. 16.9) asserts that

$$\dim H^0(X, \mathcal{O}_D) - \dim H^1(X, \mathcal{O}_D) = 1 - g + \deg(D),$$

where $\deg(D)$ is the degree of the divisor, $\sum m_i$. The theorem guarantees also that $\dim H^0$ and $\dim H^1$ are finite.

It is easy to grasp what is $H^0(X, \mathcal{O}_D)$. An element of this group is a set of functions $c_i^{(0)} \in \mathcal{O}(U_i, D)$ such that $c_j^{(0)} - c_i^{(0)} = 0$ on the intersections $U_i \cap U_j$. This means that a global function $c^{(0)} \in \mathcal{O}(X, D)$ exists such that $c_i^{(0)}$ are restrictions of $c^{(0)}$ on U_i. There are no coboundaries of the zero dimension, hence $H^{(0)}(X, \mathcal{O}_D) \cong \mathcal{O}_D(X)$.

8.2.2. We fix a point P_∞ of the Riemann surface. Let z^{-1} be a local parameter in a neighbourhood U_∞ of this point. The point itself corresponds to $z^{-1} = 0$. Without loss of generality it can be assumed that the closed domain $X_\infty : |z| \geq 1$ is contained in U_∞. Let S^1 be the curve $|z| = 1$ and $X_0 = \overline{(X \setminus X_\infty)}$. The local parameter z^{-1} enables us to identity X_∞ with the closed domain $|z| \geq 1$ of the Riemann sphere $\mathbb{C}P^1$.

Now we define an element W of the Grassmannian. This will be the set of functions on S^1, i.e. of elements of the Hilbert space H, which are boundary values of functions of $\mathcal{O}_D(X_0)$ on the Riemann surface.

8.2.3. *Proposition.* The operator $p_+|_W : W \to H_+$ is a Fredholm operator. Its index is

$$\text{ind}(W \to H_+) = \dim H^0(X, \mathcal{O}_D) - \dim H^1(X, \mathcal{O}_D) - 1.$$

Proof: First let us study the projection $W \to zH_+$. The kernel of this mapping, $\ker(W \to zH_+)$, comprises the functions on S^1 which, on the one hand, belong to W, i.e. they can be extended to X_0 as elements of $\mathcal{O}_D(X_0)$, and which, on the other hand, can be extended to X_∞ as holomorphic functions. Thus, they belong to $\mathcal{O}_D(X)$, i.e. they represent the elements of $H^0(X, \mathcal{O}_D)$.

The cokernel of this mapping, coker $(W \to zH_+) = zH_+/\text{Im}(W \to zH_+)$, consists of functions $f \in zH_+$. Two of such functions must be identified if there exist functions $\varphi_0 \in \mathcal{O}_D(X_0)$ and $\varphi_\infty \in \mathcal{O}_D(X_\infty)$ such that $f_1 - f_2 = (\varphi_0 - \varphi_\infty)|_{S^1}$. Since any function $f \in H$ is the sum of a function of zH_+ and one of $\mathcal{O}_D(X_\infty)$, it can be said that coker $(W \to zH_+)$ comprises all the functions on S^1, identified if $f_1 - f_2 = (\varphi_0 - \varphi_\infty)|_{S^1}, \varphi_0 \in \mathcal{O}_D(X_0), \varphi_\infty \in \mathcal{O}_D(X_\infty)$. This coincides with the definition of $H^1(X, \mathcal{O}_D)$. Thus,

$$\ker(W \to zH_\infty) = H^0(X, \mathcal{O}_D), \text{ coker}\,(W \to zH_\infty) = H^1(X, \mathcal{O}_D).$$

Therefore $\text{ind}(W \to zH_+) = \dim H^0 - \dim H^1$.

Now we consider the mapping $W \to H_+$. We have $1 \in W$ since constants belong to $\mathcal{O}_D(X_0)$. Then $1 \in \ker(W \to zH_+)$ but $1 \bar{\in} \ker(W \to H_+)$. We have $\dim \ker(W \to H_+) = \dim \ker(W \to zH_+) - 1$, and $\dim \text{coker}(W \to H_+) = \dim \text{coker}(W \to zH_+)$. Hence $\text{ind}\,(W \to H_+) = \text{ind}(W \to zH_+) - 1$ as required. At the same time we have obtained that the kernel and the cokernel are finite-dimensional since the corresponding cohomology groups are finite-dimensional. $\square$

8.2.4. Corollary. The index of the operator $p_+|_W : W \to H_+$ is zero if and only if $\deg(D) = g$.

Proof: See the Riemann-Roch theorem. $\square$

Solely this case will be of interest to us. More than that, we shall assume that the divisor D consists of g distinct points, and all $m_i = 1$.

8.2.5. Remark. We do not prove that the projector $p_-|_W : W \to H_-$ is compact since this is not important.

Further, the transformation $W \mapsto \exp \xi(x,z) W$ of the Grassmannian is considered, and the corresponding Baker function. However the Baker function can be constucted explicitly in terms of the θ-function (about θ-functions see the Appendix 8.A to this chapter).

8.2.6. Baker-Akhieser-Krichever lemma. The Baker function can be written in the form

$$w(x, P) = \exp\left(\sum_1^\infty x_k \left(\int_{P_0}^{P} \Omega_{P_\infty}^{(k)} - b_{k_0}\right)\right)$$

$$\frac{\theta(\mathfrak{E}(P) + \sum_1^\infty x_k \mathbf{U}_k - \mathfrak{E}(D) - \mathbb{K})\theta(\mathfrak{E}(P_\infty) - \mathfrak{E}(D) - \mathbb{K})}{\theta(\mathfrak{E}(P) - \mathfrak{E}(D) - \mathbb{K})\theta(\mathfrak{E}(P_\infty) + \sum_1^\infty x_k \mathbf{U}_k - \mathfrak{E}(D) - \mathbb{K})}.$$

Here $\Omega_{P_\infty}^{(k)}$ are the Abel differential of the second kind, see 8.A.5, $\int_{P_0}^{P} \Omega_{P_\infty}^{(k)} = z^k + \sum_0^\infty b_{kr} z^{-r}$ where z^{-1} is a local parameter in P_∞, $2\pi i\, \mathbf{U}_k$ are the vectors of β-periods of the differentials $\Omega_{P_\infty}^{(k)}$: $2\pi i U_{kj} = \int_{\beta_j} \Omega_{P_\infty}^{(k)}$, and $\mathbb{K}$ the Riemann vector (8.A.4), $\mathfrak{E}(P)$ is the Abel mapping of the point P, $\mathfrak{E}(D)$ is the Abel mapping of the divisor D.

Proof: The following facts must be verified. (i) That the function $w(x, P)$ determined by this formula is single-valued on the Riemann surface. (ii) That it belongs to $\mathcal{O}(X_0)$ in X_0. (iii) That in X_∞ it can be written as $\exp(\sum x_k z^k)(1 + \sum_1^\infty a_i z^{-c})$. All this easily follows from properties of the θ-function. When passing round the α-contours $w(x, P)$ evidently does not change. If it is the β_j contour, the exponent is multiplied by $\exp(2\pi i \sum x_k U_{kj})$ and the quotient of θ-functions is multiplied by $\exp(-2\pi i \sum x_k U_{kj})$, i.e. w does not change too. Thus, (i) is proved. This formula defines w on the whole Riemann surface except at $P = P_\infty$, the only singularities being the points of the divisor D, which are simple poles (see 8.A.4). This proves (ii). Finally, if $P \to P_\infty$ the quotient of the θ-functions tends to 1 and the exponent can be represented as $\exp \sum_1^\infty x_k z^k \left(1 + \sum_1^\infty a_i z^{-i}\right)$, as required. (In the KdV-equation theory, expressions like this with quotients of θ-functions first appeared in an article by Its [50].)

8.2.7. Proposition. The τ-function has the form

$$\tau_w(x_1, x_2, \ldots) = \exp(\sum \lambda_i x_i + \mu_{ij} x_i x_j) \theta(\alpha(P_\infty) + \sum_1^\infty x_k\, \mathbf{U}_k - \mathfrak{E}(D) - \mathbb{K}),$$

where $\{\lambda_i\}$ and $\{\mu_{ij}\}$ are constants.

Proof: Equation (7.7.2) has to be verified. Using 8.2.6 we can write $\exp(-\sum x_k z^k) w$ in X_∞ as

$$\exp\left(\sum_{r,k=1}^\infty b_{kr} z^{-k} x_k\right) \cdot \left(1 + \sum_1^\infty a_i z^{-i}\right)$$
$$\times \frac{\theta(\mathfrak{E}(P) + \sum_1^\infty x_k \mathbf{U}_k - \mathfrak{E}(D) - \mathbb{K})}{\theta(\mathfrak{E}(P_\infty) + \sum_1^\infty x_k \mathbf{U}_k - \mathfrak{E}(D) - \mathbb{K})}.$$

The first two multipliers can easily be represented as

$$\exp(\sum \lambda_i(x_i - 1/iz^l)$$
$$+ \sum \mu_{ij}(x_i - 1/iz^l)(x_j - 1/jz^j)/\exp(\sum \lambda_i x_i + \sum \mu_{ij} x_i x_j)$$

if $\{\lambda_i\}$ and $\{\mu_{ij}\}$ are properly chosen. Then we transform the θ-function in the numerator:

$$\theta(\mathfrak{E}(P_\infty) + \mathfrak{E}(P) - \mathfrak{E}(P_\infty) + \sum x_k \, \mathbb{U}_k - \mathfrak{E}(D) - \mathbb{K}).$$

According to (8.A.6) we have

$$(\mathfrak{E}(P) - \mathfrak{E}(P_\infty))_j = \int_{P_\infty}^{P} \omega_j = \int_0^{z^{-1}} \varphi_j(z^{-1}) dz^{-1}$$
$$= \sum_0^\infty z^{-(k+1)} \varphi_j^{(k)}(0)/(k+1)!$$
$$= -\sum_0^\infty z^{-k-1} \int_{\beta_i} \Omega_{P_\infty}^{(k+1)}/2\pi i(k+1)$$
$$= -\sum_0^\infty z^{-k-1} U_{k+1,j}/(k+1) = -\sum_1^\infty z^{-k} U_{k,j}/k.$$

This implies

$$\theta(\mathfrak{E}(P) + \sum_1^\infty x_k \, \mathbb{U}_k - \mathfrak{E}(D) - \mathbb{K}) = \theta(\mathfrak{E}(P_\infty) + \sum_1^\infty ((x_k - 1/kz^k) \, \mathbb{U}_k$$
$$- \mathfrak{E}(D) - \mathbb{K}).$$

The rest is clear. □

8.A. Appendix: Abel mapping and the θ-function.

8.A.1. If the genus of a Riemann surface is g this surface is homeomorphic to a sphere with g handles. Such a basic system of closed paths (or contours) $\alpha_1, \ldots, \alpha_g, \beta_1, \ldots, \beta_g$ can be chosen such that the only intersections among them are those of α_i and β_i with the same numbers i.

Let the Riemann surface be covered with charts (U_i, z_i), where z_i are local parameters in open domains U_i, the transition from z_i to z_j

in intersections $U_i \cap U_j$ being holomorphic. If in any U_i a differential $\varphi_1(z_i)dz_i$ with meromorphic $\varphi_i(z_i)$ is given and in the common parts $U_i \cap U_j, \varphi_i(z_i)dz_i = \varphi_j(z_j)dz_j$, then we say that there is an Abel differential Ω on the whole surface with restrictions $\Omega|_{v_i} = \varphi_i(z_i)dz_i$. The Abel differential is of the first kind if all the $\varphi_i(z_i)$ are holomorphic. There are exactly g linearly independent differentials of the first kind $\omega_1, \ldots, \omega_g$. They are normed if $\int_{\alpha_i} \omega_j = \delta_{ij}$, which condition determines them uniquely. We shall always assume them normed. The numbers $\int_{\beta_i} \omega_j = B_{ij}$ are called β-periods. The matrix $B = (B_{ij})$ has the following properties: 1) $B_{ij} = B_{ji}$, 2) $\tau = \operatorname{Im} B$ is a positive definite matrix.

We consider a g-dimensional vector $\mathfrak{E}(P) = \left\{ \int_{P_0}^{P} \omega_j \right\}$, where P_0 is a fixed point of the Riemann surface and P is an arbitrary point. This vector is not uniquely determined, but depends on the path of integration. If the latter is changed then a linear combination of α and β-periods with integer coefficients can be added: $(\mathfrak{E}(P))_j \mapsto (\mathfrak{E}(P))_j + \sum_1^g n_i \delta_{ij} + \sum_1^g m_i B_{ij}$, i.e. $\mathfrak{E}(P) \mapsto \mathfrak{E}(P) + \sum n_i \delta_i + \sum m_i \mathbb{B}_i$, where δ_i is the vector with coordinates δ_{ij}, $\mathbb{B}_i$ is the vector with coordinates B_{ij}. Thus $\mathfrak{E}(P)$ determines a mapping of the Riemann surface on the torus $\mathbb{J} = \mathbb{C}^g/T$, where T is the lattice generated by $2g$ vectors $\{\delta_i, \mathbb{B}_i\}$ (which are linearly independent over $\mathbb{R}$). This mapping is called the Abel mapping, and the torus $\mathbb{J}$ is the Jacobi manifold or the Jacobian or the Riemann surface.

The Abel mapping extends by linearity to the divisors: $\mathfrak{E}\left(\sum n_k P_k\right) = \sum n_k \mathfrak{E}(P_k)$.

8.A.2. Abel Theorem. Those and only those divisors go to zero of the Jacobian by the Abel mapping which are principal. The latter means that they are divisors of zeros and poles of meromorphic functions on the surface. (If P_k is a zero of the function, then $n_k > 0$ and n_k is the degree of this zero. If P_k is a pole, then $n_k < 0$ and $|n_k|$ is the degree of this pole.)

Of special interest is the case of divisors of degree g with all $n_k = 1$, i.e. of non-ordered sets of g points $P_1, \ldots, P_g$ of the Riemann surface. All the sets of such kind form the symmetrical g-th power of the Riemann surface. The Abel mapping has the form

$$\mathfrak{E}(P_1, \ldots, P_g) = \left\{ \sum_{i=1}^{g} \int_{P_0}^{P} \omega_j \right\}, j = 1, \ldots, g.$$

The symmetrical g-th power is a complex manifold of the complex dimension g and it is mapped on the Jacobian which is a manifold of the same dimension. A problem of conversion arises (the Jacobi inverse problem). It can be solved with the help of the θ-function.

For $\mathbb{p} \in \mathbb{C}^g$ arbitrary, let

$$\theta(\mathbb{p}) = \sum_{\mathbb{k} \in \mathbb{Z}^g} \exp\{\pi i (B\mathbb{k}, \mathbb{k}) + 2\pi i(\mathbb{p}, \mathbb{k})\}, (\mathbb{B}\mathbb{k}, \mathbb{k})$$
$$= \sum B_{ij} k_i k_j, (\mathbb{p}, \mathbb{k}) = \sum p_i k_i.$$

The series converges owing to the properties of the matrix B. The θ-function has the properties

(8.A.3)
$$\theta(-\mathbb{p}) = \theta(\mathbb{p}); \theta(\mathbb{p} + \delta_k) = \theta(\mathbb{p});$$
$$\theta(\mathbb{p} + \mathbb{B}_k) = \theta(\mathbb{p}) \exp\{-\pi i(B_{kk} + 2p_k)\}.$$

Note that the θ-function is not defined on the Jacobian because of the latter property.

8.A.4. *Riemann Theorem.* There are constants $\mathbb{K} = \{k_i\}, i = 1, \ldots, g$ (Riemann constants) determined by the Riemann surface such that the set of points $P_1, \ldots, P_g$ is a solution of the system of equations

$$\sum_{i=1}^{g} \int_{P_0}^{P_i} \omega_j = l_j, \mathbb{l} \in \mathbb{J}$$

if and only if $P_1, \ldots, P_g$ are the zeros of the function $\tilde{\theta}(P) = \theta(\mathfrak{E}(P) - \mathbb{l} - \mathbb{K})$ (which has exactly g zeros). Note that while the function $\tilde{\theta}(P)$ is not uniquely determined on the Riemann surface (it is multivalued) its zeros are, since distinct branches of $\tilde{\theta}(P)$ differs by exponents.

8.A.5. We define now Abel differentials of the second and of the third kind.

The Abel differential of the second kind, $\Omega_P^{(k)}, k = 1, 2, \ldots$ has the only singularity at the point P which is a pole of the order $k+1$. The differential can be represented at this point as dz_+^{-k} (holomorphic differential), z being the local parameter at this point. Such a differential is uniquely determined if it is normed: $\forall i \int_{\alpha_i} \Omega_P^{(k)} = 0$.

The Abel differential of the third kind Ω_{PQ} has only singularities which are simple poles at the points P and Q with the residues $+1$ and -1, respectively. It is uniquely determined by the same condition.

8.A.6. Proposition. If z is a local parameter in a neighbourhood of the point P and $\omega_i = \varphi_i(z)dz$ is the Abel differential of the first kind, then

$$\frac{1}{2\pi i}\int_{\beta_i}\Omega_P^{(k)} = -\frac{1}{(k-1)!}\frac{d^{k-1}}{dz^{k-1}}\varphi_i(z)\Big|_{z=0}, i = 1,\ldots,g$$

and

$$\frac{1}{2\pi i}\int_{\beta_i}\Omega_{PQ} = \int_Q^P \omega_i, i = 1,\ldots,g$$

(see e.g. Ref. [49] (6.11) and (6.12)).

CHAPTER 9. Matrix First-order Operators.

9.1. Hierarchy of equations generated by a first-order matrix differential operator.

9.1.1. An essentially new hierarchy of non-linear equations which is generated by, instead of the operator L (1.1.2) or its generalization in 5.2.1, a matrix operator

$$(9.1.2) \qquad L = -I\partial + U, \partial = \partial/\partial x$$

where I is a unity matrix $n \times n$, $U = (u_{ij})$, $u_{ii} = 0$. We shall simply write $-\partial + U$, dropping I. The elements of the matrix U will be the differential generators of a differential algebra $\mathcal{A}_u = \mathcal{A}$. More general cases can be considered if instead of the condition $u_{ii} = 0$, we take a weaker condition tr $U = 0$ or even drop this condition at all. In this chapter, however, the strictest condition will be assumed.

Any linear differential n-th order equation can be reduced to a first-order system. This enables us to consider the KdV-hierarchies as reductions of the new hierarchy (Chap. 10). In this sense the new theory is more general. The special case $n = 2$ leads to the hierarchy called the AKNS-hierarchy [52]. The general matrix case was studied by Dubrovin [53] whose work will be used here very extensively. We add to this everything concerning the Hamiltonian structures of this hierarchy and its stationary variant (Chap. 13) including integrating according to the Liouville procedure, [54].

9.1.3. Let

$$(9.1.4) \qquad \hat{L} = -\partial + U + \zeta A = L + \zeta A$$

where ζ is a parameter (similar to z earlier), A is a diagonal matrix $A = \mathrm{diag}(a_1, \ldots, a_n)$ with distinct constant elements.

Resolvents are solutions $R = \sum_{i_0}^{\infty} R_i \zeta^{-i}$ of the equation

$$(9.1.5) \qquad [\hat{L}, R] = 0 \text{ i.e. } -R' + [U + \zeta A, R] = 0.$$

We shall prove that the elements of R_i belong to $\mathcal{A}$.

9.1.6. *Proposition.* The set of all resolvents is an algebra over the field of the formal series $c(\zeta) = \sum_{i_1}^{\infty} c_i \zeta^{-i}$ with constant coefficients.

Proof: If $R^{(1)}$ and $R^{(2)}$ commute with $\hat{L}$, then so do any linear combination $c^{(1)}(\zeta) R^{(1)} + c^{(2)}(\zeta) R^{(2)}$ and $R^{(1)} R^{(2)}$. □

9.1.7. How a solution of Eq. (9.1.5) can be constructed? Without loss of generality one can take $i_0 = 0$. This can always be achieved by multiplication by a suitable power of ζ. Eq. (9.1.5) is equivalent to a recurrence relation

$$(9.1.8) \qquad -R'_i + [U, R_i] = [R_{i+1}, A], i = -1, 0, 1, 2, \ldots$$

The first of these equations, for $i = -1$, implies that R_0 is diagonal; let $R_0 = B = \mathrm{diag}(b_1, \ldots, b_n)$. We separate the second equation $-R'_0 + [U, R_0] = [R_1, A]$ into two parts: the diagonal part yields $R'_0 = 0$ i.e. $R_0 = B = \mathrm{const}$, the non-diagonal part enables us to determine the non-diagonal elements of R_1. They are uniquely determined and belong to $\mathcal{A}$. We go on with the same procedure. The non-diagonal part of the i-th equation determines the non-diagonal elements of R_{i+1} as differential polynomials in the elements of the previous matrices. The diagonal part of the equation is to determine the diagonal part of the matrix R_i, $\mathrm{diag} R_i$, but unfortunately it gives only its derivative $(\mathrm{diag} R_i)'$ expressed in terms of differential polynomials in elements of the previously determined matrices and $\{u_{ij}\}$. We shall prove that the integration of these expressions can always be performed in $\mathcal{A}$.

9.1.9. Exercise. Write expressions for R_0, R_1 and R_2.
Answer:

$$(R_0)_{jk} = b_j \delta_{jk}, (R_1)_{jk} = \frac{b_j - b_k}{a_j - a_k} u_{jk} (j \neq k), (R_1)_{jj} = 0,$$

$$(R_2)_{jk} = -\frac{b_j - b_k}{(a_j - a_k)^2} u'_{jk} - \sum_{\beta \neq j,k} \frac{u_{j\beta} u_{\beta k}}{a_j - a_k} \left(\frac{b_j - b_\beta}{a_j - a_\beta} - \frac{b_\beta - b_k}{a_\beta - a_k} \right) (j \neq k),$$

$$(R_2)_{jj} = \sum_{\beta \neq j} \frac{b_\beta - b_j}{(a_\beta - a_j)^2} u_{j\beta} u_{\beta j}.$$

9.1.10. Lemma. If R is a resolvent then tr R = const.
Proof: It suffices to take the trace of the equation (9.1.5). □

9.1.11. Proposition. All the elements of R_k belong to $\mathcal{A}$.
Proof: As it was already stated, the recurrence procedure requires integration and it is not known beforehand whether this integration can be performed within the algebra $\mathcal{A}$. To have a formal right to do this we must properly extend the algebra. In the appendix to this chapter we prove that there is such an algebra $\mathcal{A}_1 \supset \mathcal{A}$ that for every $f \in \mathcal{A}$ and $n \in \mathbb{Z}_+$ an element $g \in \mathcal{A}_1$ exists such that $g^{(n)} = f$. The elements of all R_k belong to $\mathcal{A}_1$ and we must prove that they do in fact belong to $\mathcal{A}$. We use induction. Let it be already proved that for some i the non-diagonal elements of matrices R_j for $j < i+1$, and the diagonal elements of R_j for $j < i$ all belong to $\mathcal{A}$. According to 9.1.6 R^l for any l is a resolvent. Thus, tr $R^l = c_l(\zeta) = $ const (see 9.1.10). The coefficient in ζ^{-i} yields tr $R_0^{l-1} R_i + F_l(R_j) = (c_l)_i$, where $F_l(R_j)$ is a differential polynomial in the elements of the matrices R_j with $j < i$, i.e. belongs to $\mathcal{A}$. In other words, $\sum_\beta b_\beta^{(l-1)}(R_i)_{\beta\beta} = (c_l)_i - F_l(R_j), l = 1, \ldots, n$. At first we can assume that all the elements $\{b_i\}$ are distinct. Then this system of equations determines all the diagonal elements $(R_i)_{\beta\beta}$ as elements of $\mathcal{A}$. From the non-diagonal part of Eq. (9.1.8), the non-diagonal terms of R_{i+1} can be obtained. The restriction that $\{b_i\}$ are distinct is not important as the general case can be obtained by a transition to the limit, for the elements of the matrices $\{R_i\}$ depend on $\{b_j\}$ linearly and therefore continuously. □

9.1.12. Proposition. Resolvents are uniquely determined by constants in the diagonal elements of all the matrices $\{R_i\}$.
Proof: It suffices to prove that a resolvent vanishes if there are no constants in the diagonal parts of all its terms $\{R_i\}$. We have already seen in (9.1.9) that the first non-vanishing term is always a constant diagonal matrix. □

9.1.13. Proposition. The algebra of the resolvents (see 9.1.6) is n-dimensional. As a basis of the algebra the resolvents $R^\alpha, \alpha = 1, \ldots, n$ can be chosen which start with the terms

$$R_0^\alpha = E_\alpha = \begin{pmatrix} 0 & & & & 0 \\ & \ddots & & & \\ \ldots & & 1 & & \ldots \\ & & & \ddots & \\ 0 & & & & 0 \end{pmatrix} - \alpha \text{ place}$$

and contain no constants in other terms. Therefore, any resolvent can be represented as

$$(9.1.14) \qquad R = \sum_{\alpha=1}^{n} c_\alpha(\xi) R^\alpha .$$

where R^α are spectral projectors and $c_\alpha(\xi)$ the eigenvalues of the matrix R.

Proof: It is sufficient to note that a proper choice of the series $c_\alpha(\xi)$ can provide any set of constants in the diagonal parts of all $\{R_k\}$. Further, $(R^\alpha)^2$ has the same sole constant as R^α, hence $(R^\alpha)^2 = R^\alpha$ and R^α is a projector; $R^\alpha R^\beta = 0 (\alpha \neq \beta)$ since the resolvent $R^\alpha R^\beta$ has no constants. Finally, $\sum R^\alpha = I$ because if a resolvent has $R_0 = I$ and no other constant it coincides with the resolvent $R = I$. Thus, $\{R^\alpha\}$ form a spectral decomposition of unity. We have $RR^\alpha = c_\alpha(\xi) R^\alpha$, which implies that $c_\alpha(\xi)$ is the eigenvalue corresponding to the spectral projector R^α. □

9.1.15. Proposition. Any two resolvents commute.

Proof. They have a diagonal form in one and the same basis. □

Resolvent starting from $R_0 = B$ and having no other constants will sometimes be denoted by R^B.

9.1.16. Let t be a variable. The derivatives with respect to it will be denoted by dots. We consider an equation

$$(9.1.17) \qquad \dot{U} = [A, R_{k+1}] .$$

These equations form a hierarchy which we shall study in this chapter.

Equation (9.1.17) can be written in equivalent forms. Taking (9.1.8) into account we can write

$$(9.1.18) \qquad \dot{U} - R_k' = [R_k, U]$$

An equation in this form is called the equation of the zero curvature (without parameters). It can also be written as a Lax equation

(9.1.19) $$\dot{L} = [R_k, L]$$

Denoting $V = \sum_{i=0}^{k} R_i \zeta^{k-i}$ we obtain $[-\partial + U + \zeta A, V] = -[A, R_{k+1}] = -\dot{U}$, i.e.

(9.1.20) $$-V' + (U + \zeta A)^\bullet = [V, U + \zeta A]$$

This is also the equation of the zero curvature but with a parameter ζ.

The equation with a parameter (9.1.20) combines Eq. (9.1.18) and Eq. (9.1.8) which determines the terms of a resolvent with $i < k$. Thus, Eq. (9.1.20) completely determines the system. It is often called the Zakharov-Shabat equation. However the equation for resolvents will also be useful in the other terms, $i \geq k$, to construct first integrals.

9.1.21. It is convenient to consier simultaneously all equations of the hierarchy, for all resolvents, as it has already be done for the KP hierarchy. It suffices to take the base resolvents R^α and put $V_{k\alpha} = \sum_{i=0}^{k} R_i^\alpha \zeta^{k-i} = (R^\alpha \zeta^k)_+$ (the subscript + has an obvious menaing here). Let $t_{k\alpha}$ be time variables. Then the hierarchy is a set of equations

(9.1.22) $$\partial_{k\alpha} \hat{L} = [V_{k\alpha}, \hat{L}], \partial_{k\alpha} = \partial/\partial t_{k\alpha}.$$

9.1.23. *Proposition.* By virtue of Eq. (9.1.22)

$$\partial_{k\alpha} R^\beta = [V_{k\alpha}, R^\beta].$$

Proof: We have $[\hat{L}, R^\beta] = 0$ which yields

$$0 = \partial_{k\alpha}[\hat{L}, R^\beta] = [[V_{k\alpha}, \hat{L}], R^\beta]$$
$$= -[[R^\beta, V_{k\alpha}], \hat{L}] + [\hat{L}, \partial_{k\alpha} R^\beta] = [\hat{L}, \partial_{k\alpha} R^\beta - [V_{k\alpha}, R^\beta]].$$

Hence, $\partial_{k\alpha} R^\beta - [V_{k\alpha}, R^\beta]$ is a resolvent. As it is easily seen that this resolvent does not contain constants it must vanish which completes the proof. □

9.1.24. Proposition. The equation

$$\partial_{k\alpha} V_{l\beta} - \partial_{l\beta} V_{k\alpha} = [V_{k\alpha}, V_{l\beta}]$$

holds.

Proof: Using $[R^\alpha, R^\beta] = 0$ we have:

$$\partial_{k\alpha}(\zeta^l R^\beta)_+ - \partial_{l\beta}(\zeta^k R^\alpha)_+ - [V_{k\alpha}, V_{l\beta}]$$
$$=(\zeta^l[V_{k\alpha}, R^\beta])_+ - (\zeta^k[V_{l\beta}, R^\alpha])_+ - [V_{k\alpha}, V_{l\beta}]$$
$$= -[(\zeta^k R^\alpha)_+, \zeta^l R^\beta]_+ - [(\zeta^l R^\beta)_+, \zeta^k R^\alpha]_+ - [V_{k\alpha}, V_{l\beta}]$$
$$=[(\zeta^l R^\beta)_+, (\zeta^k R^\alpha)_- - \zeta^k R^\alpha]_+ - [V_{k\alpha}, V_{l\beta}]$$
$$= -[V_{l\beta}, V_{k\alpha}] - [V_{k\alpha}, V_{l\beta}] = 0 \quad \square$$

9.1.25. Proposition. Vector fields $\partial_{k\alpha}$ and $\partial_{l\beta}$ commute.

Proof: Let us calculate the action of $[\partial_{k\alpha}, \partial_{l\beta}]$ on generators of the differential algebra $\mathcal{A}$:

$$(\partial_{k\alpha}\partial_{l\beta} - \partial_{l\beta}\partial_{k\alpha})\hat{L} = \partial_{k\alpha}[V_{l\beta}, \hat{L}) - \partial_{l\beta}[V_{k\alpha}, \hat{L}]$$
$$=[\partial_{k\alpha} V_{l\beta} - \partial_{l\beta}\partial_{k\alpha}, \hat{L}] + [V_{k\beta}, [V_{k\alpha}, \hat{L}]] - [V_{k\alpha}, [V_{l\beta}, \hat{L}]]$$
$$=[[V_{k\alpha}, V_{l\beta}], \hat{L}] + [V_{l\beta}, [V_{k\alpha}, \hat{L}]] - [V_{k\alpha}, [V_{l\beta}, \hat{L}]] = 0$$

according to the Jacobi identity. $\square$

9.1.26. Exercise. Prove the identity

$$\sum_\alpha a_\alpha \partial_{1\alpha} = \partial.$$

(Recall that in the case of the KP-hierarchy there was an identity $\partial_1 = \partial$ which enabled us to identify x_1 and x. Now there are many time variables with $k = 1 : t_{1\alpha}$. Variables $t_{1\alpha}, x$ proved to be dependent, the dependence of solutions on all the variables is the following: $f(t_{0,\alpha}, t_{1\alpha} + a_\alpha x, t_{2\alpha}, \dots)$.)

9.1.27. Exercise. Prove that flows defined by vector fields $\partial_{0\alpha}$ describe similarity transformations of the matrix U. More precisely,

$$U = \exp\left(\sum_\alpha E_\alpha t_{0\alpha}\right) U(t_{1\alpha}, t_{2\alpha}, \dots) \exp\left(-\sum_\alpha E_\alpha t_{0\alpha}\right).$$

9.1.28. The equations under consideration can be reduced to matrix submanifolds, namely, Lie subalgebras. For example, let the matrix U belong to $SU(n,\mathbb{C})$, the Lie algebra of skew symmetrical matrices with traces equal to zero. It is convenient to choose the basis

$$S_1 = \frac{i}{2}\begin{pmatrix} 1 & 0 \\ 0 & -1 \end{pmatrix}, S_2 = \frac{1}{2}\begin{pmatrix} 0 & 1 \\ -1 & 0 \end{pmatrix}, S_3 = \frac{1}{2}\begin{pmatrix} 0 & i \\ i & 0 \end{pmatrix}$$

(these are Dirac's matrices, up to a coefficient 1/2). The commutations rules are $[S_1, S_2] = S_3, [S_2, S_3] = S_1, [S_3, S_1] = S_2$. Let $A = S_1$. Resolvent R will also be of $SU(2,\mathbb{C})$. Let $R = R^{(1)}S_1 + R^{(2)}S_2 + R^{(3)}S_3$, and $U = U_2 S_2 + U_3 S_3$. Eq. (9.1.8) take the form

$$-R_k^{(1)'} + u_2 R_k^{(3)} - u_3 R_k^{(2)} = 0$$
$$-R_k^{(2)'} + u_3 R_k^{(1)} = R_{k+1}^{(3)}$$
$$-R_k^{(3)'} - u_2 R_k^{(1)} = -R_{k+1}^{(2)}$$

and Eq. (9.1.17) take the form

$$\dot{u}_2 - R_k^{(2)'} + u_3 R_k^{(1)} = 0$$
$$\dot{u}_3 - R_k^{(3)'} - u_2 R_k^{(1)} = 0$$

9.1.29. Exercise. Write these equations for $k = 2$ and $k = 3$.
Answer: if $k = 2$, then

$$\dot{u}_2 - u_3'' - \frac{1}{2}u_3(u_2^2 + u_3^2) = 0$$
$$\dot{u}_3 + u_2'' + \frac{1}{2}u_2(u_2^2 + u_3^2) = 0$$

or, letting $u = u_2 + iu_3$

(9.1.30) $$-i\dot{u} + u'' + \frac{1}{2}u|u|^2 = 0.$$

This equation is called the non-linear Schrödinger equation.

If $k = 3$ then

(9.1.31) $$\dot{u} + u''' + \frac{3}{2}|u|^2 u' = 0.$$

This equation can be restricted to the real u:

(9.1.32) $$\dot{u} + u''' + \frac{3}{2}u^2 u' = 0.$$

This is none other than the equation mKdV.

9.2. Hamiltonian structure.

9.2.1. Now we shall not assume $\{u_{ii}\} = 0$. Later on, it will be shown how to reduce the constructed structures to the submanifold $\{u_{ii}\} = 0$.

We shall construct all the elements of a Hamiltonian structure according to Sec. 2.5. To each matrix a a differentiation in $\mathcal{A}$ corresponds:

$$\partial_a = \sum a_{jk}^{(i)} \partial/\partial u_{jk}^{(i)} = \sum \operatorname{tr} a^{(j)} \partial/\partial u^{(j)}, (\partial/\partial u^{(i)})_{jk} = \partial/\partial u_{kj}^{(i)}.$$

These differentiations commute with ∂, and therefore they can be transferred to $\tilde{\mathcal{A}}$, the space of the functionals $\tilde{f} = \int f\,dx, f \in \mathcal{A}$. A relation
(9.2.2)

$$\partial_a \tilde{f} = \int \sum a_{kl} \delta f/\delta u_{kl}\, dx = \int \operatorname{tr} a\delta f/\delta u\, dx, (\delta f/\delta u)_{kl} = \delta f/\delta u_{lk}$$

holds.

Now, for the Lie algebra $\mathfrak{E}$ we take the space of all vector fields $\{\partial_a\}$, and for Ω^0 the space of the functionals $\tilde{\mathcal{A}}$; the dual space Ω^1 consists of the same matrices $X, X_{kl} \in \mathcal{A}$. The coupling between elements of $\mathfrak{E}$ and of Ω^1 is given by the formula

$$\langle \partial_a, X \rangle = \int \operatorname{tr} aX\, dx.$$

We construct the Hamilton mapping $H: \Omega^1 \to \mathfrak{E}$:

$$X \in \Omega^1 \mapsto H(X) = -X' + [U + \varsigma A, X] = [\hat{L}, X] \mapsto \partial_{[\hat{L}, X]} \in \mathfrak{E}.$$

9.2.3. Proposition.

$$[\partial_{H(X)}, \partial_{H(Y)}] = \partial_{H[X,Y] + \partial_{H(X)} Y - \partial_{H(Y)} X}.$$

Proof: $\partial_{H(X)} U = H(X)$;

$$[\partial_{[-\partial + U + \varsigma A, X]}, \partial_{[-\partial + U + \varsigma A, Y]}]$$
$$= \partial_{[[-\partial + U + \varsigma A, X], Y] - [(-\partial + U + \varsigma A, Y], X] + H(\partial_{H(X)} Y) - H(\partial_{H(Y)} X)}$$
$$= \partial_{-[X', Y] + [Y', X] + [[U + \varsigma A, X], Y] + [X, [U + \varsigma A, Y]] + H(\partial_{H(X)} Y - \partial_{H(Y)} X)}$$
$$= \partial_{[-\partial + U + \varsigma A, [X, Y]] + \partial_{H(X)} Y - \partial_{H(Y)} X}.$$

Thus, $l = \operatorname{Im} H$ is a Lie subalgebra of the algebra $\mathfrak{E}$. □

We define the form ω as usual

(9.2.4) $$\omega(\partial_{H(X)}, \partial_{H(Y)}) = \langle H(X), Y \rangle = \int \operatorname{tr} H(X) Y\, dx.$$

9.2.5. Proposition. The form ω is closed.

Proof:

$$d\omega(\partial_{H(X_1)}, \partial_{H(X_2)}, \partial_{H(X_3)}) = \partial_{H(X_1)} \int \text{tr } H(X_2) X_3 dx$$

$$+ \int \text{tr } H(X_3)([X_1, X_2] + \partial_{H(X_1)} X_2 + \partial_{H(X_2)} X_2) dx$$

$$+\text{c.p.} = \int \text{tr } [H(X_1), X_2] X_3 dx + \int \text{tr } H(X_2) \partial_{H(X_1)} X_3 dx$$

$$- \int \text{tr } H(X_3) \partial_{H(X_1)} X_2 dx + \int \text{tr } H(X_3)[X_1, X_2] dx$$

$$+ \int \text{tr } H(X_3)(\partial_{H(X_1)} X_2 - \partial_{H(X_2)} X_1) dx + \text{c.p.}$$

$$= 2 \int \text{tr } H(X_3)[X_1, X_2] dx + \text{c.p.} = 2 \int \text{tr } (-X_3' X_1 X_2 + X_3' X_2 X_1$$

$$+ [U + \varsigma A, X_3][X_1, X_2]) dx + \text{c.p.} = 2 \int \text{tr } (U + \varsigma A)[X_3, [X_1, X_2]] dx$$

$$+\text{c.p.} + 2 \int \text{tr } ((X_3 X_2 X_1)' - (X_3 X_1 X_2)') dx = 0. \ \square$$

A vector field corresponds to each functional:

(9.2.6) $\tilde{f} \mapsto \partial_{H(\delta f/\delta U)} = \partial_{[-\partial + U + \varsigma A, \delta f/\delta U]}.$

The Poisson bracket can be written as

(9.2.7) $\{\tilde{f}, \tilde{g}\} = \int \text{tr } H(\delta f/\delta U) \cdot \delta g/\delta U dx.$

with limiting cases $\varsigma = \infty$ and $\varsigma = 0$:

$$\{\tilde{f}, \tilde{g}\}^{(\infty)} = \int \text{tr } [A, \delta f/\delta U] \delta g/\delta U dx,$$

$$\{\tilde{f}, \tilde{g}\}^{(0)} = \int \text{tr } [-\partial + U, \delta f/\delta U] \delta g/\delta U dx.$$

A few words about reduction to the manifold diag $U = 0$. This reduction is possible if the mapping H has the property diag$H(X) = 0$. For $H^{(\infty)}$ this condition is automatically fulfilled. For $H^{(0)}$ it is needed that the diagonal part of the matrix X is connected with the non-diagonal one by the equation diag$(-X' + [U, X]) = 0$. The variational derivative $\delta f/\delta U$

must be understood in the following sense. The non-diagonal elements of this matrix are variational derivatives and the diagonal ones are determined by $\mathrm{diag}(-(\delta f/\delta U)'+[U,\delta f/\delta U])=0$. They are not involved in the Poisson bracket.

9.2.8. The above constructions admit a group theory interpretation in the spirit of the theory of coadjoint representation of a Lie algebra i.e. Poisson-Lie-Berezin-Kirillov-Kostant bracket (sec. 2.4). This easily can be done for H^∞. The dual space Ω^1 has a Lie algebra structure with respect to the commutator $[X,Y] = XY - YX$. The Lie algebra $\mathfrak{E}$ is considered now as a dual space to Ω^1 (a Lie algebra structure in $\mathfrak{E}$ plays no role here). The coupling is as usual, $\int \mathrm{tr}\, aX dx$. We have $ad(X)Y = [X,Y]$ and $\langle a, ad(X)Y \rangle = \int \mathrm{tr}\, a[X,Y]dx = \int \mathrm{tr}\, Y[a,x]dx$ which yield $ad^*(X)a = [a,X]$. The elements $ad^*(X)a$ are tangent to the orbit at the point $a = A$. The symplectic form is

$$\omega([A,X],[A,Y]) = \int \mathrm{tr}\,[A,X]Y\,dx$$

which is exactly of the form $\omega^{(\infty)}$ defined above.

Interpretation of the form $\omega^{(0)}$ is more difficult. For this purpose we need the notion of the central extension of Lie algebra with the help of a cocycle. Recall (see 2.5.5) the definition of the cohomologies of Lie algebras (see also 3.7). A function $F(X,Y)$, where X and Y are elements of the algebra, is called a cocycle if $F([X,Y],Z) + \mathrm{c.p.} = 0$. In our case an example of a cocycle is $F(X,Y) = -\int \mathrm{tr}\, X'Y dx$. The cocycle property can be easily verified. The central extension of the Lie algebra Ω^1 is the Lie algebra $\tilde{\Omega}^1$ whose elements are pairs $(\tilde{f},X); \tilde{f} \in \mathcal{A}, X \in \Omega^1$. The commutator is

$$[(\tilde{f},X),(\tilde{g},Y)] = \left(-\int \mathrm{tr}\, X'Y dx, [X,Y]\right).$$

The Jacobi identity is equivalent to the cocycle property. The dual space comprises pairs (λ, a), where $a \in \mathfrak{E}, \lambda \in \mathbb{R}$ or $\mathbb{C}$. The coupling is $\langle (\lambda,a),(\tilde{f},X)\rangle = \lambda \tilde{f} + \int \mathrm{tr}\, aX dx \in \tilde{\mathcal{A}}$. We have $\langle (\lambda,a),(-\int \mathrm{tr}\, X'Y dx, [X,Y])\rangle = \int \mathrm{tr}\,(-X'Y\cdot\lambda + a[X,Y])dx = \langle 0, -\lambda X' + [a,X])(\tilde{f}_1,Y)\rangle$, whence $ad^*(\tilde{f},X)(\lambda,a) = (0, -\lambda X' + [a,X])$. In particular, at the point $(1,U)$, we have

$$ad^*(\tilde{f},X)(1,U) = (0, -X' + [U,X])$$

The symplectic form is defined as

$$\omega(ad^*(\tilde{f},X)(1,U), ad^*(\tilde{g},Y)(1,U)) = \int \operatorname{tr}(-X' + [U,X])y\,dx$$

which coincides with $\omega^{(0)}$ above.

Note that in our context the group theoretical interpretation is not obligatory. This point of view is especially emphasized in the works by Gelfand and Dorfman [26–28]. On the other hand, many other authors, prefer considering Hamiltonian structures as always connected with group theory properties.

9.3. Hamiltonians of the matrix hierarchy.

9.3.1. Proposition. The formal series

$$\Phi = \sum_0^\infty \phi_i \varsigma^{-i}, \Lambda = \sum_{-1}^\infty \lambda_i \varsigma^{-i}, \phi_0 = I$$

exist, where λ_i are diagonal matrices, and the elements of matrices ϕ_i and λ_i belong to $\mathcal{A}$, for which the equation

(9.3.2) $$-\partial + U + \varsigma A = \phi(-\partial + \Lambda)\phi^{-1}$$

holds. Φ is determined up to a multiplication to the right by the series $\sum_0^\infty C_i \varsigma^{-i}$, where $\{C_i\}$ are diagonal and constant. The choice of $\{C_i\}$ can be fixed by the requirement that $(\phi_i)_{jj} = 0, i > 0, j = 1, \ldots, n$.

Proof: Eq. (9.3.2) is equivalent to $\phi(-\partial + \Lambda) = (-\partial + U + \varsigma A)\phi = -\phi\partial + (U + \varsigma A)\phi - \phi'$, i.e.

(9.3.3) $$-\phi' + (U + \varsigma A)\phi = \phi\Lambda,$$

which, in turn, is equivalent to the recurrence equation

$$-\phi'_k + U\phi_k + A\phi_{k+1} = \sum_{i=-1}^k \phi_{k-i}\lambda_i, k = -1, 0, 1, 2, \ldots, (\phi_{-1} = 0).$$

We have

$$\lambda_{-1} = A, \lambda_0 = U_d, [\phi_1, A] = U_{nd}$$

(by U_d we denote a diagonal matrix with the same diagonal as U :, $U_{nd} = U - U_d$). Further

$$\lambda_k = \left(U\phi_k - \sum_{i=0}^{k-1} \phi_{k-i}\lambda_i\right)_d,$$

$$[A, \phi_{k+1}] = \phi_k' + \left(-U\phi_k + \sum_{i=0}^{k-1} \phi_{k-i}\lambda_i\right)_{nd}.$$

We put $(\phi_k)_d = 0$ and determine in succession λ_k and the non-diagonal part of ϕ_{k+1}. □

9.3.4. Proposition. The resolvent R^B, starting with $R_0 = B$ can be represented as $R^B = \phi B \phi^{-1}$.

Proof. It is clear that this expression can be expanded into a series starting from $R^0 = B$ and without other constants. Further,

$$[-\partial + U + \varsigma A, R^B] = \phi[-\partial + \Lambda, B]\phi^{-1} = 0 .\square$$

Equation (9.3.3) and 9.3.4 give the representation of the operator $\hat{L}$ and of the resolvent by dressing. Sometimes another dressing is used: $-\partial + U + \varsigma A = \phi(-\partial + A)\phi^{-1}$. Except for an unimportant advantage that the "bare" operator $-\partial + A$ has constant coefficients, this method has an essential disadvantage: determination of the matrices ϕ_k requires integration, and in increasing amount when k rises. Our matrices ϕ_k belong the same algebra $\mathcal{A}$.

9.3.5. Denoting $\psi = \phi^{-1}$, it is easy to find that ψ satisfies the conjugate equation. Let us rewrite the equations for ϕ and ψ:

(9.3.6)   (a) $-\phi' + (U + \varsigma A)\phi = \phi\Lambda$,   (b) $\phi' + \phi(U + \varsigma A) = \Lambda\psi$.

9.3.7. Proposition. A variational relation

$$\delta \text{ tr } AR = \text{ tr } R_\varsigma \delta U + \partial \text{ tr } (\delta\phi B\psi_\varsigma - \phi_\varsigma B\delta\psi)$$

holds (the subscript ς denotes derivation with respect to ς).
Proof: We apply the operator δ to Eqs. (9.3.6)

(9.3.8)   (a) $-\delta\phi' + (U + \varsigma A)\delta\phi + \delta U \cdot \phi = \delta\phi \cdot \Lambda + \phi \cdot \delta\Lambda$
           (b) $\delta\psi'' + \delta\psi(U + \varsigma A) + \psi \cdot \delta U = \Lambda \cdot \delta\psi + \delta\Lambda \cdot \psi$

and differentiate Eqs. (9.3.6) with respect to ς:

(9.3.9)
(a) $-\phi'_\varsigma + (U + \varsigma A)\phi_\varsigma + A\phi = \phi_\varsigma \Lambda + \phi \Lambda_\varsigma$
(b) $\psi'_\varsigma + \psi_\varsigma(U + \varsigma A) + \psi A = \Lambda \psi_\varsigma + \Lambda_\varsigma \psi$.

Multiply (9.3.8a) by $B\psi_\varsigma$, add (9.3.8b) multiplied by $\phi_\varsigma B$, and take the trace. Transform some of the terms:

$$\operatorname{tr}(-\delta\phi' B\psi_\varsigma + \phi_\varsigma B\delta\psi') = -\partial \operatorname{tr} \delta\phi B\psi_\varsigma + \operatorname{tr} \delta\phi B\psi'_\varsigma$$
$$+ \partial \operatorname{tr} \phi_\varsigma B\delta\psi - \operatorname{tr} \phi'_\varsigma B\delta\psi = \partial \operatorname{tr}(-\delta\phi B\psi_\varsigma + \phi_\varsigma B\delta\psi)$$
$$- \operatorname{tr}\{\delta\phi B(\psi_\varsigma(U + \varsigma A) + \psi A - \Lambda\psi_\varsigma - \Lambda_\varsigma\psi)$$
$$+ ((U + \varsigma A)\phi_\varsigma + A\phi - \phi_\varsigma\Lambda - \phi\Lambda_\varsigma)B\delta\psi\}$$

(using 9.3.9). We then obtain

$$\partial \operatorname{tr}(-\delta\phi B\psi_\varsigma \phi_\varsigma B\delta\psi) - \operatorname{tr}(\delta\phi B\psi + \phi B\delta\psi)A$$
$$+ \operatorname{tr} \delta U(\phi B\psi_\varsigma + \phi_\varsigma B\psi) + \operatorname{tr} \Lambda_\varsigma(\psi\delta\phi + \delta\psi\phi)B$$
$$- \operatorname{tr} \delta\Lambda B(\psi_\varsigma\phi + \psi\phi_\varsigma) = 0.$$

We have $\psi \cdot \delta\phi + \delta\psi \cdot \phi = \delta(\psi\phi) = \delta I = 0$. The term with $\delta\Lambda$ vanishes for the same reason. The rest of terms yields the desired results. □

9.3.10. Corollary. The equation

$$\delta \operatorname{tr} AR/\delta U = R_\varsigma$$

holds. In particular,

$$\delta \operatorname{tr} AR_{m+2}/\delta U = -(m+1)R_{m+1}.$$

If $\operatorname{tr} U = 0$ then these equalities hold for the non-diagonal terms.

Let

(9.3.11)
$$h^B_k = \frac{1}{k}\int \operatorname{tr} AR^B_{k+1}\,dx$$

be taken as Hamiltonians. In the Hamiltonian structure given by the mapping H, vector fields $\partial_{H(\delta h_k/\delta U)}$ relate to these Hamiltonians.

9.3.12. Proposition. The equation (9.1.17) is of Hamilton type; it relates to the Hamiltonian $-h_{k+1}$ in the $H^{(\infty)}$ structure, and to h_k in the $H^{(0)}$ structure.

$$H^0(\delta h_k/\delta U) = -[-\partial + U, R_k] = [A, R_{k+1}],$$
$$H^\infty(\delta h_{k+1}/\delta U) = -[A, R_{k+1}]. \quad \Box$$

9.3.13. Proposition. The Hamiltonians (9.3.11) are in involution with respect to both the structures, $H^{(0)}$ and $H^{(\infty)}$.
Proof. We have

$$\{h_m^B, h_k^C\}^\infty = \int \operatorname{tr} [A, R_m^B] R_k^C\, dx = -\{h_{m-1}^B, h_k^C\}^0.$$

Transform

$$\alpha_{mk} \equiv \operatorname{tr} [A, R_m^B] R_k^C = -\operatorname{tr}\left(-R_{m-1}^{B'} + [U, R_{m-1}^B]\right) R_k^C$$
$$= \partial \operatorname{tr} R_{m-1}^B R_k^C + \operatorname{tr}\left(-R_k^{C'} + [U, R_k^C]\right) R_{m-1}^B$$
$$= \partial \operatorname{tr} R_{m-1}^B R_k^C - \operatorname{tr} [A, R_{k+1}^C] R_{m-1}^B$$
$$= \operatorname{tr} [A, R_{m-1}^B] R_{k+1}^C + \partial \operatorname{tr} R_{m-1}^B R_k^C = \alpha_{m-1,k+1} + \partial \operatorname{tr} R_{m-1}^B R_k^C.$$

Changing the expression by an exact derivative we managed to decrease the first index and to increase the second by 1. We can continue until the first index becomes negative and $\alpha_{m,k} = 0$. We then obtain a formula (9.3.14)

$$\operatorname{tr} [A, R_{m+1}^B] R_{k+1}^C = \partial \operatorname{tr} \left(R_m^B R_{k+1}^C + R_{m-1}^B R_{k+2}^C + \ldots + R_0^B R_{m+k+1}^C\right)$$

which proves our assertion. $\Box$

In fact, Eq. (9.3.14) is richer than the assertion of 9.3.13. Its right-hand side vanishes by integration and plays no role in 9.3.13, but is very important when we consider stationary equations in Chap. 13.

9.3.15. Corollary. All the vector fields $\partial_{[A, R_k]}$ commute. This already was proved in 9.1.25.

9.3.16. Corollary. All the Hamiltonians are first integrals in involution of Eq. (9.1.17).

9.4. Segal-Wilson style theory, soliton solutions.

9.4.1. Let H be the Hilbert space of all vector series $\sum_{-\infty}^{\infty} v_k z^k$, $v_k \in C^n$ converging in the sense of L_2 on the circle $|z| = 1$. Then we use the same decomposition $H = H_+ \oplus H_-$ as in Sec. 8.1.1.

The Grassmannian Gr is the set of all subspaces $W \subset H$ such that (i) $p_+|_W$ is one-to-one correspondence (i.e. we consider only the generic transversal case), (ii) $zW \subset W$.

Following to Wilson [112] we can prove that this Grassmannian is isomorphic to the Grassmannian $\mathrm{Gr}^{(n)}$ (see 8.14) of scalar functions; however we shall not use this fact here.

We say that a matrix belongs to Gr if all its rows do.

9.4.2. Example. Let $\{\alpha_i\}, i = 1, \ldots, Nn$ be some points inside the unit circle, $|\alpha_i| < 1$. Here N is a natural number ("soliton number"). Let $\eta_i, i = 1, \ldots, Nn$ be vector-columns in C^n which span this space. Let W be the sub-space of all elements of H having form $v(z) = \sum_{-N}^{\infty} v_k z^k$ satisfying relations

$$v(\alpha_i)\eta_i = 0.$$

For almost all sets $\{\alpha_i, \eta_i\}$ this space W belongs to Gr.

9.4.3. Let $g(z)$ be a holomorphic and non-vanishing in the circle $|z| < 1$ diagonal matrix-function. It can be represented as

$$g(t,z) = \exp\left(\sum_{m=0}^{\infty}\sum_{k=1}^{n} z^m E_\alpha t_{m\alpha}\right)$$

where $t_{m\alpha}$ are some constants (which are assumed real). Then a transformation of the Grassmannian $W \mapsto Wg^{-1}$ can be considered. (In the above example Wg^{-1} has the same form as W, the vectors η_i being replaced by $g^{-1}(t,\alpha_i)\eta_i$.)

9.4.4. For an element $W \subset \mathrm{Gr}$ a Baker function $w_W(t,z)$ is a matrix-function on $|z| = 1$ satisfying conditions: (i) for any $t = (t_{k\alpha})$ it must be $w_W \in W$ as a function of z, (ii) $p_+(w_W \cdot g^{-1}) = 1$. Both the conditions mean that $w_W(t,z)$ is the only element of W of the form

$$(9.4.5) \qquad w_W(t,z) = \left(1 + \sum_{1}^{\infty} w_i z^{-i}\right) g(t,z)$$

where 1 is a matrix unity.

9.4.6. Proposition. If $w = w_W$ is a Baker function then (i) $w'w^{-1}$ is a linear function of z : $w'w^{-1} = -Az - U$, (ii) U satisfies all the equations of the hierarchy.

Proof. Equation (9.4.5) yields that $w'w^{-1} = -Az - U + (w'w^{-1})_-$ where $U = -(w'w^{-1})_0$. Let $(w'w^{-1})_- = -Q$. One must prove that $Q = 0$. We have
$$w' + (U + Az)w = Qw.$$
The l.h.s. is an element of W. Now
$$(w' + (U + Az)w)g^{-1} = Qwg^{-1}.$$
The r.h.s. is an element of Wg^{-1} of an order $O(z^{-1})$ which implies that it vanishes. Thus $Q = 0$ and $w'w^{-1} = -Az - U$. This is the same as
$$\hat{L} = w(-\partial + Az)w^{-1}$$
(ii) We have $R^\alpha = wE_\alpha w^{-1}$, and $V_{k\alpha} = (z^k R^\alpha)_+$. Then leting $w = \hat{w}g$:
$$\partial_{k\alpha} w - V_{k\alpha} w = (\partial_{k\alpha}\hat{w})g + wz^k E_\alpha - V_{k\alpha} w$$
$$= (\partial_{k\alpha}\hat{w})g + z^k R^\alpha w - V_{k\alpha} w$$
$$= (\partial_{k\alpha}\hat{w} - (z^k R^\alpha)_-\hat{w})g.$$

The l.h.s. is an element of W. Therefore $\partial_{k\alpha}\hat{w} - (z^k R^\alpha)_-\hat{w}$ is an element of Wg^{-1} of an order $O(z^{-1})$; hence it vanishes. We obtain $\partial_{k\alpha} w = V_{k\alpha} w$, which is equivalent to the equations of the hierarchy. □

9.4.7. Example 9.4.2 presents soliton-like solutions. It is essentially the same as given by Its [58] in the special case $n = k = 2$.

9.4.8. The algebraic-geometrical solutions ([57], [58]) also can be obtained in this way.

9.A. Appendix: Extension of the algebra $\mathcal{A}$ to an algebra closed with respect to indefinite integration.

Not every element of the algebra $\mathcal{A}$ is an exact derivative. However, a differential algebra can be embedded into an algebra where all elements are derivatives. This can be made in the following way [59].

Let $\mathcal{A}_1$ be the set of all formal power series

$$f(x) = \sum_0^\infty \frac{x^k}{k!} a_k, \, a_k \in \mathcal{A}.$$

Addition and multiplication are as usual. Action of the operator ∂ is

$$\partial f(x) = \sum_0^\infty \frac{x^{k-1}}{(k-1)!} a_k = \sum_0^\infty \frac{x^k}{k!} a_{k+1}$$

i.e. the sequence of the coefficients shifts to the left: $(a_0, a_1, \ldots) \mapsto (a_1, a_2, \ldots)$. It is easy to see that this differentiation has the necessary property $\partial(fg) = f'g + fg'$. Therefore $\mathcal{A}_1$ is a differential algebra.

Any element of $\mathcal{A}_1$ has an indefinite integral $\int f(x)dx = \sum_0^\infty \frac{x^k}{k!} a_{k-1}$, where $a_{-1} \in \mathcal{A}$ is arbitrary and the coefficients shift to the right.

The embedding of $\mathcal{A}$ into $\mathcal{A}_1$ is given by the formula $f \in \mathcal{A} \mapsto f(x) = \sum_0^\infty \frac{x^k}{k!} f^{(k)} \in \mathcal{A}_1$. It is easy to check that this is, indeed, a mapping of differential algebras.

CHAPTER 10. KdV-Hierarchies as Reductions of Matrix Hierarchies.

10.1. Reduction of matrix manifolds.

10.1.1. The initial idea is very simple: a linear differential n-th order equation can be reduced to a system of n first-order equations which implies a mapping of the corresponding hierarchies of non-linear equations as well as their Hamiltonian structures. Besides of general interest these reductions are important since matrix hierarchies are in a sense simpler. The most general case, where instead of $sl(n)$ an arbitrary semi-simple group is considered, was studied by Drinfeld and Sokolov [60, 61].

10.1.2. We consider the space of matrices $gl(n)$ (sometimes we shall consider the second variant of the theory with $sl(n)$). The subspace S_+ consists of matrices with only one non-vanishing row: the last one. (In the second variant the last element of the last row vanishes too; the subspace of such matrices will be designed as S_+^0. The numeration of rows and columns will be $i,j = 0, 1, \ldots, n-1$. Let

$$U_a = \begin{pmatrix} 0, & \ldots, & 0 \\ \vdots & & \vdots \\ 0, & \ldots, & 0 \\ -a_0, & \ldots, & -a_{n-1} \end{pmatrix}, \quad J = \begin{pmatrix} 0 & 1 & 0 & \ldots & 0 \\ & \ddots & \ddots & & \\ & & \ddots & \ddots & \\ & & & \ddots & 1 \\ 0 & & \ldots & & 0 \end{pmatrix},$$

$$A = \begin{pmatrix} 0 & & \\ \vdots & 0 & \\ 1 & \cdots & 0 \end{pmatrix}$$

and the differential operators be

$$l_{U_a} = -\partial + J + U_a, \hat{l}_{U_a} = l_{U_a} + \varsigma A.$$

One of the matrices U_a will be distinguished; its elements $a_i = u_i, i = 0, \ldots, n-1$ are taken as the generators of a differential algebra $\mathcal{A}$. This matrix U_u will be denoted simply as U.

Elements of the dual to S_+ space, S_-, are matrices with a fixed last column, the rest of the elements being indifferent.

We want to transfer the theory of Chapter 9 to this case as far as possible. In particular, the Hamiltonian mapping $H : S_- \to S_+$ will be given by the formula $[\hat{l}_U, Q]$. But Q cannot be an arbitrary representive of the class of elements of S_-, for it happens that if the last column of Q is given, the rest of elements of Q can be chosen (and chosen uniquely) in such a way that $[\hat{l}_U, Q] \in S_+$.

10.1.3. We shall follow Ref. [54] and consider matrices not in an abstract n-dimensional linear space but in the space of operators $R_-/R_{(-\infty,-n-1)}$ (see Sec. 1.3), i.e. the space of operators $R_- = \left\{ \sum_0^\infty \partial^{-i-1} X_i \right\}, X_i \in \mathcal{A}$, divided by $\partial^{-n} R_-, R_-/\partial^{-n} R_-$. With any matrix Q_{ij} we connect operator-columns Q^j and operator-rows Q_i:

$$Q^j = \sum_{i=0}^{n-1} \partial^{-i-1} Q_{ij} \in R_-/\partial^{-n} R_- , Q_i = \sum_{j=0}^{n-1} Q_{ij} \partial^j \in R_{(0,n-1)}.$$

The action of the operator Q on an element $X \in R_-/\partial^{-n} R_-$ is $QX = \sum_0^{n-1} \partial^{-i-1} \text{res}(Q_i X) \in R_-/\partial^{-n} R_-$. The adjoint operator acts in $R_{(0,n-1)}$:

$$X \in R_{(0,n-1)} \mapsto Q^* X = \sum_{j=0}^{n-1} \text{res}(Q^j X) \partial^j \in R_{(0,n-1)}.$$ Multiplication of operators can be written as

$$(Q \cdot P)^j = \sum_{i=0}^{n-1} \partial^{-i-1} \text{res}(Q_i P^j), \text{or} (Q \cdot P)_i = \sum_{j=0}^{n-1} \text{res}(Q_i P^j) \partial^j.$$

10.1.4. Proposition. For $[\hat{l}_U, Q] \in S_+$, it is necessary and sufficient that the matrix Q has the following structure. If the last column (which can be an arbitrary one) is designated by X, $X = Q^{n-1} = \sum_{i=0}^{n-1} \partial^{-i-1} Q_{i,n-1}$, $Q_{i,n-1} \in \mathcal{A}$, then the rows are given by

(10.1.5)
$$Q_i = \partial^i(X\hat{L})_+ - (\partial^i X)_+ \hat{L} = -\partial^i(X\hat{L})_- + (\partial^i X)_- \hat{L}, \, i = 0,\ldots, n-1$$

where $L = \partial^n + u_{n-1}\partial^{n-1} + \ldots + u_0$, $\hat{L} = L - \varsigma$. If we want more, $[\hat{l}_U, Q] \in S_+^0$, then an additional condition on X

(10.1.6)
$$\text{res}[\hat{L}, X] = 0$$

must be satisfied. The relation

(10.1.7)
$$[\hat{l}_U, Q] = U_{H(X)}$$

holds where $H(X)$ is the Hamilton mapping 1.7.6, $\varsigma = z^n$. If $H(X) = \sum_0^{n-1} h_i \partial^i$, then $U_{H(X)}$ is the matrix

$$U_{H(X)} = \begin{pmatrix} 0, & \ldots, & 0 \\ \cdots & \cdots & \cdots \\ 0, & \ldots, & 0 \\ -h_0, & \ldots, & -h_{n-1} \end{pmatrix}$$

in particular

$$[A, Q] = U_{H^{(\infty)}(X)}, \, [-\partial + U + J, Q] = U_{H^{(0)}(X)}.$$

Proof: At first we notice that Eq. (10.1.5) are in accordance with the fact that X is the last column of Q. Indeed, the coefficient in ∂^{n-1} in the expression $Q_i = -\partial^i(X\hat{L})_- + (\partial^i X)_- \hat{L}$ is $\text{res}(\partial^i X)$, i.e. X_i, and the last column has the form $\sum_0^{n-1} \partial^{-i-1} X_i = X$. Further, the elements of the matrix $P = [\hat{l}_U, Q]$ are as follows. If $i = 0,\ldots, n-2$, then

$$P_{ij} = -Q'_{ij} - Q_{i,j-1} + Q_{i+1,j} + Q_{i,n-1}\hat{u}_j, \, (\hat{u}_j = u_j, j > 0; \hat{u}_0 = u_0 - \varsigma),$$

If $i = n - 1$, then

$$P_{n-1,j} = -Q'_{n-1,j} - Q_{n-1,j-1} + Q_{n-1,n-1}\hat{u}_j - \sum_{\alpha=0}^{n-1} \hat{u}_\alpha Q_{\alpha,j}.$$

In terms of the operators, these are

$$P_i = -\partial \circ Q_i + Q_{i+1} + Q_{i,n-1}\hat{L}, i = 0, \ldots, n-2.$$

$$P_{n-1} = -\partial \circ Q_{n-1} + Q_{n-1,n-1}\hat{L} - \sum_{\alpha=0}^{n-1} \hat{u}_\alpha Q_\alpha.$$

Vanishing of the first $n - 1$ rows yields the recurrence relation

$$Q_{i+1} = \partial \circ Q_i - Q_{i,n-1}\hat{L}, i = 0, \ldots, n-2.$$

The same equation can be written for $i = n - 1$; however Q_n is no more the n-th row (there is no such a row) but $Q_n = -\sum_{\alpha=0}^{n-1} \hat{u}_\alpha Q_\alpha - P_{n-1}$, a differential operator of order $\leq n - 1$. These equations are equivalent to

$$Q_i = \partial^i \circ Q_0 - \sum_{\alpha=0}^{i-1} \partial^{i-1-\alpha} \circ Q_{\alpha,n-1}\hat{L}, i = 1, \ldots, n.$$

In particular

$$Q_n = \partial^n \left(Q_0 - \sum_{\alpha=0}^{n-1} \partial^{-i-\alpha} Q_{\alpha,n-1} \cdot \hat{L} \right) = \partial^n (Q_0 - X \cdot \hat{L})$$

This operator has an order $\leq n - 1$ if and only if $(Q_0 - X\hat{L})_+ = 0$, i.e. $Q_0 = (X\hat{L})_+$, which implies

$$Q_i = \partial^i (XL)_+ - (\partial^i X)_+ \hat{L}, i = 1, \ldots, n - 1$$

This is exactly Eq. (10.1.5); if $i = n$ we have $-\sum_0^{n-1} u_\alpha Q_\alpha - P_{n-1} = \partial^n (X\hat{L})_+ - (\partial^n X)_+ \hat{L}$ i.e.

$$P_{n-1} = -\partial^n (X\hat{L})_+ + (\partial^n X)_+ \hat{L} - \sum_{\alpha=0}^{n-2} \hat{u}_\alpha (\partial^\alpha (X\hat{L})_+ - (\partial^\alpha X)_+ \hat{L})$$

$$= -\hat{L}(X\hat{L})_+ + (\hat{L}X)_+ \hat{L} = -H(X).$$

Equation (10.1.7) is proved. The coefficient in ∂^{n-1} in $H(X)$ is $\operatorname{res}[\hat{L}, X]$: we have obtained the condition (10.1.6), which we already met in 3.4.4. □

The formulas (10.1.5) were studied in some other context by Cherednik [105].

The matrix Q which was constructed according to Eq. (10.1.5) will be denoted as Q_X.

10.1.8. Proposition. The matrix Q_X can be written as the operator

$$Y \in R_-/R_{(-\infty,-n-1)} \mapsto Q_X(Y) = (X(\hat{L}Y)_+ - (X\hat{L})_- Y)_-$$
$$\in R_-/R_{(-\infty,-n-1)}.$$

(Recall that this expression is already familiar to us: $[X,Y]_L = Q_Y(X) - Q_X(Y)$, see 3.2.2).

Proof: Using (10.1.5) we obtain

$$Q_X(Y) = \sum_{i=0}^{n-1} \partial^{-i-1} \operatorname{res}(Q_i Y) = \sum_{i=0}^{n-1} \partial^{-i-1} \operatorname{res}(\partial^i (X\hat{L})_+ Y)$$
$$- \sum_{i=0}^{n-1} \partial^{-i-1} \operatorname{res}((\partial^i X)_+ \hat{L} Y) = ((X\hat{L})_+ Y)_- - \sum_{i=0}^{n-1} \partial^{-i-1} \operatorname{res}(\partial^i X(\hat{L}Y)_-)$$
$$= ((X\hat{L})_+ Y - X(\hat{L}Y)_-)_- . \square$$

10.1.9. Proposition. A matrix Q is a resolvent, i.e. commutes with $\hat{l}_U$, if and only if $Q = Q_X$, where X is a resolvent of the operator $L: H(X) = 0$. Proof: This directly follows from 10.1.4. □

10.2. Hamiltonian structures. Resolvent.

10.2.1. On vector fields $\left\{\partial_{[\hat{l}_U, Q_X]}\right\}$ can be defined a symplectic form ω:

$$\omega\left(\partial_{[\hat{l}_U, Q_X]}, \partial_{[\hat{l}_U, Q_Y]}\right) = \int \operatorname{tr}[\hat{l}_U, Q_X] Q_Y \, dx$$

As usual, this is a family of forms depending on a parameter ς.

10.2.2. Proposition. A relation

$$\omega\left(\partial_{[\hat{l}_U, Q_X]}, \partial_{[\hat{l}_U, Q_Y]}\right) = \int \operatorname{res} H(X) Y \, dx$$

holds which coincides with the definition of the form 3.2.3.

Proof: This is an obvious corollary of 10.1.4. □

It is not difficult to find the vector field related to a functional $\tilde{f} = \int f(x)dx$. We have $d\tilde{f}$ as an element of S_- whose last column is the operator $\delta f/\delta L$. Thus, the vector field is

$$\tilde{f} \mapsto \partial_{[\tilde{l}_U, Q_{\delta f/\delta L}]}.$$

The Poisson bracket of two functionals is

(10.2.3) $\{\tilde{f}, \tilde{g}\} = \int \mathrm{tr}[\tilde{l}_U, Q_{\delta f/\delta L}] Q_{\delta g/\delta L} dx = \int \mathrm{res} H(\delta f/\delta L) \cdot \delta g/\delta L dx.$

This Poisson bracket is the same as in Chap. 3 for $\varsigma = z^n$.

10.2.4. Till now the representation of the operator L in terms of the matrix operator l_U held no advantage. But now we shall discuss a problem where this representation is distinctly advantageous. Namely, formulas of dressing on the basis of which it becomes possible to prove the variational relation for the resolvents 3.5.3. This proof gives an explicit expression for the term with an exact derivative which we shall use further in Chap. 14 to integrate stationary equations. We know that the dressing operator for L (Sec. 7.1) can be obtained only beyond the limits of the differential algebra $\mathcal{A}$ as it requires integration. For the matrix first-order operator we managed to determine the dressing operator within the algebra $\mathcal{A}$ (see 9.3.1). Now we shall do the same for operators l_U.

At first we change the basis. Let

$$Z = \begin{pmatrix} 1 & & & 0 \\ & z & & \\ & & \ddots & \\ 0 & & & z^{n-1} \end{pmatrix}, z^n = \varsigma.$$

We put $V^{(1)} = Z^{-1}UZ, \hat{l}_{V^{(1)}} = Z^{-1}\hat{l}_U Z$, then $\hat{l}_{V^{(1)}} = -\partial + V^{(1)} + zA^{(1)}$, where

$$V^{(1)} = \begin{pmatrix} & & & 0 \\ -z^{-n+1}u_0, & -z^{-n+2}u_1, & \ldots, & -u_{n-1} \end{pmatrix},$$

$$A^{(1)} = \begin{pmatrix} 0 & 1 & & 0 \\ \vdots & \ddots & \ddots & \\ \vdots & & 0 & \ddots & 1 \\ 1 & \ldots & \ldots & 0 \end{pmatrix}.$$

The advantage of this basis is that the matrix $A^{(1)}$ is non-degenerate.

Another convenient basis can be chosen reducing the matrix $A^{(1)}$ to a diagonal form. Letting K be the matrix with elements $n^{-1/2}\varepsilon^{ij}$ where ε is a primitive root of 1, $\varepsilon^n = 1$, and putting $A^{(2)} = K^{-1}A^{(1)}K, V^{(2)} = K^{-1}V^{(1)}K, \hat{l}_{V^{(2)}} = K^{-1}\hat{l}_{V^{(1)}}K$ we have

$$A^{(2)} = \begin{pmatrix} 1 & & & 0 \\ & \varepsilon & & \\ & & \ddots & \\ 0 & & & \varepsilon^{n-1} \end{pmatrix}, V^{(2)}_{ij} = -\frac{1}{n(\varepsilon^i z)^{n-1}} \sum_{\alpha=0}^{n-1} u_\alpha(\varepsilon^j z)^\alpha,$$

$$\hat{l}_{V^{(2)}} = -\partial + V^{(2)} + zA^{(2)}.$$

In such a form the operator resembles $\hat{l}_U$ of Chap. 9. The distinction consists in the special form of the matrix $V^{(2)}$ which incorporates the parameter z in negative powers. This, however, will give us no trouble.

10.2.5. Proposition. The formal series

$$\phi^{(2)} = \sum_{i=0}^{\infty} \phi_i^{(2)} z^{-i}, \Lambda = \sum_{i=-1}^{\infty} \Lambda_i z^{-i}$$

exist, where Λ_i are diagonal matrices, elements of $\phi_i^{(2)}$ and Λ_i belong to $\mathcal{A}, \Lambda_{-1} = A^{(2)}, \phi_0 = I$ and

$$-\partial + V^{(2)} + zA^{(2)} = \phi^{(2)}(-\partial + \Lambda)(\phi^{(2)})^{-1}.$$

the choice of the dressing series can be fixed by the requirement that the constants can exist only in ϕ_0.

The proof is exactly the same as in 9.3.1. The dependence of the matrix $V^{(2)}$ on z does not matter since only negative powers of z are involved in $V^{(2)}$. □

How we return to the old basis. Let $\phi^{(1)} = K\phi^{(2)}, \phi = ZK\phi^{(2)}$ then

(10.2.6) $\qquad -\partial + U + J + \varsigma A = \phi(-\partial + \Lambda)\phi^{-1}.$

Recall that in process of the proof of 10.2.5 it turned out that $\phi^{(2)}$ and $\psi^{(2)} = (\phi^{(2)})^{-1}$ satisfied equations

(10.2.7) $\qquad \begin{aligned} -\phi^{(2)\prime} + (V^{(2)} + zA^{(2)})\phi^{(2)} &= \phi^{(2)}\Lambda, \\ \psi^{(2)\prime} + \psi^{(2)}(V^{(2)} + zA^{(2)}) &= \Lambda\psi^{(2)}, \end{aligned}$

This implies

$$-\phi' + (U + J + \varsigma A)\phi = \phi\Lambda,$$
$$\psi' + \psi(U + J + \varsigma A) = \Lambda\psi,$$

10.2.8. Proposition. Resolvents of the operator $\hat{l}_{V^{(2)}}$, i.e. the series $Q^{(2)} = \sum_{0}^{\infty} Q_j^{(2)} z^{-j}$ which commute with $\hat{l}_{V^{(2)}}$, can be expressed in terms of n basic resolvents

$$Q^{(2)\alpha} = \phi^{(2)} E_\alpha (\phi^{(2)})^{-1},$$

$$E_\alpha = \begin{pmatrix} 0 & & & \\ & \ddots & & \\ & & 1 & 0 \\ & & & \ddots \\ 0 & & & 0 \end{pmatrix}, \text{(1 on the } \alpha\text{-th place)}$$

as $Q^{(2)} = \sum_\alpha c_\alpha(z) Q^{(2)\alpha}$, where $c_\alpha(z)$ are series in z^{-1} with constant coefficients.

The proof is the same as that in 9.1.13. □

Returning to the old basis we obtain the following. The basic resolvents of the operator $\hat{l}_{V^{(1)}}$ are $Q^{(1)\alpha} = KQ^{(2)\alpha}K^{-1} = \sum_{0}^{\infty} Q_j^{(1)\alpha} z^{-j}$, where $Q_0^{(1)\alpha} = KE_\alpha K^{-1} = n^{-1}(\varepsilon^{\alpha(i-j)})$. The corresponding resolvents of the operator $\hat{l}_U$ are $Q^\alpha = ZQ^{(1)\alpha}Z^{-1}$. This can be written as $Q^\alpha = \sum_{0}^{\infty} ZQ_r^{(1)\alpha} Z^{-1} z^{-r}$, but the coefficients of $ZQ_r^{(1)\alpha}Z^{-1}$ themselves depend on z. The series can be re-expanded as

$$Q^\alpha = \sum_{i=-(n-1)}^{\infty} Q_i^\alpha z^{-i}$$

where Q_i^α are independent of z.

Dependence on α turns out to be very simple as will be seen in the following proposition.

10.2.9. Proposition. The coefficients $(Q_r^\alpha)_{ij}$ can be written as $\varepsilon^{-\alpha r} Q_{ij,r}$ where $Q_{ij,r}$ are independent of α. If $z^n = \varsigma$, then $Q^\alpha(z)$ be branches of the function $Q(\varsigma)$.

Proof: If in the series representing a resolvent z is replaced by $\varepsilon^\alpha z$, we obtain another resolvent since z is involved in the equation for resolvents only in the form of $\varsigma = z^n$. It remains to look at the constants in the coefficients Q_r^α. The constants arise only from the term $ZQ_0^{(1)\alpha}Z^{-1}$. The ij-th element of this matrix is $n^{-1}(\varepsilon^\alpha z)^{i-j}$. Thus ε and z are involved only in the form $\varepsilon^\alpha z$; hence this is correct also for the other terms. □

We consider now the resolvent $Q = \sum b_\alpha Q^\alpha$, i.e. $Q = \phi B \phi^{-1}$ where $B = \text{diag}(b_0, \ldots, b_{n-1}) = \text{const}$. For this resolvent the following holds.

10.2.10. Proposition.

$$\delta \text{tr} AQ = \text{tr}(Q_\varsigma \delta U) - \partial \text{tr}(\delta \phi B \psi_\varsigma - \phi_\varsigma B \delta \psi).$$

The proof is similar to that in 9.3.7. □

We recall from (10.1.10) that the resolvent Q is Q_x for some $X \in R_-/R_{(-\infty, -n-1)}$, which is a resolvent of the operator L. Then 10.2.10 becomes

(10.2.11) $\qquad \delta \text{res} X = -\partial_\varsigma \text{res} \delta L \cdot X - \partial \, \text{tr}(\delta \phi B \psi_\varsigma - \phi_\varsigma B \delta \psi).$

In particular, $\delta \int \text{res} X dx / \delta L = -\partial_\varsigma X$, which is exactly prop. 3.5.3. This proof is technically more complicated. However it has an enormous advantage: in Eq. (10.2.11) the term with $\partial, -\partial \, \text{tr}(\delta \varphi B \psi_\varsigma - \phi_\varsigma B \delta \psi)$, is explicitly written, which will be necessary in the study of stationary equations in Chap. 14.

10.2.12. Exercise. Prove that the matrices Q^α relate to the resolvents $T_{(\varepsilon)}$ (1.7.2), which we denote now as $T_\alpha = T_{(\varepsilon^\alpha)} \in R_-/R_{(-\infty, -n-1)}$ (ε is a primitive root), as $Q^\alpha = Q_{T_\alpha}$.

Hint: compare constants in the last columns of the matrices Q^α and Q_{T_α}.

10.3. Reduction to a factor-manifold.

10.3.1. Now we represent the results obtained above in a more natural form. Prop. 10.2.2 states that the symplectic form for scalar operators can be considered as a restriction of the form for matrix operators to the vector fields tangent to the submanifold S_+. It makes more sense to use it, not as a restriction, but as a reduction of the form. The distinction between restriction and reduction of a form is that the former refers to a submanifold while the latter refers to a factor-manifold. The submanifold S_+ happens to be a section of a factor-manifold.

Let us return to the essence of our discussion, the relation between a scalar differential operator of the n-th order and a matrix first-order operator. A matrix operator corresponds to each scalar n-th order operator, but not vice versa in general. The converse is true when the matrix operator has a triangular form. Let

$$l_q = -\partial + q + J$$

where J is as above and q is a lower triangular matrix i.e. $q_{kl} = 0$ for $l > k$. The linear manifold of such matrices will be denoted as $\mathfrak{b}$, $q \in \mathfrak{b}$. We consider a system of equations $l_q \varphi = 0$, where φ is a vector $\varphi = (\varphi_0, \ldots, \varphi_{n-1})$. This system has the form

$$\varphi_{i+1} = \varphi_i' - \sum_{\alpha=0}^{i} q_{i\alpha} \varphi_\alpha$$

Expressing in succession all the φ_i in terms of φ_0, we obtain in the end an equation $L\varphi_0 = 0$, where L is a differential n-th order operator whose coefficients are differential polynomials in $\{q_{kl}\}$. For example, if q has the form U (see 10.1.2) then $L = \partial^n + u_{n-1}\partial^{n-1} + \ldots + u_0$. The second example is $q = V$, where

$$(10.3.2) \qquad V = \begin{pmatrix} -v_0 & & & 0 \\ & \ddots & & \\ & & \ddots & \\ 0 & & & -v_{n-1} \end{pmatrix},$$

with $L = (\partial + v_{n-1}) \ldots (\partial + v_0)$.

One and the same L can correspond to distinct matrices q. For example, let us consider a group of matrices $\mathcal{N}$ which comprises matrices $g = 1 + \nu$, where ν are strictly lower triangular matrices. $\nu_{kl} = 0$ for $l \geq k$. Then $\nu^n = 0$, $(1+\nu)^{-1} = 1 - \nu + \nu^2 - \ldots \pm \nu^{n-1}$. Let us consider the transformations of operators l_q under the action of g: $l_{q_1} = g^{-1} l_q g$ (these transformations are called gauge transformations).

10.3.3. *Proposition.* If l_{q_1} and l_{q_2} belong to an orbit of the action of the group $\mathcal{N}$, i.e. there exists a matrix $g \in \mathcal{N}$ such that $l_{q_1} = g^{-1} l_{q_2} g$, then one and the same operator $L_1 = L_2 = L$ corresponds to those operators.
Proof: Let $l_{q_1} \varphi = 0$. Then $g^{-1} l_{q_2} g \varphi = 0$ and $l_{q_2} g \varphi = 0$. The zero component φ_0 of the vector φ satisfies the equation $L_1 \varphi_0 = 0$; the zero component $(g\varphi)_0$ of the vector $g\varphi$ satisfies the equation $L_2(g\varphi)_0 = 0$. We have

$\varphi_0 = (g\varphi)_0$ since $g = 1 + \nu$, φ_0 being an arbitrary solution. This means that equations $L_1\varphi_0 = 0$ and $L_2\varphi_0 = 0$ are equivalent, the highest terms of L_1 and of L_2 being ∂^n. Thus, $L_1 = L_2$. $\square$

10.3.4. Proposition. Each orbit of the action of the group $\mathcal{N}$ contains one, and only one, element of the form l_{U_a}, where $U_a \in S_+$ is defined in 10.1.2. The elements of the matrices g and U_a are differential polynomials in elements of matrices q belonging to the orbit.

Proof: The equation $g(-\partial + q + J)g^{-1} = (-\partial + U_a + J)$, i.e.

$$\nu' + [\nu, J] + (1 + \nu)q - U_a(1 + \nu) = 0$$

is to be solved with respect to matrices ν and U_a. First, we write this equation for the diagonal elements (i, i), then for the underdiagonal elements, $(i, i-1)$, then for $(i, i-2)$, etc. For (i, i),

$$\nu_{i,i-1} - \nu_{i+1,i} + q_{ii} + \delta_{i,n-1}a_{n-1} = 0, i = 0, \ldots, n-1 (\nu_{0,-1} = \nu_{n,n-1} = 0).$$

Here we have n equations with $n - 1$ unknowns $\nu_{i,i-1}$ and a_{n-1}. Summing up with respect to i we obtain $a_{n-1} = -\operatorname{tr} q$; after that all the $\nu_{i,i-1}$ can be found. For $(i, i-1)$,

$$\nu'_{i,i-1} + \nu_{i,i-2} - \nu_{i+1,i-1} + q_{i,i-1} + \nu_{i,i-1}q_{i-1,i-1} +$$
$$\delta_{i,n-1}a_{n-2} = 0, i = 1, \ldots, n-1.$$

Summing up again we find a_{n-2}, and then all $\nu_{i,i-2}$, etc. $\square$

10.3.5. Corollary. An operator l_V, where V is from (10.3.2), can be reduced by a gauge transformation to the form l_{U_a}, where $U_a \in S_+$. This yields

(10.3.6) $\qquad (\partial + v_{n-1}) \ldots (\partial + v_0) = \partial^n + a_{n-1}\partial^{n-1} + \ldots + a_0$.

This is well known to us as Miura transformation (Sec. 4.2).

10.3.7. Thus, there is a one-to-one correspondence between orbits with respect to gauge transformations and the n-th order operators. The submanifold S_+ is transversal to the orbits and intersects with each of them at one point. The choice of this section gives a coordinatization of the manifold of the orbits. The natural manifold is exactly the manifold of the orbits, i.e. the factor-manifold of the manifold $\mathfrak{b}$ under the action of the group of gauge transformations:

$$q \mapsto g'g^{-1} + gqg^{-1} + gJg^{-1} - J, g = 1 + \nu.$$

The manifold of the orbits will be denoted as $\mathcal{U}$.

10.3.8. Vector fields on $\mathcal{U}$ can be obtained from vector fields on $\mathfrak{b}$ in such a way. The vector fields preserving orbits must be taken, i.e. the vector fields which transform functions constant on orbits into functions with the same property. Then they must be factorized with respect to vector fields tangent to orbits, i.e. sending the functions constant on the orbits to zero. One representant of each class can be obtained if a field tangent to S_+ is taken and transferred to the whole orbit with the help of gauge transformations. If a vector field tangent to S_+ at a point l_U has the form $\partial_{[l_U, Q]}$ then at a point $l_q = g l_U g^{-1}$ of the orbit it has the form $\partial_{g[l_U, Q]g^{-1}} = \partial_{[l_q, gQg^{-1}]}$. The gauge transformation is $g l_q g^{-1}$, therefore a vector tangent to the orbit is $\partial_{[l_q, \nu]}$, where ν is a lower strictly triangular matrix.

10.3.9. Proposition. A symplectic form which is defined on vectors $\partial_{[l_q, Q]}$ by the formula

$$\omega\left(\partial_{[l_q, Q_1]}, \partial_{[l_q, Q_2]}\right) = \int \mathrm{tr}\, [l_q, Q_1] Q_2 \, dx$$

degenerates on vectors tangent to the orbits of gauge transformations and therefore can be transferred to the manifold of orbits.

Proof: If ν is a lower strictly triangular matrix, then

$$\omega\left(\partial_{[l_q, gQg^{-1}]}, \partial_{[l_q, \nu]}\right) = \int \mathrm{tr}\, [l_q, gQg^{-1}] \nu \, dx = 0. \square$$

The form of Chap. 3 is obtained when the points of S_+ are taken as coordinates of the orbits.

10.4. Kupershmidt-Wilson theorem.

10.4.1. The meaning of the Kupershmidt-Wilson theorem (Chap. 4) in this context is as follows. The orbits are intersected by two submanifolds of $\mathfrak{b}$: S_+ and another one comprising matrices V (10.3.2). Matrices U and V belonging to the same orbit are connected by the Miura transformation (10.3.6). The symplectic form on the orbit manifold can be transferred to these two submanifolds. This yields a relation between symplectic structures, which is exactly the Kupershmidt-Wilson theorem. Now we shall show this in a precise way.

10.4.2. Lemma. If l_U and l_V are connected by a gauge transformation

$$g l_V g^{-1} = l_U, \, g \in \mathcal{N}$$

and Q_X is a matrix constructed with the help of X according to (10.1.5), then
$$\mathrm{tr}\, g^{-1}Q_X g\delta l_V = \mathrm{tr}\, Q\delta l_U\,.$$

Proof: Apply the operator δ to the given identity:
$$\delta g l_V g^{-1} - g l_V g^{-1}\delta g g^{-1} + g\delta l_V g^{-1} = \delta l_V\,.$$

We then multiply this by Q_X and take the trace. Two of the terms are
$$\mathrm{tr}\, Q_X \delta g l_V g^{-1} - \mathrm{tr} Q_X g l_V g^{-1}\delta g g^{-1} = \mathrm{tr}\, Q_x \delta g \cdot g^{-1} l_U$$
$$- \mathrm{tr}\, Q_X l_U \delta g \cdot g^{-1} = \mathrm{tr}\, [l_U, Q_X]\delta g \cdot g^{-1}\,.$$

The matrices $[l_U, Q_X]$ and g^{-1} are triangular, the matrix δg is strictly triangular. Therefore, this expression vanishes. The rest of the terms yields the required equation. $\square$

10.4.3. Lemma. Let $X_f = \delta f/\delta L$ for a functional $\tilde{f} = \int f(u)dx$. Then the matrix $\tilde{Q}_f = g^{-1}Q_{X_f}g$ has diagonal elements $\delta f/\delta V_i$.

Proof:
$$\delta\tilde{f} = -\int \mathrm{tr}\, Q_{X_f}\delta l_U dx = -\int \mathrm{tr} g^{-1}Q_{X_f}g\delta l_V dx = -\int \mathrm{tr}\tilde{Q}_f \delta l_V dx\,.$$

Now, δl_V is a diagonal matrix with elements $-\delta v_i$. This yields the required fact. $\square$

10.4.4. Lemma. $[l_V, \tilde{Q}_f]$ is a lower triangular matrix.

Proof: $[l_V, \tilde{Q}_f] = g^{-1}[l_U, Q_{X_f}]g$, which is the product of three triangular matrices. $\square$

Now we construct a matrix $\tilde{\tilde{Q}}_f$:
$$(\tilde{\tilde{Q}}_f)_{kl} = \begin{cases} (\tilde{Q}_f)_{kl}, & l \geq k \\ 0, & l < k \end{cases}$$

(matrices $\tilde{Q}_f$ and $\tilde{\tilde{Q}}_f$ differ by a strictly triangular matrix; matrices $[l_V, \tilde{Q}_f]$ and $[l_V, \tilde{\tilde{Q}}_f]$ differ by a matrix tangent to the orbit of the group of gauge transformations).

10.4.5. Lemma. The matrix $[l_V, \tilde{\tilde{Q}}_f]$ is diagonal, with elements $-\tilde{Q}'_{ii} = -(\delta f/\delta V_i)'$ on the diagonal.

Proof: The matrices l_V and $\tilde{\tilde{Q}}_f$ are upper triangular; thus $[l_V, \tilde{\tilde{Q}}_f]$ is also upper triangular. On the other hand, this matrix differs from $[l_V, \tilde{Q}_f]$, i.e. from a lower triangular matrix by $[l_V, \nu]$ where $\nu = \tilde{\tilde{Q}}_f - \tilde{Q}_f$ is lower strictly triangular. Thus, $[l_V, \tilde{\tilde{Q}}_f]$ is lower triangular. Therefore $[l_v, \tilde{\tilde{Q}}_f]$ is diagonal. Its elements can be easily found, they are $-\tilde{Q}'_{ii} = -\tilde{Q}'_{ii}$. □

10.4.6. Lemma.

$$\mathrm{tr}[l_V, \tilde{\tilde{Q}}_1]\tilde{\tilde{Q}}_2 = \mathrm{tr}\,[l_V, \tilde{Q}_1]\tilde{Q}_2 + \partial(\ )$$

for any Q_1 and Q_2 of the form Q_X; $\partial(\ )$ means, as usual, an exact derivative of an element of $\mathcal{A}$.

Proof: Let $\nu_i = \tilde{Q}_i - \tilde{\tilde{Q}}_i, i = 1,2$. Then $\mathrm{tr}[l_V, \tilde{Q}_1]\tilde{Q}_2 = \mathrm{tr}\,[l_V, \tilde{Q}_1]\tilde{\tilde{Q}}_2$, since $\mathrm{tr}[l_V, \tilde{Q}_1]\nu_2 = 0$ as the product of a triangular and a strictly triangular matrices. Further,

$$\mathrm{tr}[l_V, \tilde{Q}_1]\tilde{\tilde{Q}}_2 = -\mathrm{tr}[l_V, \tilde{\tilde{Q}}_2]\tilde{Q}_1 - \partial(\mathrm{tr}\,\tilde{Q}_1\tilde{\tilde{Q}}_2) = -\mathrm{tr}[l_V, \tilde{\tilde{Q}}_2]\tilde{\tilde{Q}}_1 + \partial(\ )$$
$$= \mathrm{tr}[l_V, \tilde{\tilde{Q}}_1]\tilde{\tilde{Q}}_2 + \partial(\ ).\,\square$$

10.4.7. Proposition.

$$\int \sum (\delta f/\delta v_i)'(\delta g/\delta v_i)dx = \int \mathrm{res}\, H^{(0)}(\delta f/\delta L) \cdot \delta g/\delta L\, dx\,.$$

Proof: We find

$$\mathrm{tr}[l_V, \tilde{Q}_f]\tilde{\tilde{Q}}_g = -\sum (\delta f/\delta v_i)'\delta g/\delta v_i$$

and

$$\mathrm{tr}[l_V, \tilde{Q}_f]\tilde{Q}_g = \mathrm{tr}\,g^{-1}[l_V, Q_{X_f}]Q_{X_g}g = \mathrm{tr}\,[l_V, Q_{X_f}]Q_{X_g}$$
$$= \mathrm{res}\, H^{(0)}(X_f)X_g, X_f = \delta f/\delta L, X_g = \delta g/\delta L\,.\,\square$$

10.4.8. Remark. We can also easily prove this proposition if we want to restrict ourselves to the case $u_{n-1} = 0$. In this case $\mathrm{res}[L, X_f] = 0$, which yields $\mathrm{tr}\, Q_{X_f} = 0$ (see 10.1.4). Then $\mathrm{tr}\, \tilde{Q}_f = 0$, and $\sum \delta f/\delta v_i = 0$, the only condition that must be verified.

10.5. The general Drinfeld-Sokolov scheme.

10.5.1. Drinfeld and Sokolov [60,61] have shown that to each simple subalgebra of the Lie algebra $\mathrm{gl}(n)$ there is a reduction of equations and Hamilton structures, including the Kupershmidt-Wilson theorem. What has been done must be rewritten in group-theoretical terms. The theory of simple Lie algebras in detail can be found, for example, in [63].

10.5.2. Let $\mathfrak{G}$ be a semisimple matrix Lie algebra (it was $\mathrm{sl}(n)$ in the above). Let $\mathfrak{h}$ be a Cartan subalgebra, i.e. a maximum commutative subalgebra. (In the case of $\mathrm{sl}(n)$, these were diagonal matrices). The subalgebra $\mathfrak{h}$ acts in the space $\mathfrak{G}$ by the adjoint representation: $h \in \mathfrak{h} \mapsto \mathrm{ad}\,(h), \mathrm{ad}\,(h)X = [h, X]$. It can be proved that each linear transformation $\mathrm{ad}\,(X)$ can be reduced in some basis of the diagonal form. Since all of them commute, there is a common basis where they are diagonal. Let us denote this basis by $\{e_i\}$. (In the above case of $\mathrm{sl}(n)$, the basis consisted of matrices e_{ij} with an only nonzero element: the unity on the ij-th place). Eigenspaces Ce_i are called root spaces; their elements are root vectors. Corresponding eigenvalues can be considered as linear functions on $\mathfrak{h}$, $\alpha_i(h), h \in \mathfrak{h}$, i.e. as elements of the dual $\mathfrak{h}$. They are called the roots. (In the case of $\mathrm{sl}(n)$, the root corresponding to e_{ij} is $\alpha_{ij}(h) = h_i - h_j$). The basis of a system of roots is a set of roots such that each root can be uniquely represented as a linear combination of elements of this set with the integer coefficients of the same sign. (In the case of $\mathrm{sl}(n)$, a basis was $\alpha_{01}, \alpha_{12}, \ldots, \alpha_{n-2,n-1}$. The existence of a basis can be proved. A root is positive in a basis if the coefficients are positive, otherwise it is negative. The direct sum of the root spaces which correspond to negative roots is denoted by n^-, or simply n; the same for positive roots n^+. The subspaces n and n^+ are subalgebras. Subalgebras $b^+ = n^+ + \mathfrak{h}$ and $b = n + \mathfrak{h}$ are called Borel subalgebras. The relations $[b^+, b^+] \subset n^+$ and $[b, b] \subset n$ hold. (In the case of $\mathrm{sl}(n)$, the subalgebra b consists of lower triangular matrices, n of strictly triangular matrices. Correspondingly, b^+ and n^+ are upper triangular matrices.)

The sequence of subalgebras $n_0 = b \supset n_1 = n = [n, n_0]_2 \supset n_2 = [n, n_1] \supset n_3 = [n, n_2] \supset \ldots$ is finite, i.e. n is a nilpotent algebra. In every pair $n_i \supset n_{i+1}$ the subalgebra n_{i+1} is an ideal and the quotient-algebra n_i/n_{i+1} is commutative, i.e. the algebra b is solvable. Let us take one root vector for every root belonging to a basis S. These vectors will be called $\{J_k\}$. Let $J = \sum J_k$. (In the case of $\mathrm{sl}(n)$, the subalgebra n_i consists of matrices $\nu : \nu_{kl} = 0, l > k - i$. The matrix J is none other than the matrix J

in 10.1.2.) We have $[J, n_{i+1}] \subset n_i$, in particular $[J, n] \subset b$.

Let P be a projector ($P^2 = P$) of subspaces $b \to [J, n]$, and $Pn_i \subset n_i$. (For sl(n), this projector can be constructed as follows. $P : \alpha \in n_i \mapsto [J, \beta]$, where

$$\beta_{kl} = \alpha_{k-1,l} + \alpha_{k-2,l-1} + \ldots + \alpha_{k-l-1,0}, k = 1, \ldots, n-1, l < k .$$

Then $[J, \beta]_{kl} = \beta_{k+1,l} - \beta_{k,l-1} = \alpha_{kl}$ if $k < n - 1$ and $[J, \beta]_{n-1,l} = -\beta_{n-1,l-1} = -\alpha_{n-2,l-1} - \ldots - \alpha_{n-1-l,0}$.)

Let $S_+ = \ker P$ (for sl(n) and P written above, S_+ is the same as in 10.1.2). Finally, A is an element of the center of the algebra n.

10.5.3. So far all the matrices were with constant coefficients. Let $\mathcal{A}$ be a differential algebra. This can be the algebra of smooth functions on a circle (Drinfeld, Sokolov), or the algebra of differential polynomials of formal symbols $u_0, u_1, \ldots, u_{n-1}$ as we prefer, or whatever one wishes. Any Lie algebra introduced above can be multiplied by $\mathcal{A} : \widetilde{\mathfrak{G}} = \mathfrak{G} \otimes \mathcal{A}, \tilde{b} = b \otimes \mathcal{A}, \tilde{n} = n \otimes \mathcal{A}$, etc. These algebras comprise matrices whose elements are no longer numbers but elements of $\mathcal{A}$, e.g. smooth functions on a circle, in which case algebra $\widetilde{\mathfrak{G}}$ has the name "loop algebra" over the algebra $\mathfrak{G}$.

Now we have everything to formulate the above results in a general form. One only has to use instead of the words "triangular matrices", "strictly triangular matrices", "diagonal matrices", etc, words like "elements of $b, n, \mathfrak{h}$", etc.

We thus consider the differential operators

$$l_q = -\partial + q + I, q \in \tilde{b}, \tilde{l}_q = l_q + \varsigma A .$$

Let $\mathcal{N}$ be the Lie group corresponding to the algebra n. The gauge transformations are $g l_q g^{-1}, g \in \tilde{\mathcal{N}}$. Let $\tilde{A}_{\mathcal{N}}$ be the space of functionals invariant with respect to gauge transformations, i.e. constant on the orbits. Vector fields preserving orbits are those which map $\tilde{A}_{\mathcal{N}}$ onto itself. If the symplectic form defined as above (instead of the trace, the Killing scalar product must be used) is restricted to vector fields preserving orbits, then it degenerates on the vector fields tangent to the orbits. The final object on which the form will be defined is the space of vector fields preserving orbits, factorized by the vector fields tangent to orbits. The space of such vector fields can be represented by the space of vector fields tangent to the transversal to orbits in submanifold S_+.

We refer the reader to Refs. [60,61] for more details, examples and construction of Hamiltonians which is equivalent to the resolvent.

CHAPTER 11. Stationary Equations.

11.1. The ring of functions on the equation phase space.

11.1.1. This and the next three chapters deal with a new type of problems related to stationary solutions of equations or stationary equations. We have studied equations $\dot{L} = H(\delta f/\delta L)$ for a scalar differential operator L (H is a Hamiltonian mapping), and $\dot{l} = H(\delta f/\delta U)$, where $l = -\partial + U$, is a matrix operator. Now we call these equations non-stationary. If a solution L or l does not depend on t, it satisfies the equation $H(\delta f/\delta L) = 0$ ($H(\delta f/\delta U) = 0$, respectively). In terms of Lax pairs these equations are $[P, L] = 0$. We call these equations stationary ones. Stationary equations are ordinary differential equations. The general fact is that the stationary equations arising from nonstationary integrable systems are integrable Hamiltonian systems themselves. Their Hamiltonian structure will be constructed, together with sets of first integrals in involution in an amount equal to the number of degrees of freedom, i.e. to half of the dimension of the phase space. More than that, the integration procedure can be carried out effectively, and variables of the "action-angle" type can be constructed so that the system is linear and thus easily integrated. The angle variables are obtained by means of the Abel mapping of the g-th power of the Riemann surface (g is the genus of the surface) onto its Jacobi manifold. The return to the original variables can be fulfilled, after Riemann, with the aid of the θ-functions.

The importance of the stationary equations is particularly great for the reason that the finite-dimensional manifold of solutions of the equation $[P, L] = 0$ is an invariant manifold for any equation $\dot{L} = [P_1, L]$ if vector fields $\partial_{[P,L]}$ and $\partial_{[P_1,L]}$ commute, i.e. for an equation of the same hierarchy. Thus, finite-dimensional manifolds of solutions of nonstationary equations will be obtained having explicit form (as a rule, in terms of the θ-functions). These solutions are called algebraic-geometrical. We have mentioned them earlier but now we consider them from another point of view.

The discovery of the algebraic-geometrical solutions is connected, in the first place, with the names of Its, Matveev, Novikov, Dubrovin, Lax, Mac Kean, van Moerbeke, Krichever. This exposition differs from the original one by the fact that we interpret this problem as integration of a Hamiltonian system with sufficiently many first integrals in involution, according to the Liouville procedure. To this end, it is necessary (i) to construct a symplectic form and a Hamiltonian for the equation $[P, L] = 0$ (do not confuse these with the form and Hamiltonian for the nonstationary equation $\dot{L} = [P, L]$!), (ii) to construct first integrals in involution in a necessary amount, (iii) to find, according to Liouville, the angle variables corresponding to the first integrals taken as action variables, (iv) to write the formulas for the solutions in the original variables in terms of the θ-functions.

This chapter contains some general conceptions of Hamiltonian structure of variational Euler-Lagrange equations, in addition to Sec. 2.3. In the next chapter we shall consider the simplest case of the KdV-hierarchy for $n = 2$, i.e. the KdV-hierarchy in the narrow sense. Incidentally, we shall give the theory of this hierarchy independent of the method of Chap. 1, which is interesting by itself. In Chap. 13 we shall study stationary equations for matrix hierarchies, and in Chap. 14 those for the general KdV-hierarchies.

11.1.2. Stationary equations have the form $H(\delta f/\delta L) = 0$, whence $\delta f/\delta L \in$ ker H. The kernel of the operator H is always very small, and we shall see that by slightly changing f one can in every case write the equation simply as $\delta f/\delta L = 0$. Thus, we deal with the variational Euler-Lagrange equations. These equations will be of the Cauchy-Kowalevsky type, i.e. they have the form

(11.1.3) $$Q_i \equiv v_i^{(m_i)} + F_i(v) = 0, i = 1, \ldots, N,$$

where $\{v_i\}$ are some variables and the functions $\{F_i\}$ are polynomials in $\{v_i^{(j)}\}$, the order j of derivatives $v_i^{(j)}$ in all the terms being less than m_i.

Let $\mathcal{A}$ be the differential algebra of the polynomials in $\{v_i^{(j)}\}$. We shall introduce a new system of generators for it. Any element of $\mathcal{A}$ can be uniquely written as a polynomial in two kinds of variables:
(11.1.4)
   (a) $\{v_i^{(j)}\}, i = 1, \ldots, N, j < m_i$ (b) $Q_i^{(j)}, i = 1, \ldots, N, j = 0, 1, \ldots$.

The first kind of variables will be called the phase variable. The variables (b) generate a differential ideal in the algebra $\mathcal{A}: J_Q = \left\{\sum a_{ij} Q_i^{(j)}\right\}, a_{ij} \in \mathcal{A}$. Let $\mathcal{A}_Q = \mathcal{A}/J_Q$. The elements of the differential algebra $\mathcal{A}_Q$ are classes of differential polynomials which are equal modulo the differential equations (11.1.3). These elements we consider as functions in the phase space. In each class there is one polynomial which is expressed only in terms of the phase variables (a).

What is the purpose of this complicated approach? If the functions in the phase space are understood as elements of $\mathcal{A}_Q$, then the vector field corresponding to the system of equations (11.1.3), i.e. differentiation along trajectories of this system, can be written extremely simple. Indeed, this system can be written as

$$\partial v_i = v_i', \partial v_i' = v_i'', \ldots, \partial(v_i^{(m_i-2)}) = v_i^{(m_i-1)}, \partial(v_i^{(m_i-1)}) = v_i^{(m_i)},$$

where $v_i^{(m_i)}$ must be eliminated with the help of $Q_i = 0$. This series can be extended: $\partial v_i^{(m_i)} = v_i^{(m_i+1)}, \ldots$. Thus, the differentiation of any function along the trajectory is the application of the operator ∂ modulo the equation $\{Q_i^{(j)} = 0\}$, i.e. in the space $\mathcal{A}_Q$. The vector field corresponding to the system (11.1.3) is simply the operator ∂ in $\mathcal{A}_Q$.

What are other vector fields in the phase space? Any differentiation in $\mathcal{A}$ is

$$\xi = \sum a_{ij} \partial/\partial v_i^{(j)}, a_{ij} \in \mathcal{A}.$$

The space of these fields is denoted by $T\mathcal{A}$. This field can be transferred to $\mathcal{A}_Q$ if and only if it preserves the ideal $J_Q, \xi J_Q \subset J_Q$. The subspace of such fields we denote as $T\mathcal{A}_Q$. An example of such fields is a partial derivative with respect to a variable (a) when variables (b) are fixed. Another example is $\xi = \partial = \sum v_i^{(j+1)} \partial/\partial v_i^{(j)}$.

Vector fields commuting with ∂ are, as we know, $\partial_a = \sum a_i^{(j)} \partial/\partial v_i^{(j)}$. To give an example of a vector field which preserves the ideal and, at the same time, commutes with ∂ is not at all easy. This amounts to no less

Stationary Equations

than finding a symmetry of Eqs. (11.1.3). If such a field is found, the flow determined by it, $\partial v_i/\partial \tau = a_i$, transforms solutions of Eqs. (11.1.3) into other solutions. A symmetry can be found if a first integral is known. Just now we shall describe how to do this.

An element $f \in \mathcal{A}$ and its class in $\mathcal{A}_Q$ we denote by the same letter. The meaning of equality signs must be pointed out precisely: the exact equality in $\mathcal{A}$ will be denoted simply as $f = g$, while the equality in $\mathcal{A}_Q$, i.e. modulo the equation will be denoted as $f \stackrel{Q}{=} g$; this means that f and g differ by an element of the ideal J_Q.

11.2. Characteristics of first integrals.

11.2.1. This method for constructing effectively the correspondence between first integrals and vector fields was suggested in Ref. [19], [16].

Let an element of the ideal J_Q be represented as

$$(11.2.2) \qquad f = \sum a_{ij} Q_i^{(j)}.$$

This representation is of course not unique.

11.2.3. Lemma. If

$$\sum a_{ij} Q_i^{(j)} = 0$$

then all the a_{ij} belong to J_Q.

Proof: Let us express all the a_{ij} in terms of the generators (11.1.4). This representation is unique. If some $a_{ij} \notin J_Q$ then the term $a_{ij} Q_i^{(j)}$ is linear with respect to the generators of the type (b) and there are no other terms to cancel with this one. □

Thus, two distinct representations (11.2.2) of an element of the ideal have coefficients which differ by elements of the ideal. □

11.2.4. Definition. The characteristic of an element of the ideal J_Q is the set of N elements of $\mathcal{A}_Q$:

$$f \in J_Q, \chi_f = \{(\chi_f)_i\} = \left\{ \sum_{j=0}^{\infty} (-1)^{(j)} a_{ij}^{(j)} \right\}, i = 1, \ldots, N.$$

Here $a_{ij}^{(j)}$ are understood as elements of $\mathcal{A}_Q$.

The characteristic is well defined owing to lemma 11.2.3.

11.2.5. Proposition. If $f = \partial g, g \in J_Q$, then $\chi_f = 0$.

Proof: Let $g = \sum a_{ij} Q_i^{(j)}$, then $f = \sum a'_{ij} Q_i^{(j)} + \sum a_{ij} Q_i^{(j+1)}$ and

$$\chi_f = \left\{ \sum_{j=0}^{\infty} (-1)^j a_{ij}^{(j+1)} + \sum_{j=0}^{\infty} (-1)^{j+1} a_{ij}^{(j+1)} \right\} = 0. \square$$

Thus, the characteristic is a mapping $J_Q/\partial I_Q \to \mathcal{A}_Q^N$.

11.2.6. Definition. The first integral of Eq. (11.1.3) is an element $f \in \mathcal{A}_Q$ such that $\partial f = 0$ (in $\mathcal{A}_Q$).

11.2.7. Definition. Characteristic of a first integral is an element of $\mathcal{A}_Q^N$ which is constructed in such a way. Let $f \in \mathcal{A}$ be a representative of the class of the first integral. Let $\partial f = g \in J_Q$. The characteristic of the first integral is the characteristic of the element g of the ideal.

The characteristic of the first integral f will be denoted also as χ_f, i.e. $\chi_f = \chi_g$. We must remember that χ_f and χ_g have quite different meanings: the characteristics of a first integral, and that of an element of the ideal respectively.

The characteristic of a first integral is also well-defined, i.e. it is independent of the representative of the class of the first integrals. If f_1 and f_2 are two distinct representants then $f_1 - f_2 = h \in J_Q$. Hence $\partial_{f_1} - \partial_{f_2} = \partial h$ and $g_1 - g_2 = \partial h, \chi_{g_1} - \chi_{g_2} = \chi_{\partial h}$. According to 11.2.5, $\chi_{\partial h} = 0$. $\square$

11.3. Hamilton structure.

11.3.1. Recall that in our case the equations have a variational form $Q_i \equiv \delta \Lambda / \delta v_i = 0$, where Λ is a Lagrangian. If the highest orders of derivatives entering the Lagrangian, $v_i^{(m_i)}$, are m_i (and they cannot be reduced by adding to the Lagrangian exact derivatives), the equations are

$$(11.3.2) \qquad Q_i \equiv \sum_{j=1}^{N} a_{ij} v_j^{(m_i+m_j)} + F_i = 0, i = 1, \ldots, N$$

the coefficients a_{ij} and F_i containing $v_j^{(k)}$ only with $k < m_i + m_j$. We shall not consider the general case. In the examples below all the coefficients a_{ij} will be constant. Then, as it was already shown for $N = 2$ in Sec. 2.3 we can express the highest derivatives $v_i^{(2m_i)}$ in terms of the lower ones:

$$(11.3.3) \qquad v_i^{(2m_i)} = \phi_i(Q, v), i = 1, \ldots, N,$$

ϕ_i are differential polynomials in $\{Q_j^{(k)}\}$ and in $\{v_j^{(k)}\}$, $k < 2m_j$. The system is of Cauchy-Kowalevski type, we have a system of generators (11.1.4) (where $2m_i$ must be substituted for m_i), and all the definitions of the last section are valid.

11.3.4. *Differential forms are expressions*

$$\omega = \sum a_{(i)}^{(j)} \delta v_{i_1}^{(j_1)} \wedge \ldots \wedge \delta v_{i_k}^{(j_k)}, \, a_{(i)}^{(j)} \in \mathcal{A}.$$

For any vector field $\xi = \sum_{i,j} b_{ij} \partial/\partial v_i^{(j)}$, the operation $i(\xi)$ of the substitution of the field into a form is defined:

$$i(\xi)\omega = \sum a_{(i)}^{(j)}(\xi v_{i_1}^{(j_1)}) \delta v_{i_2}^{(j_2)} \wedge \ldots \wedge \delta v_{i_k}^{(j_k)} - \sum a_{(i)}^{(j)} \delta v_{i_1}^{(j_1)} (\xi v_{i_2}^{(j_2)})$$
$$\wedge \ldots \wedge \delta v_{i_k}^{(j_k)} + \ldots + \sum a_{(i)}^{(j)} \delta v_{i_1}^{(j_1)} \wedge \ldots \wedge (\xi v_{i_k}^{(j_k)}).$$

The operation $\delta : \omega \mapsto \delta\omega$ is defined as usual, δ^2 being zero. Further, the Lie derivative L_ξ is defined (see Chap. 2). In particular, for the vector field $\partial = \sum_{i,j} v_i^{(j+1)} \partial/\partial v_i^{(j)}$, the Lie derivative which is denoted simply as ∂, is

$$\partial \omega = \sum \left\{ (\partial a_{(i)}^{(j)}) \delta v_{i_1}^{(j_1)} \wedge \ldots \wedge \delta v_{i_k}^{(j_k)} + a_{(i)}^{(j)} \delta v_{i_1}^{(j_1+1)} \wedge \delta v_{i_2}^{(j_2)} \wedge \ldots , \right.$$
$$\left. + a_{(i)}^{(j)} \delta v_{(i_1)}^{(j_1)} \wedge \delta v_{(i_2)}^{(j_2+1)} \wedge \ldots + \ldots \right\},$$

i.e. this is differentiation in succession of all the multipliers including those which enter under the sign δ.

Let us restrict forms to the phase space, i.e. consider them only on vector fields belonging to $T\mathcal{A}_Q$. Some of the forms become zero. One can easily see that these are the forms which, written in terms of the variables (11.1.4), have at least one multiplier $Q_i^{(j)}$ in any position in $a_{(i)}^{(j)}$ or under the sign δ. We introduce the relation of equivalence $\omega_1 \stackrel{Q}{=} \omega_2$ for forms which coincide by the restriction to the phase space.

Some auxiliary role will be played by another notion of equivalence of forms. Two forms are equivalent in a stronger sense: $\omega_1 \stackrel{QQ}{=} \omega_2$, if their values coincide as elements of $\mathcal{A}_Q$ on all the fields from $T\mathcal{A}$ (not only from $T\mathcal{A}_Q$). It is easy to see that this means that $Q_i^{(j)}$ are identified with zero only when they are in $\mathcal{A}_{(i)}^{(j)}$, and not under the sign δ.

Example: $\delta Q_i \stackrel{Q}{=} 0$ and $\delta Q_i \stackrel{QQ}{\neq} 0$.

11.3.5. Remembering the construction of the symplectic form and the Hamiltonian of a variational Euler-Lagrange equation given in Sec. 2.3, let us rewrite $\delta\Lambda$ as

$$(11.3.6) \qquad \delta\Lambda = \sum(\delta\Lambda/\delta v_k)\delta v_k + \partial\Omega^{(1)},$$

where $\Omega^{(1)}$ is a 1-form. Then $\Omega = \delta\Omega^{(1)}$ is a symplectic form. Let

$$(11.3.7) \qquad \mathcal{H} = -\Lambda + i(\partial)\Omega^{(1)}.$$

Then

$$(11.3.8) \qquad \delta\mathcal{H} = -i(\partial)\Omega - \sum(\delta\Lambda/\delta v_k)\delta v_k$$

whence the equation $\{\delta\Lambda/\delta v_k = 0\}$ is equivalent to

$$(11.3.9) \qquad \delta\mathcal{H} = -i(\partial)\Omega.$$

This can also be expressed thus: if $\mathcal{H}$ is defined by the formula (11.3.7) as an element of $\mathcal{A}_Q$ then the identity (11.3.9) holds in $\mathcal{A}_Q$.

11.3.10. *Proposition.* The Hamiltonian $\mathcal{H}$ can be determined from the equation

$$\mathcal{H}' = -\sum v_i' \delta\Lambda/\delta v_i.$$

Proof: One must apply the operation $i(\partial)$ to both sides of the equation (11.3.8).

To every element $f \in \mathcal{A}_Q$, i.e. to every function in the phase space, a vector field ξ_f is attached with the property

$$\delta f \stackrel{Q}{=} -i(\xi_f)\Omega.$$

It happens that ξ_f can be easily found if f is the first integral. This procedure is based on the notion of the characteristic introduced in Sec. 11.2.

First we prove two simple lemmas.

11.3.11. *Lemma.* If $\delta F \stackrel{QQ}{=} 0$ then $F = \text{const} + F_1$, where $F_1 \in J_Q$ and $\chi_{F_1} = 0$.

Proof: Let us write F in terms of the variables (11.1.4). Then $F = F_1 + F_2$, where $F_1 \in J_Q$, and F_2 is expressed exceptionally in terms of phase variables (a). Let $\xi \in T\mathcal{A}_Q$. Then $\xi F = i(\xi)\delta F \in J_Q$. This yields $\xi F_2 \in J_Q$ since

$\xi F_1 \in J_Q$. Let ξ be, for example, a partial derivative with respect to a phase variable (with fixed (b)-variables). We shall see that $F_2 = c = \text{const}$. Hence $F = c + \sum a_{ij} Q_i^{(j)}$. Now we take $\xi = \partial/\partial Q_i^{(j)}$ and find that all the a_{ij} belong to J_Q and $\chi_{F_1} = 0$. □

11.3.12. Lemma. If $\sum b_i \delta v_i + \partial \omega \stackrel{QQ}{=} 0$, where $b_i \in \mathcal{A}$ and ω is a 1-form then $b_i \stackrel{Q}{=} 0$, and $\omega \stackrel{QQ}{=} 0$.

Proof: Let $\omega = \sum c_{ij} \delta v_i^{(j)}$ then $\partial \omega = \sum c_{ij} \delta v_i^{(j+1)} + \sum c'_{ij} \delta v_i^{(j)}$. Let j_i be the greatest value of j with which $\{v_i^{(j)}\}$ are involved in the form. Then $c_{ij_i} \in J_Q$. We obtain in succession $c_{i,j_i-1} \in J_Q, c_{i,j_i-2} \in J_Q, c_{i,j_i-3} \in J_Q$, etc. This gives also $b_i \in J_Q$. □

11.3.13. Proposition. Let $f \in \mathcal{A}_Q$ be a first integral with the characteristic χ_f. Then the vector field ∂_{χ_f} has the property

$$\partial_{\chi_f} \Lambda = \partial(\ ),$$

where $\partial(\ )$ is the derivative of an element of $\mathcal{A}$.

Proof: We have

$$\partial_{\chi_f} \Lambda = \sum_{i,j} (\chi_f)_i^{(j)} \partial \Lambda / \partial v_i^{(j)} = \sum_i (\chi_f)_i \delta \Lambda / \delta v_i + \partial(\ ) = \sum_{i,k} (-1)^k (\chi_f)_{i,k}^{(k)}$$
$$\times \delta \Lambda / \delta v_i + \partial(\ ) = \sum_{i,k} (\chi_f)_{i,k} (\delta \Lambda / \delta v_i)^{(k)} + \partial(\ ) = \partial F + \partial(\ ) = \partial(\ )$$

according to the definition of a characteristic. □

This can also be written as

(11.3.14) $$\partial_{\chi_F} \int \Lambda dx = 0.$$

11.3.15. Proposition. The vector field ∂_{χ_F} is a symmetry of the equation, i.e. $\partial_{\chi_f} \in T\mathcal{A}_Q$.

Proof: We act on the equation (11.3.6) with the operator ∂_{χ_f}:

$$\delta \partial_{\chi_f} \Lambda = \sum \partial_{\chi_f} (\delta \Lambda / \delta v_k) \delta v_k + \sum (\delta \Lambda / \delta v_k) \delta (\chi_f)_k + \partial (\partial_{\chi_f} \Omega).$$

According to 11.3.13, $\partial_{\chi_f} \Lambda = \partial(\ )$, hence

$$\sum \partial_{\chi_f} (\delta \Lambda / \delta v_k) \delta v_k + \sum (\delta \Lambda / \delta v_k) \delta (\chi_f)_k = \partial(\ ),$$

whence

$$\sum \partial_{\chi_f}(\delta\Lambda/\delta v_k)\delta v_k \stackrel{QQ}{=} \partial(\ ).$$

Lemma 11.3.12 yields $\partial_{\chi_f}(\delta\Lambda/\delta v_k) \in J_Q$, i.e. $\partial_{\chi_f} J_Q \subset J_Q$ as required. □

11.3.16. Proposition. For a first integral f,

$$\delta f \stackrel{Q}{=} i(\partial_{\chi_f})\Omega$$

holds. That means that the vector field $\partial_{-\chi_f}$ corresponds to the first integral f with respect to the symplectic form Ω.

Proof: Let f be the representative of the class of the first integral which is expressed in terms of only the phase variables. Then $\partial f = \sum(\chi_f)_k \delta\Lambda/\delta v_k$ (the derivatives $(\delta\Lambda/\delta v_k)^{(i)}$ with $i > 0$ are not involved). Apply the operator δ to (11.3.6): $\sum \delta(\delta\Lambda/\delta v_k) \wedge \delta v_k + \partial\Omega = 0$.

Further,

$$\partial i(\partial_{\chi_f})\Omega = i([\partial, \partial_{\chi_f}])\Omega + i(\partial_{\chi_f})\partial\Omega = i(\partial_{\chi_f})\partial\Omega,$$

(here the equality $[L_\xi, i(\eta)] = i([\xi, \eta])$ is used, (see 2.5.7), as well as the fact that L_∂ is ∂). Transform the right-hand side of this equation:

$$i(\partial_{\chi_f})\partial\Omega = -i(\partial_{\chi_f})\sum \delta(\delta\Lambda/\delta v_k) \wedge \delta v_k = -\sum \partial_{\chi_f}(\delta\Lambda/\delta v_k)\delta v_k$$
$$+ \sum(\chi_f)_k \delta(\delta\Lambda/\delta v_k) \stackrel{QQ}{=} \sum(\chi_f)_k \delta(\delta\Lambda/\delta v_k).$$

On the other hand,

$$\partial\delta f = \delta\sum(\chi_f)_k \delta\Lambda/\delta v_k \stackrel{QQ}{=} \sum(\chi_f)_k \delta(\delta\Lambda/\delta v_k).$$

Hence

$$\partial(\delta f - i(\partial_{\chi_f})\Omega) \stackrel{QQ}{=} 0.$$

Lemma 11.3.12 gives $\delta f - i(\partial_{\chi_f})\Omega \stackrel{QQ}{=} 0$. If another representative of the class f is chosen then the sign $\stackrel{QQ}{=}$ must be changed to $\stackrel{Q}{=}$. □

11.3.17. Corollary. The symmetry ∂_{χ_f} preserves the form $\Omega : \partial_{\chi_f}\Omega = 0$.
Proof: Apply δ to Eq. (11.3.16). □

Thus, a first integral generates a symmetry which preserves the Lagrangian (more exactly, the action), and the symplectic form. This is, in a sense, a conversion of the Noether theorem.

Now we shall consider our particular integrable systems.

CHAPTER 12. Stationary Equations of the KdV-Hierarchy in the Narrow Sense ($n = 2$).

12.1. Theory of the KdV-hierarchy ($n=2$) independent of Chap. 1.

12.1.1. The easiest way to explain the integration of stationary equations is to use the simple example of the KdV-hierarchy in the narrow sense. We shall introduce this hierarchy anew, independent of the fractional powers method. The method we use now (see [19]) is, despite its restricted applicability, very instructive and, as we believe, beautiful.

The differential algebra a consists now of polynomials in $u, u', u'', \ldots$. We have already written the equation (1.7.14) for the coefficient in ∂^{-1} of a resolvent:

$$(12.1.2) \qquad R''' + 4uR' + 2u'R + 4z^2 R' = 0$$

(here we use R instead of S_1). Now we change our approach. We forget the origin of this equation and start directly from it; so to say, spin it out of thin air.[a] Let us look for solutions of the form

---

[a] The origin of this equation can be explained, for example, as follows. The series R must be a formal analog of the diagonal of the kernel of a resolvent, i.e. an operator inverse to $\partial^2 + u + z^2$ (this kernel is called the Green function). The latter, as it is well known, can be constructed as the product of two solutions of the equation $(\partial^2 + u + z^2)\varphi = 0$. If φ and ψ are solutions then $R = \varphi\psi$ satisfies exactly Eq. (12.1.2).

$$R = \sum_{K_0}^{\infty} R_k z^{-k}, \quad R_k \in a$$

where the series is formal. These solutions are called resolvents.

If $R(z)$ is a resolvent then so is $R(-z)$. Therefore, the even and the odd parts of a resolvent are also resolvents. We shall consider only odd resolvents, even ones can be obtained with the help of multiplication by z. Rewrite the series as

(12.1.3) $$R = \sum_{K}^{\infty} R_{2k+1} \varsigma^{-k-1/2}, \quad \varsigma = z^2.$$

12.1.4. *Proposition.* Eq. (12.1.2) for solutions (12.1.3) is equivalent to the set of equations

(12.1.5) $$2R''R - R'^2 + 4(u+\varsigma)R^2 = c(\varsigma)$$

where $c(\varsigma)$ is a formal series in ς^{-1} with constant coefficients starting with an even power of ς.

Proof: Multiply Eq. (12.1.2) by R and integrate. □

12.1.6. *Proposition.* There is one and only one (up to the sign) solution (12.1.3) of the equation (12.1.5) for each $c(\varsigma) = \sum_{2a}^{\infty} c_k \varsigma^{-k}$. This solution can be written as $R = \sqrt{c(\varsigma)} R^{(1)}$ where $R^{(1)}$ is the solution for $c \equiv 1$; the series of $R^{(1)}$ starts with the term $(1/2) \cdot \varsigma^{-1/2}$.

Proof: Substituting the series (12.1.3) into (12.1.5) we obtain a recurrence system for determining R_{2k+1}. All R_{2k+1} are uniquely determined by the first of them which is equal to $\pm\sqrt{c_{2a}/4} \cdot \varsigma^{-a-1/2}$. The rest is clear. □

We shall use further only the basis resolvent $R^{(1)}$, denoting it simply as R. It is completely determined by the condition that it starts with the term $(1/2) \cdot \varsigma^{-1/2}$ and does not contain constants in other terms.

All the differential polynomials and series of them we use are homogeneous with respect to a grading. Namely the multiplier $u^{(j)}$ has the weight $j+2$, ς the weight 2, ∂ the weight 1, and R the weight -1. A coefficient R_{2k+1}, of weight $2k$, has a linear term $(-1/4)^k u^{(2k-2)}$; the rest of the terms contain $u^{(j)}$ only with $j < 2k - 2$.

12.1.7. *Proposition.* The equation

$$\delta R = R_\varsigma \delta u + \partial[(R_\varsigma/2R)\delta R' - (R'_\varsigma/2R)\delta R], \text{ with } R_\varsigma = \partial R/\partial \varsigma$$

holds. Proof (the method of this proof belongs to Yusin [64]):
We have

(12.1.8) $$2R''R - R'^2 + 4(u+\varsigma)R^2 = 1.$$

Let us apply the operator δ:

$$R''\delta R + R\delta R'' - R'\delta R' + 4(u+\varsigma)R\delta R + 2R^2 \delta u = 0.$$

Multiply by $R_\varsigma/2R^2$. After transformations we obtain

$$[R''R_\varsigma/2R^2 + R''_\varsigma/2R - R'R'_\varsigma/2R^2 + 2R_\varsigma(u+\varsigma)/R]\delta R + R_\varsigma \delta u$$
$$+ \partial[R_\varsigma \delta R'/2R - (R_\varsigma/2R)'\delta R - (R'R_\varsigma/2R^2)\delta R] = 0.$$

On the other hand, differentiating (12.1.8) with respect to ς, we have

$$R''R_\varsigma + R''_\varsigma R - R'R'_\varsigma + 4(u+\varsigma)RR_\varsigma + 2R^2 = 0$$

whence the coefficient in δR in the first term is -1, which yields the required equation. □

12.1.9. Corollary.

$$\delta \int R\,dx/\delta u = R_\varsigma \quad \text{i.e.} \quad \delta \int R_{2k+1}\,dx/\delta u = (-k+1/2)R_{2k-1}.$$

12.1.10. Definition. The KdV-hierarchy is the set of equations

$$\dot u = R'_{2k+1}(u).$$

(The more general equations $\dot u = \sum_0^n a_{2k+1} R'_{2k+1}$ can also be considered).

12.1.11. Exercise. To write a few equations of the hierarchy.
Answer:

$$k = 1,\, 4\dot u = -u';\ k = 2,\, 16\dot u = u''' + 6uu';$$
$$k = 3,\, 64\dot u = -(10u^3 + 10uu'' + 5u'^2 + u^{iv})';$$
$$k = 4,\, 256\dot u = (35u^4 + 70uu'^2 + 70u^2 u'' + 21u''^2$$
$$+ 28u'u''' + 14uu^{iv} + u^{vi})'.$$

For $k=2$ we have the KdV-equation. The other equations are higher KdV equations.

12.1.12. Proposition. Elements $P_{k,l} \in \mathcal{A}$ exist for any k and l such that

$$R_{2k+1} R'_{2l+1} = P'_{k,l}.$$

Proof: We use induction with respect to one of the numbers k and l. Let this be k. For $k=0$ we have $R_1 R'_{2l+1} = R'_{2l+1}/2$. i.e. $P_{0,l} = R_{2l+1}/2$ for every l. Let us write (12.1.2) as a recurrence equation:

(12.1.13) $$R'''_{2k+1} + 4u R'_{2k+1} + 2u' R_{2k+1} = -4 R'_{2k+3}.$$

Let the proposition be proved for some k. Then

$$R_{2k+3} R'_{2l+1} = (R_{2k+3} R_{2l+1})' - R'_{2k+3} R_{2l+1} = (R_{2k+3} R_{2l+1})'$$
$$+ (1/4)(R'''_{2k+1} + 4u R'_{2k+1} + 2u' R_{2k+1}) R_{2l+1} = (R_{2k+3} R_{2l+1}$$
$$+ R''_{2k+1} R_{2l+1}/4 - R'_{2k+1} R'_{2l+1}/4 + R_{2k+1} R''_{2l+1}/4 + u R_{2k+1} R_{2l+1})'$$
$$- (1/4) R_{2k+1}(R'''_{2l+1} + 4u R'_{2l+1} + 2u' R_{2l+1}) = R_{2k+1} R'_{2l+3}$$
$$+ (R_{2k+3} R_{2l+1} + R''_{2k+1} R_{2l+1}/4 - R'_{2k+1} R'_{2l+1}/4$$
$$+ R_{2k+1} R''_{2l+1}/4 + u R_{2k+1} R_{2l+1})'.$$

Thus, if $R_{2k+1} R'_{2l+3} = P'_{k,l+1}$ then $R_{2k+3} R'_{2l+1} = P'_{k+1,l}$, where

$$P_{k+1,l} = P_{k,l+1} + R_{2k+3} R_{2l+1} + R''_{2k+1} R_{2l+1}/4 - R'_{2k+1} R'_{2l+1}/4$$
$$+ R_{2k+1} R''_{2l+1}/4 + u R_{2k+1} R_{2l+1}. \quad \Box$$

12.1.14. Proposition. The quantities $J_{2l+1} = \int R_{2l+1} dx$ are first integrals of all the equations of the KdV-hierarchy.
Proof. We have

$$\dot{J}_{2l+1} = \int (\delta R_{2l+1}/\delta u)\dot{u}\, dx = (-l+1/2)\int R_{2l-1} R'_{2k+1} dx$$
$$= (-l+1/2)\int R'_{l-1,k} dx = 0.$$

12.1.15. Let us construct the Hamilton structure of the hierarchy: the symplectic form and Hamiltonians. That was already done in Chap. 3. We cannot say anything new about this and shall only briefly repeat it without proof.

We have a Lie algebra $\mathfrak{E}$ of vector fields

$$\partial_a = \sum_0^\infty a^{(i)} \partial/\partial u^{(i)}, a \in \mathcal{A}.$$

The left $\mathfrak{E}$-module $\Omega^0 = \tilde{\mathcal{A}}$ comprises functionals $\tilde{f} = \int f\,dx, f \in \mathcal{A}$. The action of ∂_a on $\tilde{f}$ is thus

$$\partial_A \tilde{f} = \int \sum a^{(i)} \partial f/\partial u^{(i)} dx = \int a\delta f/\delta u\,dx.$$

The space $\Omega^{(1)}$ of 1-forms over $\mathfrak{E}$ coincides with $\mathcal{A}$. The coupling of the elements of $\mathfrak{E}$ and of $\Omega^{(1)}$ is as follows. For $\partial_a \in \mathfrak{E}$ and $X \in \Omega^{(1)} = \mathcal{A}$, we have

$$\langle \partial_a, X \rangle = \int aX\,dx.$$

Finally, there is a mapping $H : \Omega^{(1)} \to \mathfrak{E}$

(12.1.16) $\qquad X \in \Omega^{(1)} \mapsto \partial_{H(X)}, H(X) = X''' + 4(u+\xi)X' + 2u'X$

for any fixed ξ. In Chap. 3 it is proved that the image of this mapping, $\{\partial_{H(X)}\}$ is a Lie subalgebra of the algebra $\mathfrak{E}$. A form is constructed:

$$\omega(\partial_{H(X)}, \partial_{H(Y)}) = \langle \partial_{H(X)}, Y \rangle = \int H(X)Y\,dx,$$

and this form happens to be closed.

For $\varsigma = 0$ and $\varsigma = \infty$ we have the limiting cases of this form. So, $H^{(\infty)}(X) = X'$, and $\omega^{(\infty)}(\partial_{X'}, \partial_{Y'}) = \int X'Y\,dx$.

Further, there is a mapping of functionals into vector fields

(12.1.17) $\qquad\qquad \tilde{f} \in \tilde{\mathcal{A}} \mapsto \partial_{H(\delta f/\delta u)} = \partial_{\tilde{f}},$

and Poisson bracket

$$\{\tilde{f}, \tilde{g}\} = \omega(\partial_{H(\delta f/\delta u)}, \partial_{H(\delta g/\delta u)}) = \partial_{H(\delta f/\delta u)}\tilde{g}.$$

For any two functionals a relation

(12.1.18) $\qquad\qquad \partial_{\{\tilde{f},\tilde{g}\}} = [\partial_{\tilde{f}}, \partial_{\tilde{g}}]$

holds. The standard formula of correspondence between functionals and vector fields

(12.1.19) $$\delta \tilde{f} = -i(\partial_{\tilde{f}})\omega$$

also holds.

12.1.20. Proposition. The quantities $\{J_{2k+1}\}$ (12.1.14) are involution.

Proof.

$$\{J_{2k+1}, J_{2l+1}\} = \text{const.} \int \left[R'''_{2k-1} + 4(u+\varsigma)R'_{2k-1} + 2u' R_{2k-1} \right] R_{2l-1} dx$$

$$= \text{const.} \int 4(R'_{2k+1} + \varsigma R'_{2k-1}) R_{2l-1} dx$$

$$= \text{const.} \int (P'_{k,l-1} + \varsigma P'_{k-1,l-1}) dx = 0. \square$$

12.1.21. Corollary. Vector fields $\left\{\partial_{R'_{2k+1}}\right\}$ commute, and the identity

(12.1.22) $$\partial_{R'_{2k+1}} R_{2l+1} - \partial_{R'_{2l+1}} R_{2k+1} = 0$$

holds.

Proof: The first, $[\partial_{R'_{2k+1}}, \partial_{R'_{2l+1}}] = 0$, immediately follows from the last proposition. Applying this identity to u we get

$$0 = \partial_{R'_{2k+1}} R'_{2l+1} - \partial_{R'_{2l+1}} R'_{2k+1} = \partial \left(\partial_{R'_{2k+1}} R_{2l+1} - \partial_{R'_{2l+1}} R_{2k+1} \right)$$

which yields $\partial_{R'_{2k+1}} R_{2l+1} - \partial_{R'_{2l+1}} R_{2k+1} = \text{const}$; the constant may be only zero. $\square$

12.2. Stationary equations.[b]

12.2.1. A stationary equation is $R'_{2k+1} = 0$ i.e. $R_{2k+1} = c = \text{const}$ which can also be written as $R_{2k+1} - cR_1 = 0$. A more general differential equation (see 12.1.10) is

(12.2.2) $$Q \equiv \sum_{0}^{n+1} d_{2k+1} R_{2k+1} = 0.$$

---

[b]We follow here [106].

This is an ordinary $2n$-th order differential equation.

12.2.3. Proposition. (i) If a function $u(x)$ satisfies the equation (12.2.2) then the polynomial in ς of the n-th degree

$$\hat{R}(\varsigma) = \sum_{k=1}^{n+1} d_{2k+1} \sum_{i=0}^{k-1} R_{2i+1} \varsigma^{k-1-i} = \sum_{l=0}^{n} \hat{R}_l \varsigma^l,$$

$$\left(\hat{R}_l = \sum_{k=l+1}^{n+1} d_{2k+1} R_{2(k-l)-1} \right)$$

(where $u^{(j)}(x)$ are substituted for $u^{(j)}$) satisfies the equation (12.1.2) for the resolvent.

(ii) Conversely, if for some function $u(x)$ there is a solution of Eq. (12.1.2) which is a polynomial in ς whose coefficients are differential polynomials in $u(x)$, then $u(x)$ satisfies a stationary equation of the form (12.2.2).

Proof: (i) Let us substitute $\hat{R}(\varsigma)$ for R into the left-hand side of Eq. (12.1.2):

$$\sum_{k=1}^{n+1} d_{2k+1} \sum_{i=0}^{k-1} \left[(R'''_{2i+1} + 4uR'_{2i+1} + 2u'R_{2i+1})\varsigma^{k-i-1} \right.$$

$$\left. + 4R'_{2i+1}\varsigma^{k-i} \right] = \sum_{k=1}^{n+1} d_{2k+1} \left[\sum_{i=0}^{k-1} (R'''_{2i+1} + 4uR'_{2i+1} \right.$$

$$\left. + 2u'R_{2i+1} + 4R'_{2i+3}) - 4R'_{2k+1} \right] = -4 \sum_{k=1}^{n+1} d_{2k+1} R'_{2k+1}$$

(see (12.1.13)). This expression vanishes owing to (12.2.2). (ii) Let $\hat{R}(\varsigma)$ be a solution of Eq. (12.1.2) having the form of a polynomial of n-th degree in ς. Then $\hat{R}(\varsigma)$ satisfies also the equation (12.1.5), where $c(\varsigma)$ is a $2n + 1$-degree polynomial in ς with constant coefficients. Let $\sqrt{c(\varsigma)} = d(\varsigma) = \sum_{-\infty}^{n} d_{2k+1}\varsigma^{k+1/2}$. According to 12.1.6 $\hat{R}(\varsigma) = d(\varsigma)R(\varsigma) = \sum_{k=-\infty}^{n} \sum_{i=0}^{\infty} d_{2k+1} R_{2i+1}\varsigma^{k-i}$. On the other hand this is a polynomial. Let us write the condition that the coefficient in ς^{-1} vanishes: $\sum_{0}^{n+1} d_{2i-1} R_{2i+1} = 0$. This is an equation of the type (12.2.2). □

12.2.4. Proposition. For each l there is a first integral of Eq. (12.2.2), namely

$$f_{2l+1} = \sum_{k=0}^{n+1} d_{2k+1} P_{k,l}$$

For $l = 1, \ldots, n$ these first integrals are independent.
Proof: We find

(12.2.5) $$\partial f_{2l+1} = \sum_{k=0}^{n+1} d_{2k+1} R_{2k+1} R'_{2l+1} \stackrel{Q}{=} 0.$$

Thus, all the f_{2l+1} are first integrals.

Now we shall prove the independence of the mentioned first integrals. The highest terms with respect to weights are contained in $P_{n+1,l}$, their weight being $2n + 2 + 2l$. There are quadratic terms among them which are obtained from the product of linear terms in R_{2n+3} and R'_{2l+1}, i.e. of $u^{(2n)}$ and $u^{(2l-1)}$. We find

$$u^{(2n)} u^{(2l-1)} = (u^{(2n-1)} u^{(2l-1)} - u^{(2n-2)} u^{(2l)} + \ldots \pm \frac{1}{2}(u^{(n+l-1)})^2)'.$$

Thus f_{2l+1} has the term $c \cdot (u^{(n+l-1)})^2$. The multiplier $u^{(n+l-1)}$ can be involved in other terms only in the product with $u^{(j)}$, where $j < n + l - 1$. On the other hand, in $f_{2s+1}, s < l$, the multiplier $u^{(n+l-1)}$ can only be in the product with $u^{(j)}, j < n + l - 1$. If we put all the $u^{(j)} = 0$ except for $j = n + l - 1$, then $f_{2s+1} = 0$ for all $s < l$ and $f_{2l+1} \neq 0$, which proves the independence of this quantity of the previous ones. □

12.2.6. Proposition. Eq. (12.2.2) is of the Euler-Lagrange type, with the Lagrangian

$$\Lambda = \sum_{k=0}^{n+1} d_{2k+1} R_{2k+3}/(-k - 1/2)$$

Proof: This immediately follows from 12.1.9. □

The discussion of Chap. 11 is completely applicable to this equation. In particular, a Hamiltonian form of this equation can be written.

12.2.7. Proposition. The 1-form $\Omega^{(1)}$ related to this Lagrangian Λ is

$$\Omega^{(1)} = -\sum_{k=0}^{n+1} d_{2k+1} (R' \delta R / 2R)_{k+1/2}$$

where the subscript $k+1/2$ means the coefficient in $\varsigma^{-k-1/2}$.
Proof: According to 12.1.7 we have

$$\delta\Lambda = -\sum_{k=0}^{n+1} d_{2k+1}[R_\varsigma \delta u + \partial(R_\varsigma \delta R'/2R - R'_\varsigma \delta R/2R)]_{k+3/2}/(k+1/2)$$

whence

$$\Omega^{(1)} = -\sum_{k=0}^{n+1} d_{2k+1}(R_\varsigma \delta R'/2R - R'_\varsigma \delta R/2R)_{k+3/2}/(k+1/2)\,.$$

An arbitrary differential $\delta f, f \in a$, can be added to the form $\Omega^{(1)}$ which does not change the symplectic form Ω. Then one can write

$$\Omega^{(1)} = -\sum_{k=0}^{n+1} d_{2k+1}(-\delta(R_\varsigma/2R)\cdot R' - R'_\varsigma \delta R/2R)_{k+3/2}/(k+1/2)$$

$$= \sum_{k=0}^{n+1} d_{2k+1}[\partial/\partial\varsigma(\delta R \cdot R'/2R)]_{k+3/2}/(k+1/2)$$

$$= -\sum_{k=0}^{n+1} d_{2k+1}(R' \cdot \delta R/2R)_{k+1/2}\,.\square$$

This explicit expression for $\Omega^{(1)}$ and $\Omega = \delta\Omega^{(1)}$ was obtained by Alber [65, 66].

12.2.8. Proposition. Let $\hat{w} = \hat{R}'/2\hat{R}$. Then the forms $\Omega^{(1)}$ and Ω can also be written as

$$\Omega^{(1)} = (\hat{w}\delta\hat{R})_1,\ \Omega = (\delta\hat{w}\wedge\delta\hat{R})_1,$$

where the subscript 1 denotes the coefficient in ς^{-1}.

Proof: We have $\hat{R}(\varsigma) = \sum_{k=1}^{n+1} d_{2k+1}\varsigma^{k-1/2}\cdot R + 0(\varsigma^{-1})$, whence $\hat{w} = w + 0(\varsigma^{-n-2})$ with $w = R'/2R$. Besides, $w = 0(\varsigma^{-1})$.

$$\Omega^{(1)} = -\sum_{k=0}^{n+1} d_{2k+1}\left(w\delta\sum_{i=0}^{\infty} R_{2i+1}\varsigma^{-i-1/2}\right)_{k+1/2}$$

$$= -\left(w\cdot\delta\sum_{k=0}^{n+1} d_{2k+1}\sum_{i=0}^{\infty} R_{2i+1}\varsigma^{k-i-1}\right)_1\,.$$

The expression inside the brackets will change by $0(\varsigma^{-2})$ if $\hat{w}$ is substituted for w and $\sum_{i=0}^{k-1}$ for $\sum_{i=0}^{\infty}$. This change does not affect the coefficient in ς^{-1}. Now $\Omega^{(1)} = -(\hat{w} \cdot \delta \hat{R})_1$. The second relation follows from this one. □

The advantage of this new formula over 12.2.7 is that here the form is equal to the residue of a rational function, and not to a coefficient of a formal series.

12.2.9. Proposition. The vector fields which relate to the first integrals $-f_{2l+1}$ with respect to the symplectic form Ω are $\partial_{R'_{2l+1}}$.
Proof. We have $\partial f_{2l+1} = R'_{2l+1} \cdot \delta \Lambda / \delta u$ (see 12.2.5) whence the characteristic of the first integral f_{2l+1} is R'_{2l+1}. It remains to apply 11.3.16. □

Recall the procedure of constructing of Hamiltonians (see 11.3.10). The Hamiltonian satisfies the equation $\mathcal{H}' = -u'\delta\Lambda/\delta u$.

12.2.10. Proposition. The Hamiltonian of (12.2.2) is $\mathcal{H} = 4f_3$.
Proof: We have $\mathcal{H}' = 4R'_3 \cdot \delta\Lambda/\delta u = -u'\delta\Lambda/\delta u$. □

12.2.11. Proposition. If the first integral $-f_{2l+1}$ is taken as a Hamiltonian in the phase space of the equation (12.2.2) endowed by the symplectic form Ω then the corresponding flow is the restriction of the flow given by the nonstationary equation (12.1.10) $\dot{u} = R'_{2l+1}(u)$ to the phase space of the equation (12.2.2).
Proof: Both the flows are determined by the same vector field $\partial_{R'_{2l+1}}$ restricted to the phase space. □
This fact was established by Bogoyavlenski and Novikov [67], and by Gelfand and Dickey [25].

12.2.12. Proposition. The first integrals f_{2l+1} of the stationary equation are in involution.
Proof: Their vector fields commute. □

12.2.13. Proposition. Eq. (12.2.2) is integrable in quadrature.
Proof. The equation of the n-th order has n independent integrals in involution. The integrability follows from the classical Liouville theorem. □

In the next section we shall obtain the effective formulas for such integration.

12.2.14. The first integrals will be written in another form. The polynomial $\tilde{R}(\varsigma)$ satisfies the equation (12.1.2) in virtue of Eq. (12.2.2). Then it must satisfy also the equation (12.1.5), which will be written now as

(12.2.15) $$2\hat{R}''\hat{R} - \hat{R}'^2 + 4(u+\varsigma)\hat{R}^2 = P(\varsigma)$$

where the coefficients of the polynomial $P(\varsigma)$ of the degree $2n+1$ are not absolute constants but only constants in virtue of Eq. (12.2.2), $\partial P(\varsigma) \stackrel{Q}{=} 0$, i.e. these coefficients are first integrals of the equation. However, the higher $n+2$ coefficients of $P(\varsigma)$ are absolute constants. In fact, they do not change if we replace $\hat{R}(\varsigma)$ by $\hat{R}_1(\varsigma) = \sum_{k=1}^{n+1} d_{2k+1} \sum_{i=0}^{\infty} R_{2i+1}\varsigma^{k-i-1} = \sum_{k=1}^{n+1} d_{2k+1}\varsigma^{k-1/2} R(\varsigma)$, since $\hat{R} - \hat{R}_1 = 0(\varsigma^{-1})$ and this replacement does not affect the higher $n+2$ terms. Thus, the higher $n+2$ terms of $P(\varsigma)$ coincide with those of the polynomial $\varsigma \left(\sum_{1}^{n+1} d_{2k+1}\varsigma^{k-1} \right)^2$. The only non-trivial first integrals are the n lower terms of $P(\varsigma)$. Let

$$P(\varsigma) = \sum_{0}^{2n+1} J_l \varsigma^l.$$

The new first integrals are not independent of the old ones.

12.2.16. Proposition.

$$J_l \stackrel{Q}{=} 8 \sum_{k=l+2}^{n+1} d_{2k+1} f_{2(k-l)-1}, l = 0, 1, \ldots, n-1.$$

Proof: We find (cf. the proof of 12.2.3)

$$\partial P(\varsigma) = 2(\hat{R}''' + 2u'\hat{R} + 4(u+\varsigma)\hat{R}')\hat{R}$$
$$= -8\sum_{k=1}^{n+1} d_{2k+1} R'_{2k+1} \cdot \hat{R} = -8\hat{R}Q' = -8(\hat{R}Q)' + 8\hat{R}'Q.$$

The term $(\hat{R}Q)'$ can be dropped (changing $P(\varsigma)$ by $8\hat{R}Q \stackrel{Q}{=} 0$). Therefore the characteristic of the first integral J_l is equal to the coefficient of the polynomial $8\hat{R}'$ in ς^l, i.e.

$$\chi_{J_l} = 8 \sum_{k=l+2}^{n+1} d_{2l+1} R'_{2(k-l)-1}.$$

Taking into account that $\chi_{f_{2l+1}} = R'_{2l+1}$ we obtain $J_l \stackrel{Q}{=} 8 \sum_{k=l+2}^{n+1} d_{2k+1}$ $f_{2(k-l)-1}$ as required. □

12.2.17. Proposition. The Hamiltonian of the system is

$$\mathcal{H} = \frac{1}{2d_{2n+3}} J_{n-1}.$$

Proof: This is a corollary of 12.2.10 and of 12.2.16. □

The first integrals $\{f_{2l+1}\}$ can be expressed in terms of $\{J_l\}$:

$$(12.2.18) \qquad f_{2l+1} = \sum_{k=0}^{l-1} c_{2(l-k)+1} J_{n-1-k}, l = 1, \ldots, n$$

where the matrices

$$D = \begin{pmatrix} d_{2n+3} & & & \\ d_{2n+1} & \ddots & 0 & \\ \vdots & \ddots & \ddots & \\ d_5 & \ddots & d_{2n+1} & d_{2n+3} \end{pmatrix}, C = \begin{pmatrix} c_3 & & & \\ c_5 & \ddots & 0 & \\ \vdots & \ddots & \ddots & \\ c_{2n+1} & \ddots & c_5 & c_3 \end{pmatrix}$$

are mutually reciprocal, $D = C^{-1}$.

12.3. Integration after Liouville.

12.3.1. Recall the Liouville theorem (in local form). Let $\mathcal{U}$ be a $2n$-th dimensional symplectic manifold and Ω its symplectic form. Let $J_0, \ldots, J_{n-1}$ be n independent functions in involution, $\{J_k, J_l\} = 0$ for any k and l. Then a canonically adjoint system of the functions $\varphi_0, \ldots, \varphi_{n-1}$ such that $\{\varphi_k, \varphi_l\} = 0, \{J_k, \varphi_l\} = \delta_{kl}$ can be constructed with the help of quadratures.

If there is a Hamiltonian $\mathcal{H}$ then in the variables (J, φ) the Hamilton system has the form

$$(12.3.2) \qquad \dot{\varphi}_k = \partial \mathcal{H}/\partial J_k, \dot{J}_k = -\partial \mathcal{H}/\partial \varphi_k.$$

If $\{J_k\}$ is a system of first integrals then $\partial \mathcal{H}/\partial \varphi_k = 0$, i.e. the Hamiltonian does not depend on the variables $\{\varphi_k\}$, they are called "cyclic" variables or the variables of the angle type. On each level surface $\{J_k = \text{const}\}$ the equations (12.3.2) can be easily integrated:

$$\varphi_k = (\partial \mathcal{H}/\partial J_k)t + \text{const}, J_k = \text{const}.$$

12.3.3. We describe the procedure of constructing the angle variables. Consider an n-dimensional level surface $\{J_k = C_k\}$ in the $2n$-dimensional phase space $\mathcal{U}$. The vector fields $\{\xi_k\}$ corresponding to $\{J_k\}$ are tangent to the surface, since $\xi_k J_l = \{J_k, J_l\} = 0$. The number of independent vector fields $\{\xi_k\}$ is equal to the dimension of the surface, therefore they form a basis to the tangent space in every point of the surface. These vector fields are orthogonal to each other with respect to the form Ω : $\Omega(\xi_k, \xi_l) = \{J_k, J_l\} = 0$. Thus, the symplectic form restricted to the tangent spaces at all the points of the surface vanishes. These spaces can not be extended in such a way that the restriction of the form Ω remains zero, since the form is non-degenerating, and the orthogonal completion of a subspace of the dimension $m > n$ has dimension $2n - m < n$. A n-dimensional subspace in $2n$-dimensional symplectic space is called Lagrangian if the form Ω restricted to it completely degenerates. Thus, the level surfaces $\{J_k = c_k\}$ are Lagrangian.

Let us choose as coordinates in the phase space the variables $\{J_k\}$ and some functions $\{s_k\}$ independent of $\{J_k\}$ and among themselves. Let us write the form $\Omega^{(1)}$ in these coordinates: $\Omega^{(1)} = \sum \alpha_i ds_i + \sum \beta_i dJ_i$. Being restricted to the surfaces $\{J_k = c_k\}$ the form $\Omega^{(1)}$ has the property $d\Omega^{(1)} = \Omega = 0$. Then a function $V(J, s)$ exists (locally) such that $\Omega^{(1)} = dV(J, s)|_{\{J_k = c_k\}}$, i.e. $\alpha_i = \partial V/\partial s_i$. This function can be found with the help of quadratures. Now

$$\Omega^{(1)} = \sum(\partial V/\partial s_i)ds_i + \sum \beta_i dJ_i = dV + \sum(\beta_i - \partial V/\partial J_i)dJ_i.$$

Let $\varphi_i = -\beta_i + \partial V/\partial J_i, i = 0, \ldots, n-1$. Then

$$\Omega = d\Omega^{(1)} = \sum dJ_i \wedge d\varphi_i.$$

This yields that the variables $\{\varphi_i\}$ are canonically conjugated to $\{J_i\}$. The independence of $\{J_i, \varphi_i\}$ follows from the fact that the form Ω is non-degenerating.

12.3.4. We must apply this procedure to Eq. (12.2.2). In addition to the variables $J_0, \ldots, J_{n-1}$, coordinates $\{s_i\}$, arbitrary but convenient as far as possible must be chosen. The transition from the coordinates $(u) = (u, u', \ldots, u^{(2n-1)})$ in the phase space to the coordinates (J, s) will be carried out gradually. First we introduce the coordinate system $(\hat{R}, \hat{R}') = (\hat{R}_0, \ldots, \hat{R}_{n-1}, \hat{R}'_0, \ldots, \hat{R}'_{n-1})$. The higher derivative $u^{(j)}$ entering $\hat{R}_l$ is $u^{(2n-2l-2)}$: and it is involved linearly. In $\hat{R}'$ it is $u^{(2n-2l-1)}$, respectively. Thus, two coordinate systems (u) and $(\hat{R}, \hat{R}')$ can be expressed, one of them in terms of the other, polynomially and triangularly. So, u is proportional to $\hat{R}_{n-1}$. u' is a polynomial in $\hat{R}_{n-1}$ and $\hat{R}'_{n-1}$ which is linear which respect to $\hat{R}'_{n-1}$; u'' is a polynomial in $\hat{R}_{n-1}, \hat{R}'_{n-1}$ and $\hat{R}_{n-2}$ etc.

The next coordinate system is $(\hat{R}, \hat{w}) = (\hat{R}_0, \ldots, \hat{R}_{n-1}, \hat{w}_1, \ldots, \hat{w}_n)$ where $\hat{w} = \hat{R}'/2\hat{R} = \sum_{1}^{\infty} \hat{w}_i \varsigma^{-i}$. It is connected with the previous one also by a triangular relation. This coordinate system is convenient since the symplectic form is written in it canonically, (see 12.2.8).

Finally, following Dubrovin [68] we introduce the coordinates (J, s). When $(J_0, \ldots, J_{n-1})$ are fixed, the variables (s) are coordinates on the corresponding n-dimensional surface. If (J) are fixed then the polynomial $P(\varsigma)$ (12.2.15) is also fixed. The two-valued function $\sqrt{P(\varsigma)}$ is defined on a two-foliated Riemann surface Γ_J. Let $\varsigma_1^*, \ldots, \varsigma_n^*$ be some points of the Riemann surface Γ_J over $\varsigma_1, \ldots, \varsigma_n \in CP^1$ where $\varsigma_1, \ldots, \varsigma_n$ are roots of the polynomial $\hat{R}(\varsigma)$. In other words, $\varsigma_1^*, \ldots, \varsigma_n^*$ are the points $\varsigma_1, \ldots, \varsigma_n$ together with the indication as to which of two roots $\pm\sqrt{P(\varsigma_i)}$ should be taken. These points (ς_i^*) are taken as the additional coordinates s_i. How to express the coordinates $(\hat{R}, \hat{R}')$ in terms of (J, ς^*) and vice versa? We have

$$(12.3.5) \qquad \hat{R}(\varsigma) = (d_{2n+3}/2) \prod_{j=1}^{n} (\varsigma - \varsigma_i).$$

That means that $\hat{R}_0, \ldots, \hat{R}_{n-1}$ are the elementary symmetrical functions of $\{\varsigma_j\}$. In particular, $\hat{R}_{n-1}/\hat{R}_n = -(\varsigma_1 + \ldots + \varsigma_n)$. Taking into account that $\hat{R}_n = d_{2n+3}/2, \hat{R}_{n-1} = d_{2n+3}R_3 + d_{2n+1}R_1 = -u d_{2n+3}/4 + d_{2n+1}/2$ we obtain

$$(12.3.6) \qquad u = 2 d_{2n+1}/d_{2n+3} + 2(\varsigma_1 + \ldots + \varsigma_n).$$

The expression of $\{\tilde{R}'_l\}$ in terms of the new variables is given by the lemma below.

12.3.7. Lemma.
$$\hat{R}'(\varsigma_l) = i\sqrt{P(\varsigma_l)}, l = 1, \ldots, n.$$

Proof: One must substitute $\varsigma = \varsigma_l$ into (12.2.15). □

Conversely, if the values of the coordinates $(\hat{R}, \hat{R}')$ are given they determine the values of $\{J_l\}$ (which are polynomials in these coordinates). Solving the algebraic equation $\hat{R}(\varsigma) = 0$ one obtains $\{\varsigma_l\}$, and $\hat{R}'(\varsigma_l)$ determines the choice of the sign of the root $\sqrt{P(\varsigma_l)}$, i.e. the point on the Riemann surface $\varsigma_l^* \in \Gamma_J$.

12.3.8. Proposition. (Dubrovin). The system (12.2.2) in the coordinates (J, ς^*) has the form

$$\partial J_l = 0, l = 0, 1, \ldots, n-1; \partial \varsigma_l(x) = -(2i/d_{2n+3})$$
$$\times \sqrt{P(\varsigma_l)}/\prod_{j \neq l}(\varsigma_l - \varsigma_j), l = 1, \ldots, n$$

Proof: The quantities $\{J_l\}$ are first integrals, $\partial J_l = 0$. Further, $\partial \hat{R}_l = \hat{R}'_l$. It remains to use (12.3.5) and 12.3.7. □

12.3.9. Proposition. The form $\Omega^{(1)}$ in the coordinates (J, ς^*) is

$$\Omega^{(1)} = (i/2)\sum_{j=1}^n \sqrt{P(\varsigma_j)}\delta\varsigma_j.$$

Proof: According to 12.2.8, $\Omega^{(1)} = (\hat{R}'/2\hat{R} \cdot \delta\hat{R})_1$. This is the residue in $\varsigma = \infty$ of the differential $(\hat{R}'/2\hat{R} \cdot \delta\hat{R})d\varsigma$. It is equal to the sum of the residues of the differential in the finite part of the Riemann sphere with the opposite sign. The poles are $\varsigma = \varsigma_j$, and

$$\Omega^{(1)} = \sum_{j=1}^n \left[\hat{R}'(\varsigma_j)/2\prod_{i \neq j}(\varsigma_j - \varsigma_i)\right] \cdot \prod_{i \neq j}(\varsigma_j - \varsigma_i)$$
$$\delta\varsigma_j = (i/2)\sum_{j=1}^n \sqrt{P(\varsigma_j)}\delta\varsigma_j. \square$$

12.3.10. Proposition. The variables $\{\varphi_l\}, l = 0, \ldots, n-1$ canonically adjoint to the variables $\{J_l\}$ are

$$\varphi_l = (i/4)\sum_{m=1}^n \int_{\varsigma_a^*}^{\varsigma_m^*} \left(\varsigma^l/\sqrt{P(\varsigma)}\right)d\varsigma, l = 0, \ldots, n-1$$

where ς_a^* is a fixed point on the Riemann surface.

Proof: According to Liouville's procedure one must find a function $V(J,\varsigma^*)$ such that $\delta V = \Omega^{(1)}$ on the surfaces $\{J_k = c_k\}$. The form $\Omega^{(1)}$ from 12.3.9 enables one to perform this integration very easily since the variables are separated (this is the reason for the convenience of using the Dubrovin variables, although the paper [68] does not deal with differential forms). We have

$$V = (i/2) \sum_{m=1}^{n} \int_{\varsigma_a^*}^{\varsigma_m^*} \sqrt{P(\varsigma)} d\varsigma.$$

Further, $\varphi_l = \partial V/\partial J_l$ (see 12.3.3). From $P(\varsigma) = \sum_{0}^{2n+1} J_l \varsigma^l$ we find

$$\partial V/\partial J_l = i/2 \sum_{m=1}^{n} \int_{\varsigma_a^*}^{\varsigma_m^*} \left(\varsigma^l/\sqrt{P(\varsigma)}\right) d\varsigma.$$

It is convenient to take $\varsigma_a^* = \infty$. □

12.3.11. Thus we have constructed a system of variables in which the equation (12.2.2) can be integrated as linear functions of x. This integration can be performed either directly, in which case using 12.3.8 we have

$$\partial \varphi_l = (i/4) \sum_{k=1}^{n} \left(\varsigma_k^l/\sqrt{P(\varsigma_k^*)}\right) \cdot \varsigma_k' = (i/4) \sum_{k=1}^{n} \varsigma_k^l$$

$$(-2i/d_{2n+3})/\prod_{j \neq k}(\varsigma_k - \varsigma_j) = (1/2d_{2n+3}) \sum_{k=1}^{n} \text{res}_{\varsigma=\varsigma_k} \frac{\varsigma^l}{\prod_j(\varsigma - \varsigma_j)}$$

$$= (1/2d_{2n+3}) \cdot 2\pi i \oint \varsigma^l \cdot \prod_j (\varsigma - \varsigma_j)^{-1} d\varsigma = \begin{cases} 0, & l < n-1 \\ 1/2d_{2n+3}, & l = n-1, \end{cases}$$

whence

(12.3.12) $\quad \varphi_l = \varphi_l^{(0)} + \delta_{l,n-1} \cdot (1/2d_{2n+3}) x, J_l = J_l^0; \varphi_l^{(0)}, J_l^{(0)} = \text{const}$

or by writing one of the canonic equations (12.3.2) as $\varphi_l' = \partial \mathcal{H}/\partial J_l$ and taking the Hamiltonian from 12.2.17.

12.3.13. The finite-dimensional manifold of the solutions of Eq. (12.2.2) is invariant with respect to the flow generated by each of the equations (2.1.10), i.e. the solutions transform to solutions under the action of this flow. Let t be the parameter of this flow, then, the point in (12.1.10) denotes differentiation with respect to t. Let us look for the functions $u(x,t)$ which satisfy the equation (12.2.2) with respect to x and the equation (12.1.10) with respect to x and t. We find these solutions first in the coordinates (J, φ). The constants in (12.3.12) will depend now on the parameter t. The vector field $\partial_{R'_{2k+1}}$ relating to equation (12.2.10) in the phase space of the equation (12.2.2) is a Hamiltonian one, with the Hamiltonian $-f_{2k+1}$ (see 12.2.9). We express this Hamiltonian in the coordinates (J, φ). According to (12.2.18), $-f_{2k+1} = -\sum_{l=0}^{k-1} c_{2(k-l)+1} J_{n-l-1}$. It remains to write the canonical system with this Hamiltonian:

$$\dot{J}_i = 0, \dot{\varphi}_i = -\partial f_{2k+1}/\partial J_i = -c_{2(k+i-n)+3}(i \geq n-k), \dot{\varphi}_i = 0 \ (i < n-k).$$

This system can be integrated and we finally obtain
(12.3.14)
$$J_i = J_i^0, \varphi_i = \varphi_i^0 + \delta_{n-1,i}(1/2 d_{2n+3})x - c_{2(k+i-n)+3}t, i = 0, \ldots, n-1$$

where $c_i = 0$ if $i < 3$, J_i^0 and φ_i^0 are constants. Note that a common solution of several equations (12.1.10) could be considered, depending on several parameters, "times", as it was done in Chap. 5 for the KP-hierarchy.

12.4. Return to original variables.

12.4.1. Now we have to return from the variables (J, φ) to the variables (u). In 12.3.10 the variables $\{\varphi_l\}$ are obtained by integrating the differentials

(12.4.2) $$\widetilde{\omega_j} = \left(\varsigma^i/\sqrt{P(\varsigma)}\right) d\varsigma, j = 0, 1, \ldots, n-1$$

over paths on the Riemann surface.

12.4.3. *Proposition.* The differentials (12.4.2) are Abel differentials of the first kind (see 8A).

Proof: The genus of a Riemann surface can be calculated with the help of the Riemann-Hurwitz formula (see [47])

(12.4.4) $$g = \sum \nu_r/2 - N + 1$$

where N is the number of sheets of the surface, ν_r is the degree of a branch point (the number of sheets joining at this point minus one).

In our case the Riemann surface of the function $\sqrt{P(\varsigma)}$ has 2 sheets, $N = 2$, the branch points are, in a generic case, simple, $\nu_r = 1$, and their number is $2n+2$ since the polynomial $P(\varsigma)$ is of the $(2n+1)$-th degree. Thus, $g = n$. The differentials $\widetilde{\omega_j}$ could have singularities only in branch points. However at a branch point $\varsigma_a \neq \infty$ the substitution $\varsigma - \varsigma_a = z^2$ transforms this differential into $\varphi(z)dz$ where $\varphi(z)$ is a holomorphic function. In $\varsigma = \infty$ so does the substitution $\varsigma = z^{-2}$ (the condition $j < n$ is important here). The Abel differentials $\widetilde{\omega_j}$ are not normed. □

The Abel differentials should be normalized:

$$(12.4.5) \qquad \omega_j = (i/4) \sum_{l=0}^{n-1} r_{jl} \widetilde{\omega}_l , j = 1, \ldots, n$$

where $\int_{\alpha_i} \omega_j = \delta_{ij}$, α_i are α-contours (see 8A).
Then

$$\psi_j = \sum_{l=0}^{n-1} r_{jl} \varphi_l = \sum_m \int_\infty^{\varsigma_m^*} \omega_j = \sum_{l=0}^{n-1} r_{jl}(\varphi_l^0 + \delta_{n-1,l}(1/2d_{2n-3})x \\ - c_{2(k+l-n)+3} t).$$

This is exactly the Abel mapping of the g-th symmetrical power of the Riemann surface into its Jacobi manifold. We know how $\{\psi_j\}$ depend on x and t. If the Abel mapping is converted we shall know the dependence of $\{\varsigma_m^*\}$ on x and t. After that we find the dependence of u on x and t from (12.3.6), which is our aim.

We use the Riemann theorem (see 8.A.4). If

$$\sum_{m=1}^{n} \int_\infty^{\varsigma_m^*} \omega_j = \psi_j$$

then $\{\varsigma_m^*\}$ are the zeros of the Riemann theta-function

$$\tilde{\vartheta}(\varsigma^*) = \theta(\mathfrak{E}(\varsigma^*) - \psi - \mathbb{K}), \psi = (\psi_1, \ldots, \psi_n),$$

To find u we must calculate the sum of zeros of $\tilde{\vartheta}(\varsigma^*)$. The symmetrical function of the zeros can be found by integration over a contour. Let us cut the Riemann surface along all the contours α_i and β_i. Instead of handles on the surface holes are formed (see Fig. 1a,b). The edge of a hole consists

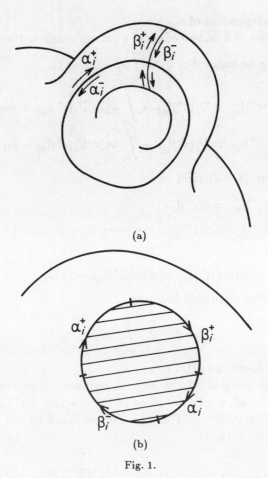

Fig. 1.

of four edges of the cuts: $\alpha_i^+, \beta_i^+, \alpha_i^-$ and β_i^-, which are being passed in the positive direction (the orientation on the Riemann surface is induced by the orientation on the complex plain ς). We consider the positive directions of the contours α_i and β_i coinciding with α_i^+ and β_i^+. Let γ be the contour consisting of all edges of the holes. The function $\tilde{\vartheta}(\varsigma^*)$ is single-valued on the cut surface.

12.4.6. *Lemma.* The value of the integral

$$\frac{1}{2\pi i} \oint_\gamma \varsigma d\ln \tilde{\vartheta}(\vartheta^*) = a$$

is a constant independent of ψ.

Proof: The values of $\tilde{\vartheta}$ on two edges of a cut are connected with each other by the following formulas. For $U_j = \int\limits_\infty^{\varsigma^*} \omega_j$

$$U_j(\varsigma^*)|_{\alpha_i^-} = U_j(\varsigma^*)|_{\alpha_i^+} + \int_{\beta_i} \omega_j = U_j(\varsigma^*)|_{\alpha_i^+} + B_{ij}$$

$$U_j(\varsigma^*)|_{\beta_i^-} = U_j(\varsigma^*)|_{\beta_i^+} - \int_{\alpha_i} \omega_j = U_j(\varsigma^*)|_{\beta_i^+} - \delta_{ij}$$

(see Fig. 1). Eqs. (8.A.3) yield

$$\tilde{\vartheta}(\varsigma^*)|_{\beta_i^-} = \tilde{\vartheta}(\varsigma^*)|_{\beta_i^+},$$

$$\tilde{\vartheta}(\varsigma^*)|_{\alpha_i^-} = \tilde{\vartheta}(\varsigma^*)|_{\alpha_i^+} \cdot \exp\{-\pi i(B_{ii} + 2U_i(\varsigma^*))\}.$$

Finally,

$$d\ln\tilde{\vartheta}(\varsigma^*)|_{\alpha_i^-} = d\ln\tilde{\vartheta}(\varsigma^*)|_{\alpha_i^+} - 2\pi i\omega_i.$$

The integral under discussion becomes

(12.4.7)
$$a = \sum_i \int_{\alpha_i} \varsigma\omega_i$$

which is independent of $\{\psi_i\}$. $\square$

Replace this integral by the sum of residues. The residues are at the points $\varsigma_1^*, \ldots, \varsigma_n^*$, which are zeros of the function $\tilde{\vartheta}(\varsigma^*)$ and equal to $\varsigma_1, \ldots, \varsigma_n$, and at $\varsigma = \infty$. This one can be found by the substitution $\varsigma = z^{-2}$. Then ($\bar{r}_l = \{r_{jl}\}$)

$$\tilde{\vartheta}(\varsigma^*) = \theta(-\psi + \mathbb{K} + (i/4)\int_0^z \bar{r}_l z^{-2l}/\sqrt{P(z^{-2})} \cdot (-2z^{-3})dz)$$

$$= \theta(\psi - \mathbb{K} + (i/2d_{2n+3})\bar{r}_{n-1} \cdot z + 0(z^3))$$

since $P(z^{-2}) = d_{2n+3}^2 z^{-2(2n+1)} + 0(z^{-2 \cdot 2n})$. Now

$$\vartheta(\varsigma^*) = \theta(\psi - \mathbb{K}) + \sum_j \partial\theta/\partial p_j \cdot (i/2d_{2n+3})r_{j,n-1}z$$

$$+ (1/2)\sum_{i,j} \partial^2\theta/\partial p_i \partial p_j \cdot (i/2d_{2n+3})^2 r_{i,n-1} \cdot r_{j,n-1} z^2 + 0(z^3).$$

$$= \theta(\psi - \mathbb{K}) + i\frac{\partial}{\partial x}\theta(\psi - \mathbb{K}) \cdot z - \frac{\partial^2}{\partial x^2}\theta(\psi - \mathbb{K})z^2 + 0(z^3)$$

The constant $\mathbb{K}$ is not important, it can be included into arbitrary constants $\varphi_l^{(0)}$. We have

$$\mathrm{res}_{z=0} z^{-2} \frac{d}{dz} \ln \tilde{\theta}(\varsigma^*) = \frac{(\partial \theta(\psi)/\partial x)^2 - \theta(\psi) \partial^2 \theta(\psi)/\partial x^2}{\theta^2(\psi)} = -\frac{\partial^2 \ln \theta(\psi)}{\partial x^2},$$

and

$$a = \varsigma_1 + \ldots + \varsigma_g - (\partial^2/\partial x^2) \ln \theta(\psi(x,t)).$$

Substituting this into (12.3.6) we obtain

(12.4.8) $\qquad u(x,t) = 2 d_{2n+1}/d_{2n+3} + 2a + (\partial^2/\partial x^2) \ln \theta(\psi(x,t)).$

This explicit formula for the solutions was suggested by Its and Matveev [69].

CHAPTER 13. Stationary Equations of the Matrix Hierarchy.

13.1. Hamilton structure. First integrals.

13.1.1. In Chap. 9 the matrix hierarchy was introduced (9.1.1). Let diag$U = 0$. The corresponding stationary equations have the form

(13.1.2) $$Q \equiv [A, R_{m+1}] = 0.$$

Here $R_{m+1} = R^B_{m+1}$, where R^B is the resolvent whose expansion starts from the constant diagonal matrix B, other terms containing no constants. (More general equations $\left[A, \sum_0^{m+1} c_k R\right]$ can also be considered).

Equation (13.1.2) is a system of $n(n-1)$ ordinary differential equations of the m-th order; the order of the system is $mn(n-1)$.

The basic theory of this equation was worked out by Dubrovin [53]. We shall expound this theory from the point of view of the Hamiltonian mechanics (after [54]), we shall see that the process of integration of this equation is no other than the Liouville procedure, similar to that in the previous chapter.

13.1.3. *Proposition.* Eq. (13.1.2) is a variational one, with the Lagrangian

(13.1.4) $$\Lambda = -\mathrm{tr}\, A R_{m+2}/(m+1).$$

Proof: Eq. (13.1.2) means vanishing of the nondiagonal elements of the matrix R_{m+1}, nondiag $R_{m+1} = 0$. From 9.3.10 we have $\delta\Lambda/\delta V = R_{m+1}$ for the nondiagonal elements. Thus, Eq. (13.1.2) is equivalent to $\delta\Lambda/\delta U = 0$. □

This implies that Eq. (13.1.2) is of the Hamiltonian type

13.1.5. Proposition. The quantities

$$F_k^\alpha = -\mathrm{tr}(R_m R_k^\alpha + R_{m-1} R_{k+1}^\alpha + \ldots + R_0 R_{m+k}^\alpha),$$

where R^α are basic resolvents (see 9.1.13), are first integrals of Eq. (13.1.2). The corresponding vector fields are $\partial_{[A, R_k^\alpha]}$.

Proof: Eq. (9.3.14) implies $\partial F_k^\alpha = -\mathrm{tr}[A, R_k^\alpha] R_{m+1} \stackrel{Q}{=} 0$, and that the characteristic of the first integral F_k^α is $-[A, R_k^\alpha]$. The rest follows from 11.3.16. □

Thus, the flows which are restrictions of the flows given by the equations of the hierarchy (9.1.17) to the phase space of Eq. (13.1.2) correspond to the first integrals 13.1.5.

13.1.6. Proposition. The first integrals 13.1.5 are in involution.
Proof: The vector fields $\{\partial_{[A, R_k^\alpha]}\}$ commute. □

It is clear that not all of the first integrals F_k^α are independent since the phase space is finite-dimensional.

We suggest now a second way of constructing the first integrals. Let

$$\tilde{R} = \sum_{k=0}^m R_k \varsigma^{m-k}.$$

(In Chap. 9 we used another notation, V.)

13.1.7. Lemma. $\tilde{R}$ satisfies the equation

$$-\tilde{R}' + [U + \varsigma A, \tilde{R}] = -[A, R_{m+1}].$$

Proof: Let us multiply the recurrence relation (9.1.8) by ς^{m-i} and sum up from $i = 0$ to $i = m$. Taking into account $[A, R_0] = 0$ we get the required. □

13.1.8. Corollary. Eq. (13.1.2) is equivalent to the fact that the polynomial in $\varsigma: \tilde{R}(\varsigma)$ satisfies the resolvent equation

(13.1.9) $$-\tilde{R}' + [U + \varsigma A, \tilde{R}] = 0.$$

13.1.10. Proposition. The coefficients of polynomials in ς

$$\operatorname{tr}\tilde{R}^k, k = 1, \ldots, n$$

are first integrals of Eq. (13.1.2). There are $mn(n-1)/2$ nontrivial first integrals of this kind.

Proof: If $\tilde{R}$ is a resolvent (by virtue of Eq. (13.1.2)), then so does $\tilde{R}^k$. This implies $\partial \operatorname{tr}\tilde{R}^k \stackrel{Q}{=} \operatorname{tr}[U + \varsigma A, \tilde{R}^k] = 0$. Hence $\operatorname{tr}\tilde{R}^k$ are first integral. The quantity $\operatorname{tr}\tilde{R}^k$ is a polynomial in ς of degree mk. The higher $m + 1$ terms of this polynomial coincide with those of the series $\operatorname{tr}(\varsigma^m R)^k$, while $\operatorname{tr} R^k$ are absolute constants (not only by virtue of the equation). Therefore the higher $m + 1$ terms of $\operatorname{tr}\tilde{R}^k$ are trivial first integrals. There remain $mk+1-(m+1) = m(k-1)$ nontrivial terms. In all, there are $\sum_{k=1}^{n} m(k-1) = mn(n-1)/2$ nontrivial first integrals. $\square$

Below it will be proved that these first integrals can be expressed in terms of $\{F_k^\alpha\}$ which implies that they are in involution. More or less cumbersome calculations can prove that they are independent. Then the Liouville theorem guarantees that Eq. (13.1.2) is integrable. This integration will be performed effectively.

Instead of $\operatorname{tr} \tilde{R}^k$, the coefficients of the characteristic polynomials

$$(13.1.11) \quad f(w,\varsigma) = \det(\tilde{R} - w \cdot 1) = \sum_{l=0}^{n} J_l(\varsigma) w^l = \sum_{l=0}^{n} \sum_{k=0}^{(n-l)m} J_{lk} w^l \varsigma^k$$

can be considered. Here also for each fixed l the higher $m + 1$ coefficients yield trivial first integrals; thus, the non-trivial ones are obtained for each of $l = 0, \ldots, n-2$ when $k = 0, \ldots, (n-l-1)m - 1$. In all, there are $\sum_{l=0}^{n-2}[(n-l)m - m] = mn(n-1)/2$ of them. As to the higher terms they coincide with those of the polynomial

$$f_0(w,\varsigma) = \det(\varsigma^m R - w1) = \det(\varsigma^m R_0 - w1) = \prod_{\beta=1}^{n}(\varsigma^m b_\alpha - w).$$

Let a point of the phase space $U, U^1, \ldots, U^{(m-1)}$ be fixed. Then the equation $f(w,\varsigma) = 0$ determines an algebraic function $w(\varsigma)$. To distinct points of the phase space but belonging to the same level surface

$\{J_{lk} = \text{const}\}$, one and the same algebraic function $w(\varsigma)$ corresponds. The Riemann surface of this function has n sheets. The points of the Riemann surface will be denoted by the letter, $P, P = (\varsigma, w(\varsigma))$.

13.1.12. Proposition. If $\varsigma \to \infty$ then n branches of $w(\varsigma)$ satisfy the asymptotic relations
$$w(\varsigma) = \varsigma^m b_\alpha + 0(\varsigma^{-1}), \alpha = 1, \ldots, n$$
where b_α are elements of the diagonal of $R_0 = B$.

Proof: Let us compare the equations: a) $\det(\tilde{R} - w1) = 0$ and b) $\det(\varsigma^m R - w \cdot 1) = 0$. The higher $m+1$ terms of the functions $\tilde{R}$ and $\varsigma^m R$ coincide, hence in the asymptotics of $w(\varsigma)$ for $\varsigma \to \infty$ the higher $m+1$ terms coincide too. As to the equation b), the eigenvalues of the matrix $\varsigma^m R$ are exactly $\varsigma^m b_\alpha$. $\square$

We denote the infinite point on the sheet where $w(\varsigma) \sim b_\alpha \varsigma^m$ by $\{\alpha\}$.

A 1-dimensional eigenspace corresponds to the eigenvalue $w(\varsigma)$ of the matrix $\tilde{R}(\varsigma)$. Let $g(P)$ be the spectral projector onto this space.

13.1.13. Proposition. The projector $g(P)$ can be written as
$$g(P) = f_w^{-1} \sum_{l=1}^{n} J_l(\varsigma) \sum_{k=0}^{l-1} w^k \tilde{R}^{l-1-k}$$
where $f_w = \partial f / \partial w$.

Proof: Let $g(P)$ be a matrix defined by this equation. If ξ is the eigenvector corresponding to the eigenvalue $w(\varsigma)$ then
$$g(P)\xi = f_w^{-1} \sum_{l=1}^{n} J_l(\varsigma) \sum_{k=0}^{l-1} w^k \cdot w^{l-1-k} \xi$$
$$= f_w^{-1} \sum_{l=1}^{n} J_l(\varsigma) l w^{l-1} \xi = f_w^{-1} \cdot f_w \xi = \xi.$$

If ξ is an eigenvector for another eigenvalue $w_1 \neq w$ then
$$g(P)\xi = f_w^{-1} \sum_{l=1}^{n} J_l(\varsigma) \sum_{k=0}^{l-1} w^k \cdot w_1^{l-1-k} \xi = f_w^{-1} \cdot \sum_{l=1}^{n} J_l(\varsigma) \cdot \frac{w^l - w_1^l}{w - w_1} \xi$$
$$= f_w^{-1}[f(w, \varsigma) - f(w_1, \varsigma)]/(w - w_1) \cdot \xi = 0. \square$$

13.1.14. Corollary. The relations
$$g^2(P) = g(P), \text{tr} g^k(P) = 1$$

hold.

13.1.15. Proposition. $g(P)$ is a resolvent by virtue of Eq. (13.1.2).
Proof: $\tilde{R}$ is a resolvent. The product of resolvents is a resolvent, so does the sum. The rest is from 13.1.13. □

13.1.16. Proposition. The asymptotics

$$g(P) = R^\alpha + 0(\varsigma^{-\infty})$$

holds, by virtue of the Eq. (13.1.2) (the latter means that the higher derivatives $U^{(m)}, U^{(m+1)}, \ldots$ should be excluded with the help of (13.1.2)).
Proof: The fact that the higher $m+1$ terms of $g(P)$ and R^α coincide can be established in the same way as in 13.1.22: they are spectral projectors of matrices $\tilde{R}$ and $\varsigma^m R$, asymptotically coinciding in $m+1$ terms. The fact that this coincidence holds also for rest of the terms follows from the equalities $g^k = g, (R^\alpha)^k = R^\alpha$ for any k, and from a reasoning of the type of 9.1.11. □

As is seen from 13.1.13 the elements of the matrices $g(P) : g_{ij}(P)$ are rational functions of ς and w, i.e. rational functions on the Riemann surface. The poles are branch points, $f_w = 0$.

13.1.17. Proposition. If $g_{ij}(P) = 0$ at a point P of the Riemann surface, then at this point all the elements either of the same row, $g_{ik}(P) = 0, k = 1, \ldots, n$ or of the same column, $g_{kj} = 0, k = 1, \ldots, n$ also vanish. Therefore the divisor of zeros of the element $g_{ik}(P)$ divides into two divisors: $d_i^{(1)} + d_j^{(2)}$, the divisor of the zeros of the i-th row and the divisor of the zeros of the j-th column.
Proof: The projector $g(P)$ onto a one-dimensional subspace is a matrix of rank one. Locally, in a neighbourhood of a zero of $g_{ij}(P)$ a decomposition $g_{ij} = \varphi_i \psi_j$ holds. The rest is clear. □

13.1.18. Proposition. The number of all branch points, taking into account their multiplicities, is $mn(n-1)$.
Proof: The discriminant of the polynomial $f(w,\varsigma)$ is $\Delta = \prod_{i \neq j}(w_i(\varsigma) - w_j(\varsigma))$. The discriminant vanishes at the points ς where two (or more) roots coincide, i.e. at ς where the Riemann surface has a branch point. The discriminant is a symmetrical function of the roots, hence it is a polynomial in coefficients $J_l(\varsigma)$, and a polynomial in ς. To each of its zero a branch point relates, the multiplicity of the zero being equal to that of the branch point. Thus, the number of the branch points is equal to the degree of Δ as

a polynomial in ς. This can be easily calculated. The number of multipliers is $n(n-1)$, each of which is ς^m at infinity, therefore the degree is $mn(n-1)$. □

13.1.19. Corollary. The genus of the Riemann surface of the function $w(\varsigma)$ is $g = mn(n-1)/2 - n + 1$. (The genus g and the projector $g(P)$ are denoted by the same letter g; we hope that there will be no confusion).

Proof: This follows from the Riemann-Hurwitz formula (12.4.4). □

13.1.20. Proposition. Each element g_{ik} has $2n-2$ zeros at infinity.

Proof: If $P \to \{\alpha\}$ then $g(P) = R^\alpha + 0(\varsigma^{-\alpha})$. A few first terms of the asymptotics can be found from 9.1.9 letting $b_i = \delta_{i\alpha}$:

$$(R_0^\alpha)_{ij} = \delta_{i\alpha}\delta_{j\alpha}, (R_1^\alpha)_{ij} = \delta_{i\alpha}(a_\alpha - a_j)^{-1}u_{\alpha j} - \delta_{\alpha j}(a_i - a_\alpha)^{-1}u_{i\alpha}.$$

This implies the following. If $i \neq j$ and α coincides with one of i or j, then at the point $\{\alpha\}$ there is a simple zero. If $\alpha \neq i, j$ then at $\{\alpha\}$ there is a double zero. We have on all sheets $2 + 2(n-2) = 2n-2$ zeros at infinity. If $i = j$, then there is no zero in $\{\alpha\}$ for $\alpha = i = j$ and a double zero for $\alpha \neq i$. In all we have $2(n-1) = 2n-2$ zeros at infinity also in this case. □

13.1.21. Corollary. Functions g_{ij} have $mn(n-1) - 2n + 2$ zeros in the finite part of the Riemann surface.

Proof: The meromorphic function g_{ij} has the number of zeros equal to that of poles. The latter is $mn(n-1)$, see (13.1.18). From this number $2n-2$ zeros at infinity must be subtracted. □

13.1.22. Proposition. The degrees of the divisors $d_i^{(1)}$ and $d_j^{(2)}$ are

$$\deg(d_i^{(1)}) = \deg(d_j^{(2)}) = mn(n-1)/2 - (n-1)$$

(We recall that the degree of a divisor is the number of its points taking into account their multiplicities).

Proof. First of all, $\deg(d_i^{(1)})$ does not depend on i (and $\deg(d_j^{(2)})$ on j). This follows from the fact that the meromorphic function g_{ij}/g_{kj} has equal numbers of zeros and poles. At infinity the numbers of zeros and poles coincide too. Therefore they are equal also in the finite part of the Riemann surface.

How the relation $\deg(d_i^{(1)}) = \deg(d_j^{(2)})$ reflects the fact that rows and columns are equal in rights which is more or less evident. If the reader is convinced by this reasoning he can consider the proof as finished. The strict deduction can be carried out for example, as follows.

Pay our attention to the dependence of $\tilde{R}(\varsigma), w(\varsigma)$ and $g(P)$ on the point of the phase space $(U, U', \ldots, U^{(m-1)})$. This dependence is continuous. Therefore, if the point moves on the level surface $\{J_{lk} = \text{const}\}$ the roots of g_{ij} change continuously, they cannot appear or disappear (if a zero does not meet a pole which can be avoided). A zero also cannot pass from the divisor $d_i^{(1)}$ to divisor $d_j^{(2)}$. Thus $\deg(d_i^{(1)}) = \text{const}$, i.e. is independent of the point of the phase space. On the other hand it is easy to see that the resolvent equation (13.1.9) admits of the following transformation (an inversion) $\tilde{R}(U(x)) \mapsto \tilde{R}^t(U^t(-x))$ where the superscript t denotes the transpose, $U(-x)$ denotes changing the sign of odd derivatives: $U, -U', U'', -U''', \ldots$. This inversion does not change R_0, therefore $\tilde{R}(U(x)) = \tilde{R}^t(U^t(-x))$, and $\tilde{R}(U^t(-x)) = \tilde{R}^t(U(x)), g(U^t(-x)) = g^t(U(x))$. The point $U^t(-x) = (U^t, -U'^t, \ldots)$ of the phase space belongs to the same level surface $\{J_{lk} = \text{const}\}$ as $U(x)$. Thus, $\deg(d_i^{(1)}(U(x)) = \deg(d_i^{(1)}(U^t(-x)))$. The equation $g(U^t(-x)) = g^t(U(x))$ implies $\deg(d_i^{(1)}(U^t(-x)) = \deg(d_i^{(2)}(U(x)))$. We have obtained $\deg(d_i^{(1)}) = \deg(d_j^{(2)})$, and this is a half of the full number of zeros (see 13.1.21). □

13.1.23. Proposition. The connection between the two kinds of first integrals, $\{F_k^\alpha\}$ and $\{J_{lk}\}$, is given by the formula

$$F_k^\alpha = -w_{\alpha,k}$$

where $w_{\alpha,k}$ is the coefficient in ς^{-k} in the asymptotics of $w(P)$ when $P \to \{\alpha\}$. (These coefficients can be easily expressed in terms of the coefficients $\{J_{lk}\}$).

Proof: We have

$$F_k^\alpha = -\text{tr}(R_m R_k^\alpha + \ldots + R_0 R_{m+k}^\alpha) = -\text{tr}(\tilde{R} R^\alpha)_k$$

(the subscript k denotes the coefficient in ς^{-k}). In this relation R^α can be replaced by the expansion of $g(P)$ when $P \to \{\alpha\}$, see 13.1.16, then $F_k^\alpha = -\text{tr}(\tilde{R}g(P))_k, P \to \{\alpha\}$. The matrix $g(P)$ is the spectral projector of $\tilde{R}$, hence $\text{tr} g \tilde{R}$ is equal to the eigenvalue $w(P)$, and this gives the required relation. □

13.1.24. We want to give an explanation for the purpose of introducing two kinds of first integrals and establishing a connection between them. As a rule, $\{J_{lk}\}$ are more convenient to deal with. They determine a Riemann surface, algebraic function $w(\varsigma)$ etc. In return, the first integrals F_k^α are

Hamiltonians of restrictions of the phase flows given by the equations of the hierarchy (9.1.17) to the finite-dimensional phase space of the stationary equation (13.1.2). Owing to this fact we can write explicit formulas of solutions for this flows.

13.2. Hamilton structure of stationary equations

13.2.1. Proposition. The Hamiltonian of the stationary equation (13.1.2) is

$$\mathcal{H} = -\sum_{\alpha=1}^{n} a_\alpha w_{\alpha,2}$$

(a_α are the diagonal elements of the matrix A).

Proof. One must verify the equality $\mathcal{H}' = -\operatorname{tr} U' \delta \Lambda / \delta U$ where Λ is the Lagrangian (13.1.4) (see 11.3.10). Let $R^A = \sum a_\alpha R^\alpha$ be the resolvent starting with $R_0 = A$. We have

(a) $-R_0^{A'} + [U, R_0^A] + [A, R_1^A] = 0$, (b) $-R_1^{A'} + [U, R_1^A] + [A, R_2^A] = 0$,

being the first two of the recurrence relations. Eq. (a) implies that the nondiagonal elements in R_1^A are equal to those in U. Then (b) yields $U' = [A, R_2^A]$. Now

$$-\operatorname{tr} U' \delta \Lambda / \delta U = -\operatorname{tr} U' R_{m+1} = -\operatorname{tr} [A, R_2^A] R_{m+1}$$
$$= -\operatorname{tr} \sum a_\alpha [A, R_2^\alpha] R_{m+1} = \sum a_\alpha \partial F_2^\alpha = -\partial \sum a_\alpha w_{\alpha,2} = \partial \mathcal{H}.$$

Let us express the Hamiltonian in terms of $\{J_{lk}\}$.

13.2.2. Proposition. The second and the third of the nonvanishing coefficients of the asymptotics of $w(P)$ when $P \to \{\alpha\}$:

$$w(P) = b_\alpha \varsigma^m + p_\alpha \varsigma^{-1} + q_\alpha \varsigma^{-2} + 0(\varsigma^{-3})$$

are

$$p_\alpha = \sum_{l=0}^{n-2} J_{l,(n-l-1)m-1} b_\alpha^l / \prod_{\beta \neq \alpha} (b_\alpha - b_\beta),$$

$$q_\alpha = \sum_{l=0}^{n-2} J_{l,(n-l-1)m-2} b_\alpha^l / \prod_{\beta \neq \alpha} (b_\alpha - b_\beta).$$

Proof: Let us expand $f(w,\varsigma)$ (see 13.1.11) into a sum

$$f(w,\varsigma) = f_0(w,\varsigma) + f_{m+1}(w,\varsigma) + f_{m+2}(w,\varsigma) + \ldots,$$

gathering together terms with the highest power ς^k for every l. Then with the next power and so on:

$$f_0(w,\varsigma) = \prod_{\alpha=1}^{n}(b_\alpha\varsigma^m - w) = \sum_{l=0}^{n} J_{l,(n-l)m} w^l \varsigma^{(n-l)m},$$

$$f_1(w,\varsigma) = \sum_{l=0}^{n-2} J_{l,(n-l-1)m-1} w^l \varsigma^{(n-l-1)m-1},$$

$$f_2(w,\varsigma) = \sum_{l=0}^{n-2} J_{l,(n-l-1)m-2} w^l \varsigma^{(n-l-1)m-2}.$$

The series $w = b_\alpha \varsigma^m + p_\alpha \varsigma^{-1} + q_\alpha \varsigma^{-2} + \ldots$ must be substituted into the equation $f(w,\varsigma) = 0$. We have $f_0(b_\alpha \varsigma^m, \varsigma) = 0$, i.e. all the terms with the power ς^{mn} vanish. The next power is $\varsigma^{mn-(m+1)}$, this yields

$$p_\alpha \partial f_0(b_\alpha, 1)/\partial w + f_{m+1}(b_\alpha, 1) = 0$$

and the next, $\varsigma^{mn-(m+2)}$:

$$q_\alpha \partial f_0(b_\alpha, 1)/\partial w + f_{m+2}(b_\alpha, 1) = 0$$

whence

$$p_\alpha = -f_{m+1}(b_\alpha,1)/(\partial f_0(b_\alpha,1)/\partial w) = \sum_{l=0}^{n-2} J_{l,(n-l-1)m-1} b_\alpha^l / \prod_{\beta \neq \alpha} (b_\alpha - b_\beta),$$

similarly the second formula. □

13.2.3. Corollary. The Hamiltonian is

$$\mathcal{H} = -\sum_{\alpha=1}^{n} a_\alpha \sum_{l=0}^{n-2} J_{l,(n-l-1)m-2} b_\alpha^l / \prod_{\beta \neq \alpha} (b_\alpha - b_\beta).$$

We pass now to the symplectic form.

13.2.4. Proposition. The 1-form relating to the Lagrangian Λ is

$$\Omega^{(1)} = -\sum_{\alpha=1}^{n} b_\alpha (\delta R^\alpha \cdot R^\alpha)_{ij} / R^{(\alpha)}_{ij} |_{m+1}$$

where i and j are any two values of the matrix indices (if another pair is taken $\Omega^{(1)}$ will be changed by a nonessential differential).

Proof: We use the proposition 9.3.7. From $\delta\Lambda = \text{tr}\,\delta U \cdot \delta\Lambda/\delta U + \partial\Omega^{(1)}$ and (13.1.4) we find

$$\Omega^{(1)} = -\frac{1}{m+1}\text{tr}(\delta\phi B\psi_\varsigma - \phi_\varsigma B\delta\psi)|_{m+2}$$

Adding to a 1-form any differential δf does not change it significantly since the form $\Omega = \delta\Omega^{(1)}$ is not changed. The term $(m+1)^{-1}\delta\text{tr}\,\phi B\psi_\varsigma$ can be added to the last expression:

$$\Omega^{(1)} = \frac{1}{m+1}\text{tr}\,(\phi B\delta\psi_\varsigma + \phi_\varsigma B\delta\psi)|_{m+2} = \frac{1}{m+1}\text{tr}\frac{\partial}{\partial\varsigma}\phi B\delta\psi|_{m+2}$$

$$= -\text{tr}\,\phi B\delta\psi|_{m+1} = -\sum_{l,\alpha}\phi_{l\alpha}b_\alpha\delta\psi_{\alpha l}|_{m+1} = -\sum_{l,\alpha}\frac{\phi_{l\alpha}\psi_{\alpha j}}{\psi_{\alpha j}}$$

$$b_\alpha\frac{\delta(\phi_{i\alpha}\psi_{\alpha l})}{\phi_{i\alpha}} + \sum_{l,\alpha}\phi_{l\alpha}\psi_{\alpha l}\delta\phi_{i\alpha}/\phi_{i\alpha}.$$

We have $\sum_l \psi_{\alpha l}\phi_{l\alpha} = 1$, therefore the second term is a complete differential and can be dropped. With $R^\alpha_{ij} = \phi_{i\alpha}\psi_{\alpha j}$ we obtain the required expression. □

13.2.5. Besides the above-mentioned involution transforming solutions of the stationary equation to other ones, there is another obvious group of symmetries depending on $n-1$ parameters. Namely,

(13.2.6) $\qquad U \mapsto KUK^{-1}, K = \text{diag}\,(k_1,\ldots,k_n) = \text{const}, k_i \neq 0\,.$

This transformation sends a solution to another. The related infinitesimal transformations are

(13.2.7) $\qquad U \mapsto U + [C,U], C = \text{diag}(C_1,\ldots,C_n) = \text{const}\,.$

Thus, the vector fields $\{\partial_{[C,U]}\}$ correspond to the transformation group (13.2.6). Let us find the Hamiltonians of these vector fields.

13.2.8. Proposition. The Hamiltonian of a vector field $\partial_{[C,U]}$ is

$$h_C = \operatorname{tr} CR_{m+1} = \sum_{\alpha=1}^{n} C_\alpha F_1^\alpha = -\sum_{\alpha=1}^{n} C_\alpha w_{\alpha,1}$$

$$= \sum_{\alpha=1}^{n} C_\alpha \sum_{l=0}^{n-2} J_{l,(n-l-1)m-1} b_\alpha^l / \prod_{\beta \neq \alpha} (b_\alpha - b_\beta).$$

(we draw readers attention to that $\operatorname{tr} CR_{m+1}$ does not vanish by virtue of the stationary equation as it may seem to at first sight; this expression involves only diagonal elements of R_{m+1}, while the stationary equation means vanishing of the nondiagonal elements).

Proof: At first we take the first expression for h_C, $h_C = \operatorname{tr} CR_{m+1}$. We have

$$\partial h_C = \operatorname{tr} CR'_{m+1} = -\operatorname{tr} C[U, R_{m+1}] - \operatorname{tr} C[A, R_{m+2}]$$
$$= -\operatorname{tr} C[U, R_{m+1}] = -\operatorname{tr} R_{m+1}[C, U].$$

which implies that h_C is a first integral with the characteristic $-[C, U]$; the corresponding vector field is $\partial_{[C,U]}$ as required. Further, according to 13.1.5,

$$\sum_{\alpha=1}^{n} C_\alpha F_1^\alpha = -\operatorname{tr} \sum_{\alpha=1}^{n} c_\alpha (R_m R_1^\alpha + \ldots + R_0 R_{m+1}^\alpha)$$

$$= -\operatorname{tr} \sum_{\alpha=1}^{n} C_\alpha (R_{m+1} R_0^\alpha + \ldots + R_0 R_{m+1}^\alpha) + \operatorname{tr} \sum_{\alpha=1}^{n} C_\alpha R_{m+1} R_0^\alpha$$

$$= -\operatorname{tr} R^B R^C|_{m+1} + \operatorname{tr} CR_{m+1} = \operatorname{tr} CR_{m+1} = h_C$$

i.e. the second expression for h_C is obtained. The third and the fourth follow from 13.1.23 and 13.2.2. □

13.2.9. We summarize the results obtained. From all the quantities $J_{l,k}$, for each l the higher $m+1$ ones $J_{l,(n-l)m}, J_{l,(n-l)m-1}, \ldots, J_{l,(n-l-1)m}$ are absolute constants; the next, $J_{l,(n-l-1)m-1}$ is the Hamiltonian of a trivial symmetry; from the next ones: $\{J_{l,(n-l-1)m-2}\}$ the Hamiltonian of the equation, $\mathcal{H}$, is combined (13.2.3). All the others yield non-trivial symmetries of the stationary equation. These are flows given by the equations of the matrix hierarchy (9.1.17) restricted to the finite-dimensional phase space of the stationary equation (13.1.2). The flows commute; the first integrals are involutive.

13.2.10. Proposition. Vector fields $\partial_{[C,U]}$ are tangent to the manifold $\mathcal{U}_C$:

(13.2.11) $$\{J_{l,(n-l-1)m-1} = C_l = \text{const}\}, l = 0, 1, \ldots, n-2.$$

Proof: This follows from the fact that the first integrals are in involution. □

From now on the stationary equation as well as the symplectic form and the Hamiltonian will be considered as restricted to the invariant submanifold $\mathcal{U}_C$ of the phase space which has the codimension $n-1$.

13.2.12. Proposition. On the submanifold $\mathcal{U}_C$ the form Ω degenerates on the vector fields corresponding to trivial symmetries.

Proof: The Hamiltonians of these vector fields are linear combinations of $\{J_{l,(n-l-1)m-1}\}$, therefore they are constant on $\mathcal{U}_C$. □

To restore the non-degeneracy of the form we must factorize the manifold $\mathcal{U}_C$ with respect to the trajectories of the trivial symmetries. In other words, points of a new manifold will be the sets $(U, U', \ldots, U^{(m-1)})$ satisfying Eq. (13.2.11); two of such sets being identified if

$$U_1^{(i)} = KU^{(i)}K^{-1}, i = 0, \ldots, m-1$$

where K is a constant diagonal matrix. Evidently,

$$\dim \mathcal{U}_C = mn(n-1) - (n-1), \dim \mathcal{U}'_C = mn(n-1) - 2(n-1).$$

The form Ω can be transferred to the factor-manifold $\mathcal{U}'_C$ since it degenerates just on the vector tangent to the trajectories along which the identification is performed. On $\mathcal{U}'_C$ the form is non-degenerate.

13.2.13. The author must confess that he does not understand the profound meaning of the fact that one has to pass from the whole phase space to the submanifold $\mathcal{U}_C$ (further passage to $\mathcal{U}'_C$ is forced by the necessity to have a non-degenerate form). This restriction happens to be important several times in what follows.

The dimension of the new manifold $\mathcal{U}'$ is $mn(n-1) - 2(n-1)$. The number of the remaining non-trivial first integrals is $mn(n-1)/2 - (n-1)$, i.e. half of the dimension, as it was earlier.

13.3. Action-angle variables.

13.3.1. Proposition. On the manifold $\mathcal{U}_C'$ the form $\Omega^{(1)}$ can be written as

$$(13.3.2) \qquad \Omega^{(1)} = -\sum_\alpha w(P) \frac{(\delta g \cdot g)_{ij}}{g_{ij}}\Big|_1 = -\sum_{\alpha,k} w(P) \frac{\delta g_{ik} \cdot g_{kj}}{g_{ij}}\Big|_1$$

(the subscript 1 means, as usual, the coefficient in ς^{-1} of the asymptotic expansion $P \to \{\alpha\}$; the sum is over all the sheets $\alpha = 1,\ldots, n$).

Proof: It must be shown that up to a complete differential this expression coincides with 13.2.4. We have

$$-\sum_\alpha w(P) \frac{(\delta g \cdot g)_{ij}}{g_{ij}}\Big|_1 = -\sum_\alpha w(P)|_{-m} \frac{(\delta g \cdot g)_{ij}}{g_{ij}}\Big|_{m+1}$$
$$-\sum_\alpha w(P)|_1 \cdot \frac{(\delta g \cdot g)_{ij}}{g_{ij}}\Big|_0$$

(the subscript $-m$ denotes the coefficient in ς^m). The first term coincides with the expression 13.2.4, owing to the asymptotics 13.2.2 and 13.1.16. As to the second term, we have $w(P)|_1 = p_\alpha$ which is constant on $\mathcal{U}_C$, and $\sum_{k\alpha} \delta g_{ik} \cdot g_{kj}/g_{ij}|_0$ is a complete differential (see 13.1.16 and the expressions for R^α in 13.1.20). □

In this proof the restriction on $\mathcal{U}_C$ is essential and we have used the fact that p_α is constant.

We have another representation of $\Omega^{(1)}$ given below.

13.3.3. Proposition. The form $\Omega^{(1)}$ can be written as

$$\Omega^{(1)} = -\sum_{P^* \in d_i^{(1)}} w(P^*)\delta\varsigma_{P^*} + \sum_{l,k} a_{lk}(J)\delta J_{lk}$$

where the first sum is over all the points P^* of the divisor $d_i^{(1)}$, the coefficients a_{lk} of the second sum depend only on the values of the first integrals $\{J_{lk}\}$, ς_{P^*} is the projection of a point P^* on the complex plain ς. (An explanation: if a point of the phase space experiences a small displacement $\{\delta U^{(i)}\}$, then the first integrals also change by δJ_{lk} and the points of the divisor $d_i^{(1)}$ are displaced by $\delta\varsigma_{P^*}$).

Proof: The right-hand side of Eq. (13.3.2) can be written as the sum of residues of the differential

$$(13.3.4) \qquad \Omega^{(1)} = -\sum_\alpha \text{res}_{\{\alpha\}} w(P) \frac{(\delta g \cdot g)_{ij}}{g_{ij}} d\varsigma$$

Replace this by the sum of residues in the finite part of the Riemann surface with the opposite sign. There are poles at two kinds of points: at the points of the divisor $d_i^{(1)}$ (but not of $d_i^{(2)}$, the zeros here being cancelled out) and at the branch points. Let $P^* \in d_i^{(1)}$. In the neighbourhood of this point, g_{ij} can be represented as $\varphi_i \psi_j$ where $\sum \varphi_i \psi_i = 1$. We have $\varphi_1 = (\varsigma - \varsigma_{P^*})\tilde{\varphi}_i$ and $\delta \varphi_i = -\delta \varsigma_{P^*} \tilde{\varphi}_i + (\varsigma - \varsigma_{P^*})\delta \tilde{\varphi}_i$. The residue is

$$\sum_k w(P^*)\delta\varsigma_{P^*} \cdot \frac{\tilde{\varphi}_i \psi_k \varphi_k \psi_j}{\tilde{\varphi}_i \psi_j} = w(P^*)\delta\varsigma_{P^*}.$$

Now we consider the residue at a branch point. The coordinates of a branch point (w_0, ς_0) depend solely on the coefficients of the polynomial $f(w,\varsigma)$ i.e. on $\{J_{lk}\}$. Therefore the residue at the branch point can be only of the form $\sum a_{lk}(J)\delta J_{lk}$, which completes the proof. (More details about the coefficients a_{lk}: Let $P_0(w_0, \varsigma_0)$ be a branch point. The expansion of $f(w,\varsigma)$ in its neighbourhood is $b(\varsigma-\varsigma_0)+d(\varsigma-\varsigma_0)(w-w_0)+e(w-w_0)^2+\ldots$

For simplicity we consider now the generic case, $b \neq 0, e \neq 0$. Then $w - w_0 = \sqrt{\varsigma - \varsigma_0} \cdot \varphi(\sqrt{\varsigma - \varsigma_0})$, where φ is holomorphic, $\varphi(0) \neq 0$, and $f_w = \sqrt{\varsigma - \varsigma_0} \cdot \psi(\sqrt{\varsigma - \varsigma_0}), \psi(0) \neq 0$. Let $g(P) = f_w^{-1}\tilde{g}(P)$. Then $\mathrm{res}_{P_0} w(P)(\delta g(P) \cdot g(P))_{ij}/g_{ij}(P)d\varsigma = -\mathrm{res}\, w(\delta f_w/f_w)(g \cdot g)_{ij}/g_{ij}d\varsigma +$ res $wf_w^{-1}(\delta\tilde{g} \cdot \tilde{g})_{ij}/\tilde{g}_{ij}d\varsigma$. The first term is $-\mathrm{res}\, w(\delta f_w/f_w)d\varsigma = -(1/2)w$ $(P_0)\delta\varsigma_0$, and the second term vanishes since $\mathrm{res}_{\varsigma_0}(\varsigma - \varsigma_0)^{-1/2}d\varsigma = 0$, which can be verified by substitution $\varsigma-\varsigma_0 = z^2$. It remains to express $\delta\varsigma_0$ in terms of $\{\delta J_{kl}\}$. Let us vary the equation $f(w,\varsigma) = 0 : \sum w^l \varsigma^k \delta J_{lk} + f_w \delta w + f_\varsigma \delta \varsigma = 0$. At the point P_0 we have $\delta\varsigma_{p_0} = -f_\varsigma^{-1}(w_0,\varsigma_0)\sum_{l,k} w_0^l \varsigma_0^k \delta J_{l,k}$, whence

$a_{lk} = (1/2)f_\varsigma^{-1}(w_0, \varsigma_0)w_0^{l+1}\varsigma_0^k.$) □

13.3.5. We consider now the Lagrange manifolds $\{J_{lk} = C_{lk} = \mathrm{const}\}$. On each of these manifolds coordinates have to be chosen. The number of coordinates must be equal to the dimension of the manifold i.e. to $nn(n-1)/2$ $(n-1)$. For all the points of such a manifold the function $f(w,\varsigma)$ is the same, so is the Riemann surface. It is convenient to choose the coordiantes in which the form $\Omega^{(1)}$ is written with separated variables; according to 13.3.3 these are the points of a divisor $d_i^{(1)}$ with an arbitrary but fixed i.

On each submanifold $\{J = C\}$, the form $\Omega^{(1)}$ can be easily integrated:

$$\Omega^{(1)} = \delta V(d_i^{(1)}, J), V(d_i^{(1)}, J) = -\sum_{P^* \in d_i^{(1)}} \int_{P_0}^{P^*} w(P)d\varsigma.$$

13.3.6. Proposition. The angle variables adjoint to the action variables J_{lk}, $k \leq (n-l-1)m-2$ can be obtained by the Abel mapping of the divisor $d_i^{(1)}$.

Proof: The angle variables adjoint to the action variables J_{lk} are (see Chap. 12) $\theta_{lk} = -\partial V/\partial J_{lk} + a_{lk}$. We have
(13.3.7)
$$\theta_{lk} = \sum \int_{P_0}^{P^*} \partial w(P)/\partial J_{lk} d\varsigma + a_{lk} = - \sum_{P^* \in d_i^{(1)}} \int_{P_0}^{P^*} f_w^{-1} w^l \varsigma^k d\varsigma + a_{lk}.$$

It remains to note that here are written integrals of Abel differentials of the first kind, i.e. of the holomorphic differentials. Indeed, at the branch points $f_w = 0$ the differential is holomorphic because it has the form $(\varsigma - \varsigma_0)^{1/\nu} d\varsigma$, where ν is the multiplicity of the branch point. At infinity we have $w \sim \varsigma^m$, $f_w \sim w^{n-1} \sim \varsigma^{m(n-1)}$, therefore $f_w^{-1} w^l \varsigma^k$ grows as $\varsigma^{k+m(l-n+1)}$. Taking into account that $k \leq (n-l-1)m-2$ we conclude that at infinity the differential is $\varsigma^{-2} d\varsigma$. □

Note that the above written differentials of the first kind are not normalized.

13.3.8. Proposition. In the variables (J, θ) the equation (13.1.2) can be integrated:

$$J_{lk} = J_{lk}^0,$$

$$\theta_{lk} = \begin{cases} \theta_{lk}^0 - \sum_{\alpha=1}^n a_\alpha b_\alpha^l \prod_{\beta \neq \alpha} (b_\alpha - b_\beta)^{-1} \cdot x, & k = (n-l-1)m-2 \\ \theta_{lk}^0, & \text{otherwise}. \end{cases}$$

Proof: This follows from $\theta_{lk} = (\partial \mathcal{H}/\partial J_{lk})x + \theta_{lk}^0$, see (12.3.2), and 13.2.3. □

The constants a_{lk} can be included into initial constants θ_{lk}^0, therefore they play no part here.

13.3.9. Recall that in the phase space the vector fields $\sum c_\alpha \partial_{[A,R_r^\alpha]}$ commute with the vector field of the stationary equation. If t is a parameter along the integral curve of such a field then we can find the dependence of the point of the phase space on this parameter i.e. integrate the non-stationary equation in the finite-dimensional manifold of solutions of the stationary equation.

13.3.10. Proposition. In the variables (J,θ) the dependence on x and t is given by the formulas

$$J_{lk} = J_{lk}^0, \theta_{lk} = \theta_{lk}^0 - \delta_{k,(n-l-1)m-2} \cdot \sum a_\alpha b_\alpha^l \cdot \prod_{\beta \neq \alpha}(b_\alpha - b_\beta)^{-1} \cdot x$$

$$- \sum c_\alpha \partial w_{\alpha,r}/\partial J_{lk} \cdot t$$

where θ_{kl}^0 are constants independent of x and t.

Proof: It suffices to remember that the Hamiltonian of the vector field $\sum c_\alpha \partial_{[A,R_r^\alpha]}$ is $\sum c_r F_r^\alpha = -\sum c_\alpha w_{\alpha,r}$. □

13.3.11. To obtain the inverse formulas it is more convenient to use the normalized Abel differentials of the first kind. If a system of contours α_i and β_i is chosen (see 8.A.1), linear combinations

$$\omega_i = \sum_{l=0}^{n-2}\sum_{k=0}^{(n-l-1)m-2} c_{i,lk} f_w^{-1} w^l \varsigma^k d\varsigma, i = 1, \ldots, g$$

can be found such that $\int_{\alpha_i} \omega_j = \delta_{ij}$. Instead of θ_{lk} we must consider variables

(13.3.12) $$\xi_i = \sum_{l,k} c_{i,lk} \theta_{lk}$$

which depend on x and t linearly, too.

13.3.13. The points of the divisor $d_i^{(1)}$ were taken as coordinates. The points of another divisor $d_j^{(1)}$ or even $d_j^{(2)}$ could be taken. This yields the same formula 13.3.10 but with other constants. To find a relation between the constants one can use the Abel theorem (8.A.2). The Abel mapping of the divisor of the rational on the Riemann surface function g_{ij} must be zero (on the Jacobian i.e. modulo the lattice periods). If D is the divisor of branch points, and $S = \sum\{i\}$ then

(13.3.14) $$\mathfrak{E}(d_i^{(1)} + d_j^{(2)} - D + 2S - \{i\} - \{j\}) = 0.$$

If we denote $\eta_i^{(1,2)} = \mathfrak{E}(d_i^{(1,2)})$ then $\eta_j^{(2)} = -\eta_i^{(1)} + C_{ij}$, where C_{ij} is a constant vector on the lattice of periods.

CHAPTER 14. Stationary Equations of the Matrix Hierarchy (continuation).

14.1. Baker function. Return to original variables.

14.1.1. In the previous chapter the variables were obtained in which solutions of the stationary equation were linear in x and t. The aim of this chapter is to return to the variables $(U, U', \ldots, U^{(m-1)})$. We follow here almost literally the work by Dubrovin [53].

So let us have a solution $U(x)$ of the stationary equation (which can also depend on the parameter t; we shall not write this parameter explicitly). $\tilde{R}$ becomes a function of x, so does the projector $g(x, P)$, P is a point on the Riemann surface. The divisors $d_i^{(1)}$ and $d_j^{(2)}$ depend on x, too.

14.1.2. The one-dimensional projector g can be represented as $g_{ij}(x, P) = \varphi_i(x, P)\psi_j(x, P)$. Substituting this into the equation $-g' + [U + \varsigma A, g] = 0$ we obtain

$$(-\varphi_i' + \sum_\alpha u_{i\alpha}\varphi_\alpha + \varsigma a_i\varphi_i)\psi_j + \varphi_i(-\psi_j' - \sum_\alpha \psi_\alpha u_{\alpha j} - \varsigma a_j \psi_j) = 0$$

whence

$$-\varphi_i' + \sum u_{i\alpha}\varphi_\alpha + \varsigma a_i\varphi_i = \lambda\varphi_i, \psi_j' + \sum \psi_\alpha u_{\alpha j} + \varsigma a_j\psi_j = \lambda\psi_j$$

λ depends on x and ς but not on i and j. By the substitution $\varphi_i \mapsto \exp \int \lambda dx \cdot \varphi_i, \psi_j \mapsto \exp(-\int \lambda dx)\psi_j$ the terms with λ can be removed without violating the property $g_{ij} = \varphi_i \psi_j$. Thus, the vector-column $\varphi = (\varphi_1, \ldots, \varphi_n)^t$ and the vector-row $\psi = (\psi_1, \ldots, \psi_n)$ satisfy the equations

$$(14.1.3) \qquad -\varphi' + (U + \varsigma A)\varphi = 0, \gamma \psi' + \psi(U + \varsigma A) = 0.$$

Since g is the projector to the eigenspaces of the matrix $\tilde{R}$ we have

$$(14.1.4) \qquad \tilde{R}\varphi = w\varphi, \psi\tilde{R} = w\psi.$$

The functions φ and ψ satisfying both the equations (14.1.3) and (14.1.4) are called the Baker and the adjoint Baker functions. We have $g = \varphi\psi, \psi\varphi = \operatorname{tr} g = 1$.

The functions φ and ψ are not expressed in terms of g locally. However some of their combinations are. These are, firstly, $g = \varphi\psi$, and secondly

$$(14.1.5) \qquad \chi_i^{(1)} = \partial \ln \varphi_i, \chi_j^{(2)} = \partial \ln \psi_i$$

(cf. 7.4.4). Indeed,

$$(14.1.6) \qquad \begin{aligned} \chi_i^{(1)} &= \partial \ln \varphi_i = \varphi_i'/\varphi_i = (\sum u_{i\alpha}\varphi_\alpha + \varsigma a_i \varphi_i)/\varphi_i \\ &= (\sum u_{i\alpha}\varphi_\alpha \psi_j + \varsigma a_i \varphi_i \psi_j)/\varphi_i \psi_j = ((U + \varsigma A)g)_{ij}/g_{ij}, \forall j \,; \\ \chi_j^{(2)} &= -(g(U + \varsigma A))_{ij}/g_{ij}, \forall i\,. \end{aligned}$$

Evidently, the relation

$$(14.1.7) \qquad \chi_i^{(1)} + \chi_j^{(2)} = \partial \ln g_{ij}$$

holds. Now let

$$\mu_i^{(1)}(x,y,p) = \varphi_i(x,P)/\varphi_i(y,P) = \exp \int_y^x \chi_i^{(1)}(s)ds$$

$$\mu_j^{(2)}(x,y,P) = \psi_j(y,P)/\psi_j(x,P) = \exp \left(-\int_y^x \chi_j^{(2)}(s)ds\right).$$

The behaviour of these functions on the Riemann surface will be studied now.

14.1.8. *Lemma*. The divisors of the functions $\mu_i^{(1)}$ and $\mu_j^{(2)}$ in the finite part of the Riemann surface are, respectively, $d_i^{(1)}(x) - d_i^{(1)}(y)$ and $d_j^{(2)}(y) - d_j^{(2)}(x)$.

Proof: This immediately follows from the definition of these functions. □

14.1.9. Proposition. If $P \to \{k\}$ then

$$\chi_i^{(1)} = \varsigma a_k + 0(1), k \neq i, \chi_k^{(1)} = \varsigma a_k + 0(\varsigma^{-1}),$$
$$\chi_i^{(2)} = -\varsigma a_k + 0(1), k \neq i, \chi_k^{(2)} = -\varsigma a_k + 0(\varsigma^{-1}).$$

Proof: This follows from (14.1.6) and the asymptotics 13.1.16:

$$\chi_i^{(1)} \sim u_{ik} g_{kk}/g_{ik} + \varsigma a_i \sim -\varsigma u_{ik}(a_i - a_k)/u_{ik} + \varsigma a_i = \varsigma a_k, i \neq k$$
$$\chi_k^{(1)} \sim (Ug)_{kk}/g_{kk} + \varsigma a_k = 0(\varsigma^{-1}) + a_k \varsigma,$$

etc. □

14.1.10. Corollary. If $P \to \{k\}$ then

$$\mu_i^{(1,2)}(x, y, P) = \exp[a_k(x-y)\varsigma] \cdot (1 + 0(\varsigma^{-1}))$$

holds.

14.1.11. We consider the following Abel differentials (see 8.A): 1) the normalized Abel differentials of the first kind ω_i, as above; 2) the Abel differentials of the second kind $\Omega_{\{k\}}$, the only singularity of such a differential is the point $\{k\}$ where it can be represented in a local parameter as $d(z^{-1})+$ (holomorphic dif.); 3) the Abel differentials of the third kind Ω_{PQ} having two singularities, simple pole in P with the residue $+1$, and in Q with the residue -1. The differentials $\Omega_{\{k\}}$ and Ω_{PQ} are unique if they are normalized by the condition $\int_{\alpha_i} \Omega_{\{k\}} = \int_{\alpha_i} \Omega_{PQ} = 0\, \forall i$. For simplicity we shall write $\Omega_k = \Omega_{\{k\}}$. We recall the proposition 8.A.6:
(14.1.12)

$$\int_{\beta_k} \Omega_{PQ} = 2\pi i \int_Q^P \omega_k, \int_{\beta_k} \Omega_{\{i\}} = -2\pi i \varphi_k(\{i\}) \text{ where } \omega_k = \varphi_k dz \text{ in}\{i\}$$

Let, as above, $\eta_i^{(1)}(x) = \mathfrak{E}(d_i^{(1)})$ be the Abel mapping of the divisor $d_i^{(1)}(x)$.

14.1.13. Akhieser's lemma.

$$(\eta_i^{(1)}(x) - \eta_i^{(1)}(y))_r = \sum_k a_k(x-y) \int_{\beta_r} \Omega_k / 2\pi i$$

where the subscript r symbolizes the r-th component of the vector.

Proof: Let us consider the Abel differential $d_\varsigma \ln \mu_i^{(1)}(x,y)$, with $d_\varsigma f = \partial f/\partial \varsigma d\varsigma$. It has poles at the points of the divisor $d_i^{(1)}(x)$ with residue $+1$ and at the points of the divisor $d_i^{(1)}(y)$ with residue -1. Besides, it has poles of the second order at all the points $\{k\}$ where it can be represented as $a_k(x-y)d\varsigma+$ (a holomorphic differential). This is why the differential can be written as

$$d_\varsigma \ln \mu_i^{(1)}(x,z) = \sum a_k(x-y)\Omega_k + \sum \Omega_{P_k,Q_k} + \sum h_k \omega_k,$$

where $P_k \in d_i^{(1)}(x), Q_k \in d_i^{(1)}(y)$. We integrate this formula, at first, over the contours $\{\alpha_r\}$. On the one hand $\int_{\alpha_r} d_\varsigma \ln \mu_i^{(1)} = h_r$, on the other hand $\mu_i^{(1)}$ is a single valued function which implies that $\int_{\alpha_r} d_\varsigma \ln \mu_i^{(1)} = 2\pi i \cdot m_r, m_r \in \mathbb{Z}$. Thus $h_r = 2\pi i m_r$. Then we integrate this formula over the contours $\{\beta_r\}$:

$$\int_{\beta_r} d_\varsigma \ln \mu_i^{(1)} = \sum a_k(x-y)\int_{\beta_r} \Omega_k + 2\pi i \sum \int_{P_k(z)}^{P_k(x)} \omega_r + \sum h_k B_{rk},$$

where $\{B_{rk}\}$ is the matrix of β-periods of the differentials $\{\omega_i\}$ (see 8.A.1). On the left-hand side we have $2\pi i n_r, n_r \in \mathbb{Z}$. Thus,

$$\mathfrak{E}(d_i^{(1)}(x) - d_i^{(1)}(y)) = \sum a_k(x-y)\int_{\beta_r} \Omega_k/2\pi i.$$

The equality holds on J i.e. modulo the lattice of periods. □

The Akhieser lemma yields another proof of the linear dependence of the Abel mapping of the divisor $d_i^{(1)}$ on x.

The next proposition contains the main result.

14.1.14. Proposition. The relation

$$\mu_i^{(1)}(x,y,P) = \exp\left[(x-y)\sum a_k \left(\int_{P_0}^{P} \Omega_k - \xi_{ki}\right)\right]$$

$$\times \frac{\theta(\mathfrak{E}(P) - \eta_i^{(1)}(x) - \mathbb{K})\theta([i] - \eta_i^{(1)}(y) - \mathbb{K})}{\theta(\mathfrak{E}(P) - \eta_i^{(1)}(y) - \mathbb{K})\theta([i] - \eta_i^{(1)}(x) - \mathbb{K})}$$

holds where $\xi_{ki} = \int_{P_0}^{\{i\}} \Omega_k (k \neq i), \xi_{ii} = \lim_{P \to \{i\}} (\int_{P_0}^{P} \Omega_i - \varsigma)$, $\mathbb{K}$ is the Riemann constant vector (see 8.A.4), and $[i] = \mathfrak{E}(\{i\})$.[a]

Proof: At first we prove that the expression is a single-valued function on the Riemann surface. Provisionally we denote this expression as $F(x, y, P)$. If the point P passes around a contour α_r then the vector $\delta_r = (0, \ldots, 1, \ldots, 0)$ is added to $\mathfrak{E}(P)$, the function θ not being changed, nor the integral $\int \Omega_k$ in the exponent. If the point P passed around a contour β_r the vector $\mathbb{B}_r = (B_{r1}, \ldots, B_{rg})$ is added to $\mathfrak{E}(P)$. Recall (8.A.3): $\theta(p_1 + B_{r1}, \ldots, p_g + B_{rg}) = \exp[-2\pi i (B_{rr}/2 + p_r)]\theta(p_1, \ldots, p_g)$. The quotient of θ-functions is multiplied by $\exp\{2\pi i[\eta_i^{(1)}(x) - \eta_i^{(1)}(y)]\} = \exp\left[-\sum_k a_k(x-y) \int_{\beta_r} \Omega_k \cdot\right]$. The exponential function is multiplied by its reciprocal, and they cancel out.

The divisors of $\mu_i^{(1)}(x, y, P)$ and $F(x, y, P)$ coincide. Indeed, according to the Riemann theorem, 8.A.4, the divisor of the zeros of the function $\theta(\mathfrak{E}(P) - \eta_i^{(1)}(x) - \mathbb{K})$ is $d_i^{(1)}(x)$ and that of $\theta(\mathfrak{E}(P) - \eta_i^{(1)}(y) - \mathbb{K})$ is $d_i^{(1)}(y)$.

Let $P \to \{l\}$. We have seen (14.1.10) that $\mu_i^{(1)}(x, y, P) = \exp[a_l(x - y)\varsigma](1 + 0(\varsigma^{-1}))$. The asymptotics of $F(x, y, P)$ is the following. If $l = i$ then

$$\int_{P_0}^{P} \Omega_k - \xi_{ki} = \int_{P_0}^{P} \Omega_k - \int_{P_0}^{\{i\}} \Omega_k = \int_{\{i\}}^{P} \Omega_k = 0(\varsigma^{-1}), k \neq i;$$

$$\int_{P_0}^{P} \Omega_i - \xi_{ii} = \varsigma + 0(\varsigma^{-1})$$

Therefore if $P \to \{i\}$ then $\mu_i^{(1)}(x, y, P)/F(x, y, P) \to 1$. If $P \to \{l\} \neq \{i\}$, then for $k \neq i, k \neq l$

$$\int_{P_0}^{P} \Omega_k - \xi_{ki} = \int_{P_0}^{P} \Omega_k - \int_{P_0}^{\{i\}} \Omega_k = \int_{\{i\}}^{P} \Omega_k \to \int_{\{i\}}^{\{l\}} \Omega_k = \text{const};$$

---

[a] The definition of the right-hand side of the formula is not quite obvious. θ-functions and the integral are not uniquely determined on the Riemann surface but they are single-valued on the cut along the α_i, β_i-contours Riemann surface (see Fig. 1 on p. 190). For the time being we understand this formula in this sense; later on, when proving this proposition we shall make sure that the expression on the right-hand side is single-valued on the entire surface.

for $k = l$, $\int_{P_0}^{P} \Omega_l - \xi_{li} = \varsigma + 0(\varsigma^{-1})$, and for $k = i$, $\int_{P_0}^{P} \Omega_i - \xi_{ii} \to \int_{P_0}^{\{l\}} \Omega_i - \xi_{ii} =$ const. Thus, $F(x, y, P) = \exp[(x - y)a_l\varsigma] \cdot [c + 0(\varsigma^{-1})]$ where $c \neq 0$ and depends only on x and t but not on ς. We have obtained that the quotient $\mu_i^{(1)}(x, y, P)/F(x, y, P)$ is a bounded holomorphic function on the closed Riemann surface i.e. is a constant. The asymptotics for $P \to \{i\}$ implies that this constant is 1. □

14.1.15. Proposition.

$$\mu_j^{(2)}(x, y, P) = \exp\left[(x - y)\sum a_k\left(\int_{P_0}^{P} \Omega_k - \xi_{kj}\right)\right] \cdot$$

$$\frac{\theta(\mathfrak{E}(P) - \eta_j^{(2)}(y) - \mathbb{K})\theta([j] - \eta_j^{(2)}(x) - \mathbb{K})}{\theta(\mathfrak{E}(P) - \eta_j^{(2)}(x) - \mathbb{K})\theta([j] - \eta_j^{(2)}(y) - \mathbb{K})}.$$

The proof is the same. □

14.1.16. Proposition. For the elements u_{ij} of the matrix U the formula

$$u_{ij}(x) = u_{ij}(y) \exp\left[(x - y)\sum a_k(\xi_{kj} - \xi_{ki})\right]$$

$$\times \frac{\theta([j] - \eta_i^{(1)}(x) - \mathbb{K})\theta([i] - \eta_i^{(1)}(y) - \mathbb{K})}{\theta([j] - \eta_i^{(1)}(y) - \mathbb{K})\theta([i] - \eta_i^{(1)}(x) - \mathbb{K})}$$

holds.

Proof: We have

$$\mu_i^{(1)}/\mu_j^{(2)} = g_{ij}(x)/g_{ij}(y)$$

If $P \to \{j\}$ this becomes $u_{ij}(x)/u_{ij}(y)$. It remains to substitute the expressions for $\mu_i^{(1)}$ and $\mu_j^{(2)}$ from 14.1.14 and 14.1.15. The value y can be considered as an initial value for x, e.g. $y = 0$. The formula can be written in a simpler form

$$(14.1.17) \quad u_{ij}(x) = k_{ij} \exp\left[x \sum a_k(\xi_{kj} - \xi_{ki})\right] \cdot \frac{\theta([j] - \eta_i^{(1)}(x) - \mathbb{K})}{\theta([i] - \eta_i^{(1)}(x) - \mathbb{K})}.$$

We recall that all the divisors $d_i^{(1)}$ can be expressed in terms of one of them, see 13.3.13. The coefficients k_{ij} do not depend on x.

Unfortunately, this formula cannot be considered as the formula for a general solution with arbitrary constants k_{ij}; they are not independent since their number is too great. Let us count how many independent constants must be involved in a general solution. The order of the system is $mn(n-1)$. Some of the constants determining a solution are already fixed: $mn(n-1)/2$ values of the first integrals $\{J_{lk}\}$, and $mn(n-1)/2 - (n-1)$ arbitrary constants in the linear functions θ_{lk}. There are $mn(n-1) - mn(n-1)/2 - [mn(n-1)/2 - (n-1)] = n-1$ constants which remain arbitrary. Thus, all the k_{ij} cannot be independent. In accordance with the group (13.2.6), with respect to which we have factorized the system, arbitrary multipliers can only be of the type $k_i k_j^{-1}$. Dubrovin has found the constraints for k_{ij}; they are however very complicated. It is better to change the point of view and to present the process of integration in the following succession. 1) The equation (13.1.2) and initial conditions $U = U_0, U' = U_1, \ldots, U^{(m-1)} = U_{m-1}$ are given. 2) One finds the coefficients of the characteristic polynomial $f(w, \varsigma)$ by substituting into $\det(\tilde{R} - wI)$ the initial conditions. 3) On the Riemann surface which is determined by the equation $f(w, \varsigma) = 0$ the normalized differentials of the first and of the second kind as well as ξ_{kj} must be found, 4) From the formula 13.1.13 for the projector g, one must calculate the points of the divisor $d_i^{(1)}$ for the initial values of $\{U^{(i)}\}$. This is probably the most difficult part of work. The Abel transformation of this divisor is taken for $\eta_i^{(1)}(0)$. Now the formula 14.1.16 will determine the solution. Here $\eta_i = \sum c_{i,lk} \theta_{lk}$ and θ_{lk} should be taken from 13.3.10.

14.2. Rotation of an n-dimensional rigid body.

14.2.1. An interesting application of the above theory concerns an n-dimensional rigid body with a fixed point. If $n = 3$ then the motion of such a body in the absence of external forces is governed by the well-known Euler equation. In the works by Arnold [70,71] it was shown that this equation is one of a series of "Euler equations" which can be written for any Lie algebra. The infinite-dimensional generalization includes also the Euler equation for the ideal liquid. Mishchenko [72] considered the "Euler equation" in the algebra $SO(n)$ which can be called the equation of rotation of the n-dimensional rigid body. He showed that besides the obvious first integrals, energy and momentum, there is a series of first integrals. In a note of the author [73] it was shown that these first integrals are in involution with respect to the Poisson bracket, which always exists for the Euler-Arnold equations and is none other than the Lie-Poisson-Berezin-Kirillov-Kostant bracket (see 2.4). However, the number of Mishchenko's first integrals is insufficient for integrability. The decisive step was made by Manakov [74] who invented the trick of multiplying the set of first integrals by introducing a parameter into the equation. He also noticed that after that procedure the equation of rotation of the rigid body become a particular case of the stationary equation (13.1.2).

14.2.2. Recall the Euler equation for the rotation of a 3-dimensional rigid body around a fixed point. We have a vector $\boldsymbol{\omega}(t)$ which is called the angular velocity (in a frame attached to the body). There is a symmetrical positive operator $J : \mathbb{R}^3 \to \mathbb{R}^3$ called the inertia tensor. The vector $\mathbf{m} = J\boldsymbol{\omega}$ is the kinetic momentum. The Euler equation is

$$d\mathbf{m}/dt = \boldsymbol{\omega} \times \mathbf{m}.$$

It is convenient to choose a coordinate system such that the matrix J is diagonal: $J = \text{diag}(J_1, J_2, J_3)$ (J_i are the principal inertia momenta). Then $m_i = J_i\omega_i$. To generalize this equation to the n-dimensional case we can write it as a matrix one using the isomorphism between the Lie algebra of vectors of $\mathbb{R}^3$ (with respect to the vector multiplication) and the algebra of skew symmetrical matrices, SO(3). This isomorphism is given by the equality $\omega_{ij} = \sum \varepsilon_{ijk}\omega_k$ where ε_{ijk} is skew symmetric, and $\varepsilon_{123} = 1$. The vector $\mathbf{m}$ is related to the matrix $m_{ij} = \sum_k \varepsilon_{ijk}J_k\omega_k$. Let $J_1 = I_2 + I_3$, $J_2 = I_3 + I_1$ and $J_3 = I_1 + I_2$, then $m_{ij} = \sum_k (I_i + I_j)\varepsilon_{ijk}\omega_k = (I_i + I_j)\omega_{ij}$ i.e.

(14.2.3) $$m = I\omega + \omega I, I = \text{diag}(I_1, I_2, I_3),$$

and the Euler equation takes the form

(14.2.4) $$\dot{m} = [\omega, m].$$

Now one can generalize this equation. Let ω and m belong to SO(n), the Lie algebra of skew symmetric matrices. Let $I = \text{diag}(I_1, \ldots, I_n) = \text{const}$. The equations (14.2.3) and (14.2.4) will remain unchanged.

14.2.5. The equation (14.2.4) is Hamiltonian in the following structure. With the aid of the scalar product $\langle X, Y \rangle = \text{tr}XY$, algebra so($n$) can be identified with its dual (coalgebra). Let us consider orbits of the coadjoint representation of the group SO(n) : $g \in \text{SO}(n) \mapsto T_g$: so(n) $\to$

so(n); $T_g X = g^{-1} X g$, ($X \in$ so(n)). Infinitesimal operators form the coadjoint representation of the algebra so(n) on itself: $Y \in$ so(n) $\mapsto T_Y$: so(n) $\to$ so(n); $T_Y X = [X, Y]$.

The quantities tr m^k are first integrals of Eq. (14.1.4). The values of these first integrals determine an orbit $\{g^{-1} m g\}$ of the coadjoint representation. Therefore the orbit is invariant with respect to the equation. The codimension of an orbit is equal to the number of the first integrals tr m^k. k must take only even integers. If n is even then the codimension is $n/2$. If n is odd then the codimension is $(n-1)/2$, i.e. it is always $[n/2]$. The dimension of the orbit is $n(n-1)/2 - n/2 = n(n-2)/2(-)$ if n is even and $n(n-1)/2 - (n-1)/2 = (n-1)^2/2$ if n is odd.

A symplectic structure on the orbits is introduced. This was already done in Sec. 2.4. Briefly recall the construction now. The general form of a vector tangent to the orbit at a point m is $\xi_x(m) = \mathrm{ad}^*(X) m = [m, X]$, where $X \in$ so(n). The tangent spaces to the orbit are canonically embedded into so(n), thus $\xi_x(m) \in$ so(n). (The commutator of $\xi_x(m)$ and $\xi_Y(m)$ as vector fields has nothing in common with their commutator as elements of so(n). To distinguish the first of them we denote it as $[[\xi_x(m), \xi_Y(m)]]$.)

14.2.6. *Lemma.*

$$[[\xi_x(m), \xi_y(m)]] = \xi_{[X,Y]}$$

where X and Y are fixed.
Proof: See (2.4.4). □

Thus, the vector fields $\xi_x(m)$ yield the right representation of so(n). The form Ω is defined as

(14.2.7) $$\Omega(\xi_X, \xi_Y) = \langle \xi_X, Y \rangle = \mathrm{tr}[m, X] Y .$$

In Sec. 2.4 it is proved that this form is a symplectic one.

If $f(m)$ is a function on an orbit then $df(m)$ is an element of the dual to the tangent space to the orbit $T_m \mathcal{U}$. This dual can be identified with so(n) modulo $X \in$ so(n) such that $[m, X] = 0$. It is not difficult to find $df(m)$ if $f(m)$ is a linear or a quadratic function.

If $f(m) = \mathrm{tr} A m, A \in$ so(n), then $df(m)$ is an element of so(n) such that for any $\xi_x \in T_m M$ the relation $\langle \xi_x, df(m) \rangle = \xi_x f(m) = \mathrm{tr} A \xi_x$ holds, which implies $df(m) = A$.

If $f(m) = \mathrm{tr} J(m) m / 2$ where J is a symmetrical operator then

$$\langle \xi_x, f(m) \rangle = \xi_x f(m) = \mathrm{tr} J(m) \xi_x ; df(m) = J(m) .$$

To every function $f(m)$ taken as a Hamiltonian, a vector field $\xi_{df} = [m, df]$ is appointed.

14.2.8. Proposition. The equation (14.2.4) is of the Hamilton type with the Hamiltonian
$$\mathcal{H} = \frac{1}{2}\langle m, \omega \rangle = \frac{1}{2}\operatorname{tr} m\omega.$$

Proof: Taking into account that ω is linearly expressed in terms of m we have $d\mathcal{H} = \omega$. Therefore the vector field related to the Hamiltonian is $[m, \omega]$. □

The Hamiltonian is a quadratic first integral of the system. Mishchenko has suggested a series of quadratic first integrals. All of them are contained in the set of the more general first integrals constructed by Manakov.

14.2.9. Proposition. Eq. (14.2.4) is equivalent to
$$\frac{d}{dt}(m + \varsigma I^2) = [\omega + \varsigma I, m + \varsigma I^2],$$
where ς is a parameter, and the equation must be fulfilled identically in ς.
Proof: One must only make sure that $[I, m] + [\omega, I^2] = 0$. But this immediately follows from (14.2.3). □

14.2.10. Corollary. The coefficients of the expansion in powers of ς of the quantities
$$\operatorname{tr}(m + \varsigma I^2)^k, k = 1, 2, \ldots, n$$
are first integrals.

Let us count the number of these first integrals. If k is even then only even powers of ς give non-vanishing first integrals. Besides, $\operatorname{tr} m^k$ are trivial on the orbit. Only ς^l with $l = 2, 4, \ldots, k-2$ remain i.e. there are $(k-2)/2$ first integrals. If k is odd then the same calculation yields $(k-1)/2$ first integrals. Altogether we have: for n even $1 + 1 + 2 + 2 + \ldots + (n-2)/2 + (n-2)/2 = n(n-2)/4$ first integrals, and for n odd $1 + 1 + \ldots + (n-3)/2 + (n-3)/2 + (n-1)/2 = (n-1)^2/4$ first integrals. In all the cases, the number of first integrals is equal to half of the dimension of an orbit i.e. half of the differential order of the system.

14.2.11. Proposition. The constructed first integrals are in involution.
Proof: Let $f_k = \operatorname{tr}(m + \varsigma I^2)^k$. The vector field ξ_{df_k} related to this first integral is $\xi_{df_k} = [m, df_k] = k[m, (m + \varsigma I^2)^{k-1}]$. One must prove that the Poisson bracket $\{f_k, f_l\} = kl \operatorname{tr}[m, (m + \varsigma I^2)^{k-1}](m + \varsigma_1 I^2)^{l-1}$ vanishes

(more precisely this is a generator of Poisson brackets which can be obtained by expansion into a double series in ς and ς_1).

We calculate

$$\mathrm{tr}[m,(m+\varsigma I^2)^k](m+\varsigma_1 I^2)^l = (\varsigma_1-\varsigma)^{-1}\mathrm{tr}[\varsigma_1(m+\varsigma I^2)$$
$$-\varsigma(m+\varsigma_1 I^2),(m+\varsigma I^2)^k](m+\varsigma_1 I^2)^l$$
$$= (\varsigma_1-\varsigma)^{-1}\{\varsigma_1\,\mathrm{tr}[(m+\varsigma I^2),(m+\varsigma I^2)^k](m+\varsigma_1 I^2)^l$$
$$+\varsigma\,\mathrm{tr}[(m+\varsigma_1 I^2),(m+\varsigma I^2)^l](m+\varsigma I^2)^k]\} = 0.\ \square$$

14.2.12. Corollary. Eq. (14.2.4) can be integrated in quadratures.

The procedure of integration can be reduced to that in the previous section with the aid of the following proposition.

14.2.13. Proposition. Eq. (14.2.9) coincides with the stationary equation (13.1.2) where $m=1, A=I$ and $B=I^2$, restricted to the subalgebra so(n).

Proof: First of all one can see that equation (13.1.2) admits of a restriction to the algebra so(n). The recurrence equations (9.1.8) imply that if $U \in$ so(n) then all R_{2k} are symmetric and all R_{2k+1} are skew symmetric. Therefore $[A, R_2]$ is skew symmetric, and if the initial values for U are skew symmetric they remain so for all the values of x.

The equation $U' = [A, R_2]$ is, as we know, equivalent to $-\tilde{R}'_1 = [U + \varsigma A, \tilde{R}_1]$, where $\tilde{R}_1 = B\varsigma + R_1$. If we put $I = A, B = I^2$ and $U = \omega$ then $(R_1)_{jk} = [(I_j^2 - I_k^2)/(I_j - I_k)]U_{jk} = (I_j + I_k)\omega_{jk}$ (see 9.1.9) i.e. $R_1 = m$, and the equation takes the form $-(m+I^2\varsigma)' = [\omega + \varsigma I, m+I^2\varsigma]$ which coincides with 14.2.9 if $t = -x$.

14.2.14. We draw the reader's attention to the fact that the Hamiltonian structure of the equation under consideration constructed in this section does not coincide with those from Chap. 13; this can be seen e.g. from the fact that the Hamiltonian is now $\frac{1}{2}\langle m,\omega\rangle$, which is equal to tr AR_2. In the theory of the equation (13.1.2) this expression was not the Hamiltonian of the given equation but of a trivial symmetry. This also means that the symplectic forms are different, too. Now this form is defined only on orbits. It would be interesting to discuss the relation between two structures.

CHAPTER 15. Stationary Equations of the KdV-Hierarchies.

15.1. Reduction to stationary equations of the matrix hierarchy.

15.1.1 Return to Chap. 1. Let $T(z) = \sum_0^{n-1} \partial^{-i-1} T_i(z) \in R_- / R_{(-\infty, -n-1)}$ be a resolvent of the operator L. Let $T_i(z) = \sum_{-\infty}^{\infty} T_{i,r} z^{-r}$ be the expansion of $T_i(z)$ into a series in z; the series is one-way infinite, $r \to \infty$. The resolvent satisfies the equation $\hat{L}(T\hat{L})_+ - (\hat{L}T)_+ \hat{L} = 0$.

The stationary equation has the form $T_{i,r} = 0, i = 0, 1, \ldots, n-1$. This is a system of n ordinary differential equations with n unknown quantities $u_0, u_1, \ldots, u_{n-1}$. By virtue of 3.5.3, it has a variational form $\delta\Lambda/\delta u_i = 0$. We shall always assume that $u_{n-1} = 0$. Then $n - 1$ equations remain:

(15.1.2) $$T_{i,r} = 0; i = 0, 1, \ldots, n-2$$

with the unknown quantities $u_0, u_1, \ldots, u_{n-2}$. ($T_{n-1,r}$ is expressed in terms of $T_{i,r}, i < n - 1$ with the aid of the equation res $[L, T_r] = 0$ or 3.4.5).

As it will be shown in 15.1.12 the system will satisfy the criterion of non-degeneration (see 2.3). This implies that the system is of the Hamilton type. We shall find a set of first integrals in involution, and their amount will be sufficient for integrability. To actually perform this integration one must find the symplectic form, and here one encounters the first difficulties. As it is already known in order to find the symplectic form one must represent $\delta\Lambda$ as $\sum(\delta\Lambda/\delta u_i)\delta u_i + \partial\Omega^{(1)}$. The method used in Sec. 3.5 does not give any explicit expression for the term $\partial\Omega^{(1)}$; more precisely, it gives a sum of a large number of awkward terms. The experience of Chap. 9 teaches us that a good expression for $\partial\Omega^{(1)}$ follows from a representation of L in terms of "dressing" operators ϕ and $\phi^{-1} = \psi$. But this time the situation is more complicated: in contrast to the case of the matrix hierarchy in Chap. 9, now the dressing functions do not belong to the differential algebra $\mathcal{A}$ and their construction requires integration, i.e., extension of the differential algebra $\mathcal{A}$. If an expression is an exact derivative in the extended algebra, it is not obliged to be such in the original algebra. (It seems very interesting to solve the problem in this way using e.g. the fact that some combinations of the dressing functions, or the Baker functions φ and φ^* belong to $\mathcal{A}$, see 7.4.)

We do not overcome, but rather circumvent, the diffialties as follows: in Chap. 10 we have learned how to embed KdV-hierarchies into the matrix ones. Now we shall reduce the stationary equation (15.1.2) to the corresponding matrix equation. Further, one can follow exactly the plan of Chap. 13 and 14.

We repeat what was already said, that the importance of stationary equations is enhanced by the fact that manifolds of their solutions are invariant with respect to all nonstationary equations of the same hierarchy. This yields finite-dimensional classes of solutions of the nonstationary equations (algebraic-geometrical solutions, in particular, solitons). We have already dealt with them in Chap. 8 (Krichever's solutions). Now we have another method and another form of these solutions which are difficult to compare with the first ones. We follow our paper [62].

15.1.3. We touch only briefly the facts which essentially repeat those of Chap. 13 and dwell on the distinctions.

We have basic resolvents (see 10.2.8) $Q^\alpha = \sum_{-n+1}^{\infty} Q_r^\alpha z^{-r}$ which satisfy the equation

$$[-\partial + U + J + \varsigma A, Q^\alpha] = 0.$$

Recall 10.2.9: $(Q_r^\alpha)_{ij} = \varepsilon^{-\alpha r} Q_{ij,r}$. If b_α are arbitrary, $\alpha = 0, \ldots, n-1$, then the resolvent $Q^B = \sum b_\alpha Q^\alpha$ can be formed, i.e. $Q_{ij,r}^B = \sum b_\alpha \varepsilon^{-\alpha r} Q_{ij,r}$.

The stationary equation comes from the Lagrangian

(15.1.4) $$\Lambda = -(m+n)^{-1} \operatorname{tr} A Q_{m+2n}^B.$$

The variational equation $\delta\Lambda/\delta U = 0$, on account of 10.2.10, is

(15.1.5) $\qquad Q^B_{i,n-1,m+n} = 0, \quad i = 0,\ldots, n-2.$

Note that this equation does not depend on B i.e. on the coefficients $\{b_\alpha\}$ since the factor $\sum b_\alpha \varepsilon^{-\alpha(m+n)}$ cancels. Hence Q^B can be considered as one of the basic resolvents, e.g. Q^0, i.e. $b_0 = 1; b_i = 0, i > 0$.

15.1.6. Let us write Eq. (15.1.5) in terms of the KdV-hierarchy. According to 10.1.9, if Q is a resolvent, then Q has the form $Q = Q_X$, where $X = \sum_0^{n-1} \partial^{-i-1} X_i, X_i = Q_{i,n-1}$ is a resolvent of the operator $L = \sum_0^n u_i \partial^i, u_n = 1, u_{n-1} = 0$. If Q is one of the basic resolvents Q^α, then X is uniquely determined by the condition that $X_i = \sum_{-n+1+i}^{\infty} X_{i,r} z^{-r}, X_{i,-n+1+i} = \varepsilon^{\alpha(i-n+1)}$ and there are no other constants.

The Lagrangian takes the form

(15.1.7) $\qquad \Lambda = -(m+n)^{-1}\operatorname{res} X|_{m+2n},$

and the equations (15.1.5) the form

(15.1.8) $\qquad X_i|_{m+n} = 0, i = 0, \ldots, n-2.$

i.e. Eq. (15.1.2).

15.1.9. *Proposition.* The system (15.1.5) is equivalent to

$$[A, Q_{m+n}] = 0.$$

Proof: From the equality written here it follows that the last (non-complete) column of the matrix Q_{m+n} vanishes, i.e. (15.1.5). Conversely, let the last column without the last element vanish, $Q_{i,n-1;n+m} = 0, i \leq n-2$. This implies $Q_{n-1,n-1;n+m} = 0$ (see 3.4.5). Let us show that the matrix Q_{m+n} is strictly lower triangular. Equation (10.1.5) implies for the operator-row

$$Q_{i;m+n} = \partial^i(X_{m+n}L)_+ - (\partial^i X_{m+n})_+ L + (\partial^i X_{m+2n})_+ = (\partial^i X_{m+2n})_+$$

(where $X = \sum_0^{n-1} \partial^{-i-1} Q_{i,n-1}$). This differential operator has an order $< i$ which proves that the matrix Q_{m+n} is strictly triangular. This yields $[A, Q_{m+n}] = 0$.

15.1.10. There is a grading in the differential algebra $\mathcal{A}$. The weight of a factor $u_i^{(j)}$ is $n-1+j$. The weight of ∂ is 1 and that of z also 1. By multiplication of two expressions their weights are summed. The operator $\hat{L}$ is homogeneous with respect to the weights, and all the formulas we write are also homogeneous.

15.1.11. Proposition. The weight of the element $Q_{ij,r}$ is equal to $i-j+r$.
Proof: The series $\phi^{(2)}$ and Λ have zero weight since they are homogeneous and start with 1. Therefore, the series $Q^{(2)\alpha}$ (see 10.2.8) have also zero weights. Further, $Q^\alpha = ZKQ^{(2)\alpha}K^{-1}Z^{-1}$. Hence the weight of the element Q_{ij}^α is $i-j$, and that of the coefficient $Q_{ij,r}^\alpha$ is $i-j+r$. $\square$

15.1.12. Proposition. (i) The system (15.1.5) has the following structure

$$0 = Q_{0,n-1;m+n} \equiv a_{00}u_0^{(m-n+1)} + a_{01}u_1^{(m-n+2)}$$
$$+ \ldots + a_{0,n-2}u_{n-2}^{(m-1)} + f_0,$$

$$\ldots\ldots\ldots\ldots\ldots\ldots\ldots$$

$$0 = Q_{n-2,n-1;m+n} \equiv a_{n-2,0}u_0^{(m-1)} + a_{n-2,1}u_1^{(m)}$$
$$+ \ldots + a_{n-2,n-2}u_{n-2}^{(m+n-3)} + f_{n-2},$$

where $f_0, \ldots, f_{n-2}$ are nonlinear terms containing only lower derivatives.
(ii) If m and n are mutually simple then $|\det(a_{ij})| = n^{-2n-m+3} \neq 0$.
Proof: (i) follows from homogeneity and the proposition 15.1.11 (ii) requires some calculations (this proposition was conjectured in [16] and proved by Veselov [75]; we produce here another proof).

The linear terms of the dressing matrices $\phi^{(2)}$ and $\psi^{(2)} = (\phi^{(2)})^{-1}$ can be found. All the equalities will be written up to non-linear terms. The matrices $\phi^{(2)}$ and $\psi^{(2)}$ satisfy Eqs. (10.2.7). Expanding the equation in z^{-1} and dropping the non-linear terms we get for $i \neq j$

$$-\phi_{ij,r-1}^{(2)'} - \frac{1}{n\varepsilon^{i(n-1)}} \sum_{\beta=0}^{n-1}\sum_{\alpha=0}^{n-2} \varepsilon^{\beta\alpha} u_\alpha \phi_{\beta j,r-n+\alpha}^{(2)} + (\varepsilon^i - \varepsilon^j)\phi_{ij,r}^{(2)} = 0.$$

The middle term is linear only if $r-n+\alpha = 0$ and $\phi_{\beta j,0}^{(2)} = \delta_{\beta j}$. Thus,

$$-\phi_{ij,r-1}^{(2)'} - \frac{1}{n\varepsilon^{i(n-1)}}\varepsilon^{j(n-r)}u_{n-r} + (\varepsilon^i - \varepsilon^j)\phi_{ij,r}^{(2)} = 0$$

($u_{n-r} = 0$ if $r > n$). Taking into account that $\varepsilon^n = 1$ this implies

$$\phi_{ij,r}^{(2)} = \frac{1}{n} \sum_{\beta=n-r}^{n-2} \varepsilon^{i+j\beta} (\varepsilon^i - \varepsilon^j)^{-r-\beta+n-1} u_\beta^{(r+\beta-n)}$$

$$= \frac{1}{n} \sum_{\beta=n-r}^{n-2} \varepsilon^{i-jr-j} (\varepsilon^{i-j} - 1)^{-r-\beta+n-1} u_\beta^{(r+\beta-n)}.$$

For the inverse matrix $\psi_{ij,r}^{(2)} = -\phi_{ij,r}^{(2)}$. Now $Q_{ij,r}^{(2)\alpha} = [\phi_{ij,r}^{(2)}, E_\alpha] = \phi_{i\alpha,r}^{(2)} \delta_{\alpha j} - \delta_{i\alpha} \phi_{\alpha j,r}^{(2)}$. We take $\alpha = 0$, $Q_{ij,r}^{(2)0} = \phi_{i0,r}^{(2)} \delta_{0j} - \delta_{i0} \phi_{0j,r}^{(2)}$. Further, $Q^{(1)0} = KQ^{(2)0}K^{-1}$ whence

$$Q_{ij,r}^{(1)0} = \frac{1}{n} \sum_{\gamma=1}^{n-1} (\varepsilon^{i\gamma} \phi_{\gamma 0,r}^{(2)} - \phi_{0\gamma,r}^{(2)} \varepsilon^{-\gamma i}) = \frac{1}{n^2} \sum_{\gamma=1}^{n-1} \sum_{\beta=n-r}^{n-2}$$

$$(\varepsilon^{i\gamma+\gamma} (\varepsilon^\gamma - 1)^{-r-\beta+n-1} - \varepsilon^{-\gamma(r+1+j)} (\varepsilon^{-\gamma} - 1)^{-r-\beta+n-1})$$

$$u_\beta^{(r+\beta-n)} = \frac{1}{n^2} \sum_{\gamma=1}^{n-1} \sum_{\beta=n-r}^{n-2} (\varepsilon^{(i+1)\gamma} - \varepsilon^{(r+j+1)\gamma})$$

$$(\varepsilon^\gamma - 1)^{-r-\beta+n-1} u_\beta^{(r+\beta-n)}$$

(in the second term we have made the substitution $\gamma \mapsto n - \gamma$ and have taken into account that $\varepsilon^n = 1$). Finally, $Q = ZQ^{(1)0}Z^{-1}$, that is, $Q_{ij} = z^{i-j} Q_{ij}^{(1)0}$, whence

$$Q_{ij,r} = Q_{ij,r+i-j}^{(1)0} = \frac{1}{n^2} \sum_{\gamma=1}^{n-1} \sum_{\beta=n-r-i+j}^{n-2}$$

$$(\varepsilon^{(i+1)\gamma} - \varepsilon^{(m+i+1)\gamma})(\varepsilon^\gamma - 1)^{-r-\beta+n-1-i+j} u_\beta^{(r+i-j+\beta-n)}.$$

Now

$$Q_{i,n-1,m+n} = \frac{1}{n^2} \sum_{\gamma=1}^{n-1} \sum_{\beta=n-m-i-1}^{n-2} (\varepsilon^{(i+1)\gamma} - \varepsilon^{(m+i+1)\gamma})$$

$$(\varepsilon^\gamma - 1)^{-m-\beta+n-2-i} u_\beta^{(m+\beta-n+1+i)} = \sum_{\beta=n-m-i-1}^{n-2} a_{i\beta} u_\beta^{(m+\beta-n+1+i)}$$

We make a note about the case $n-m-i-1>0$. The coefficient $a_{i\beta} = n^{-2}\sum_{\gamma=1}^{n-1}(1-\varepsilon^{m\gamma})\varepsilon^{(i+1)\gamma}(\varepsilon^\gamma-1)^{-m-\beta+n-2-i}$ vanishes if $0\leq\beta<n-m-i-1$. Then $0\leq -m-i-\beta+n-2\leq -i+n-m-2$ and $a_{i\beta}$ is the sum of terms of the type $\sum_{\gamma=0}^{n-1}(1-\varepsilon^{m\gamma})\varepsilon^{p\gamma}$, where $0<p\leq n-m+1$. It is easy to see that this expression vanishes. Therefore one can write

$$Q_{i,n-1;m+n} = \sum_{\beta=0}^{n-2} a_{i\beta} u_\beta^{(m+\beta-n+1+i)}$$

where $u_\beta^{(q)}$ for $q<0$ can be arbitrary; its coefficient is zero.

The coefficients a_{ij} are found. They look rather unpleasant, but it is surprising that the determinant can nevertheless be calculated. The matrix

$$b_{ij} = \sum_{\alpha=0}^{i}(-1)^{i-\alpha}\binom{i}{\alpha}a_{\alpha j} = \frac{1}{n^2}\sum_{\gamma=1}^{n-1}$$
$$\left(\sum_{\alpha=0}^{i}(-1)^{i-\alpha}\binom{i}{\alpha}\varepsilon^{\alpha\gamma}(\varepsilon^\gamma-1)^{-\alpha}\right)$$
$$\times(\varepsilon^\gamma-\varepsilon^{(m+1)\gamma})(\varepsilon^\gamma-1)^{-m-j-n+2}$$
$$= \frac{1}{n^2}\sum_{\gamma=1}^{n-1}\varepsilon^\gamma(1-\varepsilon^{m\gamma})(\varepsilon^\gamma-1)^{-i-j+m+n-2}$$

has the same determinant. Let us consider the quadratic form

$$\sum_{i,j=0}^{n-2} b_{ij}\xi^i\xi^j = \frac{1}{n^2}\sum_{\gamma=1}^{n-1}\varepsilon^\gamma(1-\varepsilon^{m\gamma})(\varepsilon^\gamma-1)^{-m+n-2}$$
$$\left(\sum_{i=0}^{n-2}(\varepsilon^\gamma-1)^{-i}\xi_i\right)^2 = \sum_{\gamma=1}^{n-1} c_\gamma \eta_\gamma^2.$$

We have changed the variables $\xi \mapsto \eta$ reducing the form to a sum of quadrats. The determinant of the matrix of this change does not vanish since this is a Vandermond's determinant

$$\prod_{\alpha>\delta}((\varepsilon^\gamma-1)^{-1} - (\varepsilon^\delta-1)^{-1}) = \prod_{\gamma>\delta}(\varepsilon^\delta-\varepsilon^\gamma)\bigg/\prod_{\gamma=1}^{n-1}(\varepsilon^\gamma-1)^{n-2}.$$

As a matter of fact we could stop at this place: all c_γ are evidently non-zero, the form is reduced to the sum of $n-1$ independent quadrats, i.e. $\det(a_{ij}) \neq 0$, which the only fact we need. However, let us calculate this determinant to the end.

$$\det(a_{ij}) = \det(b_{ij}) = n^{-2(n-1)} \prod_{\gamma=1}^{n-1} \varepsilon^\gamma (1-\varepsilon^{m\gamma})(\varepsilon^\gamma - 1)^{-m+n-2}$$

$$\cdot \prod_{\gamma > \delta}^{n-1} (\varepsilon^\delta - \varepsilon^\gamma)^2 / \prod_{\gamma=1}^{n-1} (\varepsilon^\gamma - 1)^{2(n-2)}$$

We have

$$\prod_{\gamma > \delta} (\varepsilon^\delta - \varepsilon^\gamma)^2 = (-1)^{(n-1)(n-2)/2} \prod_{\gamma,\delta=1}^{n-1} (\varepsilon^\delta - \varepsilon^\gamma)$$

$$= (-1)^{(n-1)(n-2)/2 + n - 1} \prod_{\substack{\gamma,\delta=0 \\ \gamma \neq \delta}}^{n-1} (\varepsilon^\delta - \varepsilon^\gamma) / \prod_{\gamma=1}^{n-1} (1 - \varepsilon^\gamma)^2$$

The quantity $\Delta = \prod_{\gamma \neq \delta} (\varepsilon^\delta - \varepsilon^\gamma)$ is the discriminant of the polynomial $f = z^n - 1$, hence $\Delta = \prod_{1}^{n-1} f'(\varepsilon^\gamma) = n^n \prod_{0}^{n-1} \varepsilon^{(n-1)\gamma}$. Further, $\prod_{0}^{n-1} (1 - \varepsilon^\gamma) = f'(1) = n$. In all we have $\det(a_{ij}) = \pm n^{-m-2n+3}$. □

15.1.13. Corollary. There is a sequence of non-vanishing minors inserted one into another

$$a_{0j_0} \neq 0, \begin{vmatrix} a_{0j_0} & a_{0j_1} \\ a_{1j_0} & a_{1j_1} \end{vmatrix} \neq 0, \ldots, \begin{vmatrix} a_{0j_0} & \cdots & a_{0j_{n-2}} \\ \vdots & & \vdots \\ a_{n-2,j_0} & \cdots & a_{n-2,j_{n-2}} \end{vmatrix} \neq 0$$

where $j_0, \ldots, j_{n-2}$ is a permutation of the numbers $0, 1, \ldots, n-2$.

15.1.14. Proposition. All $u_j^{(i)}$ can be uniquely expressed as polynomials in $\left\{ Q_{i,n-1;m+n}^{(j)} \right\}$ and in "phase variables"

$$u_{j_0}^{(p)}, p \leq m-n+j_0; u_{j_1}^{(p)}, p \leq m-n+j_1+1; \ldots$$
$$\ldots; u_{j_{n-2}}^{(p)}, p \leq m-n+j_{n-2}+(n-2).$$

where $m-n+j_0+1 \geq 0, m-n+j_1+2 \geq 0, \ldots, m-n+j_{n-2}+n+1 \geq 0$.
Proof: The first of Eqs. 15.1.2 yields an expression for $u_{j_0}^{(m-n+j_0+1)}$ in terms of the phase variables and Q. After that, the system of the differentiated first equation and of the nondifferentiated one enables us to express $u_{j_0}^{(m-n+j_0+2)}$ and $u_{j_1}^{(m-n+j_1+2)}$. Then the system of the first equation differentiated twice, the differentiated second equation, and the nondifferentiated third equation yields expressions for $u_{j_0}^{(m-n+j_0+3)}, u_{j_1}^{(m-n+j_1+3)}$ and $u_{j_2}^{(m-n+j_2+3)}$ etc.

Make another note about the case $n-m-1 > 0$. Let us show that all the numbers $m-n+j_0+1, m-n+j_1+2, \ldots$ are non-negative. If $m-n+j_0 < 0$ then, as we have already seen, the coefficient a_{0j_0} in $u^{(m-n+j_0+1)}$ vanishes; by assumption this is not the case. If $m-n+j_1+2 < 0$, then the coefficient of $u_{j_1}^{(m-n+j_1+2)}$ in the second equation, i.e. a_{1j_1}, and those of $u_{j_1}^{(m-n+j_1+1)}$ in the first equation (i.e. a_{0j_1}) vanish. Then

$$\begin{vmatrix} a_{0j_0}, & a_{0j_1} \\ a_{1j_1}, & a_{1j_1} \end{vmatrix} = 0$$

which contradicts the assumption etc. Thus, in the described process, we express the really existing $u_j^{(p)}$, with $p \geq 0$ in terms of $Q_{i,n-1;m+n}^{(j)}$ and the phase variables. □

Further, m and n will always be mutually simple.

15.1.15. *Corollary.* The order of the system of equations (15.1.5) is $(m-1)(n-1)$.

Proof. The number of the phase variables is $(m-n+j_0+1)+(m-n+j_1+2)+\ldots+(m-n+j_{n-2}+n-1) = (m-n)(n-1)+n(n-1)/2+(n-1)(n-2)/2 = (n-1)(m-1)$. □

15.2. First integrals.

15.2.1. Just as in Chap. 13, two kinds of first integrals can be constructed. The first construction uses the proposition 3.6.3. We have $\{\tilde{T}_{m+2n}, \tilde{T}_{k+n}^{(1)}\}^{(\infty)} = 0$ where $\tilde{T}_r = \int \operatorname{res} T_r dx$. $T^{(1)}$ is another resolvent: $T_r^{(1)}$ and T_r are propositional. This means that $\int \operatorname{res} H^{(\infty)}(T_k^{(1)}) T_{m+n} dx = 0$. In other words, an element $F_k^{(1)} \in \mathcal{A}$ exists such that $\operatorname{res} H^{(\infty)}(T_k^{(1)}) T_{m+n} = \partial F_k^{(1)}$. This implies that $\partial F_k^{(1)} = 0$ by virtue of the stationary equation $T_{m+n} = 0$ i.e. $F_k^{(1)}$ is a first integral with characteristic $H^\infty(T_k^{(1)})$. The corresponding vector field is $-\partial_{H^{(\infty)}(T_k^{(1)})}$. Exactly the same vector field relates to the first integral $\tilde{T}^{(1)}_{k+n}$ in the non-stationary theory. All these vector fields commute since the functionals $\{\tilde{T}^{(1)}_{k+n}\}$ are in involution (see 3.6.3). Thus, all the first integrals $\{F^{(1)}_k\}$ are in involution in the stationary theory.

Now the same will be written in terms of the matrix hierarchy using the correspondence $X \mapsto Q_X$. We transform:

$$\partial F^{(1)}_k = \operatorname{res} H^{(\infty)}(T^{(1)}_k) T_{m+n} = \operatorname{res}[L, T^{(1)}_k]_+ T_{m+n}$$
$$= \operatorname{res}[L, T^{(1)}_k] T_{m+n} = \operatorname{res}(LT^{(1)}_k)_+ T_{m+n} - \operatorname{res} T^{(1)}_k (LT_{m+n})_+ .$$

Recall that $(LT^{(1)}_k)_+$ is the first operator-row of $Q_{T^{(1)}_k} = Q^{(1)}_k$ (see (10.1.6)), similarly $(LT_{m+n})_+$. Therefore $\operatorname{res}(LT_k)_+ T_{m+n} = \operatorname{tr} AQ^{(1)}_k Q_{m+n}$, $\operatorname{res} T^{(1)}_k (LT_{m+n})_+ = \operatorname{tr} Q^{(1)}_k A Q_{m+n}$ and

$$(15.2.2) \qquad \partial F^{(1)}_k = \operatorname{tr}[A, Q^{(1)}_k] Q_{m+n} .$$

It is not difficult to give for $F^{(1)}_k$ an explicit expression similar to (9.3.14):

$$(15.2.3) \qquad F^{(1)}_k = \operatorname{tr}(Q_m Q^{(1)}_k + Q_{m-n} Q^{(1)}_{k+n} + Q_{m-2n} Q^{(1)}_{k+2n} + \cdots)$$

In fact, the series is finite since $Q_r = 0, r \leq -n$. The resolvent equation is equivalent to the recurrence relation

$$(15.2.4) \qquad -Q'_r + [U + J, Q_r] + [A, Q_{r+n}] = 0 .$$

The matrices Q_r are connected by this relation not in succession but in n. Therefore if a chain

$$Q^* = Q_{-\mu} z^\mu + Q_{-\mu+n} z^{\mu-n} + Q_{-\mu+2n} z^{\mu-2n} + \cdots, \mu \geq 0$$

is picked out from the series $\sum\limits_{-n+1}^{\infty} Q_r z^{-r}$ then this chain is also a resolvent. We take μ such that this chain contains a term $Q_m z^{-m}$ i.e. $-\mu = m \pmod{n}$. In other words there is a number S_1 such that $-\mu + s_1 n = m$. The resolvent Q^* can be expressed in terms of the basic resolvents using 10.2.9:

$$(15.2.5) \qquad Q^* = \frac{1}{n} \sum_{\alpha=0}^{n-1} \varepsilon^{-\alpha\mu} Q^\alpha .$$

Let
$$\tilde{Q} = z^m(Q_{-\mu}z^\mu + Q_{-\mu+n}z^{\mu-n} + \ldots + Q_m z^{-m})$$
$$= Q_{-\mu}\varsigma^{S_1} + Q_{-\mu+n}\varsigma^{S_1-1} + \ldots + Q_m.$$

15.2.6. Proposition. Eq. (15.1.5) is equivalent to the fact that $\tilde{Q}$ is also a resolvent, i.e. $-\tilde{Q}' + [U + J + \varsigma A, \tilde{Q}] = 0$.

Proof. Eq. (15.2.4) implies that

$$-\tilde{Q}' + [U + J + \varsigma A, \tilde{Q}] = -[A, Q_{m+n}]$$

It remains to recall 15.1.9. □

15.2.7. Proposition. The coefficients of the polynomials in z:

$$\mathrm{tr}\,\tilde{Q}^k, k = 1, \ldots, n$$

are first integrals. There are $(m-1)(n-1)/2$ such non-trivial integrals.

Proof: The first assertion is obvious (cf. 13.1.10). The computation of the number of the first integrals is much more complicated.

The series $z^m Q^*$ is a resolvent, hence $\mathrm{tr}\,(z^m Q^*)^k$ is a series with constant coefficients. In contrast to this, $\mathrm{tr}\,\tilde{Q}^k$ are constant by virtue by Eq. (15.1.5), i.e. they are first integrals. However a few coefficients in higher powers of z in these series coincide, i.e. a few first integrals are trivial (cf. 13.1.10). It should be established how many coefficients of $\mathrm{tr}\,\tilde{Q}^k$ are distinct from the corresponding coefficients of $\mathrm{tr}(z^m Q^*)^k$.

Note that the matrices $Q_{-\mu}$ and $Q_{-\mu+n}$ have the structures

$$Q_{-\mu} = \overset{(\mu)}{\begin{pmatrix} 0 & & & & & \\ 0 & \ddots & & 0 & & \\ \vdots & & \ddots & & & \\ * & & & \ddots & & \\ \vdots & \ddots & & & \ddots & \\ * & \cdots & * & \cdots & 0 & 0 \end{pmatrix}},$$

$$Q_{-\mu+n} = \begin{pmatrix} * & \cdots & \overset{(n-\mu)}{*} & 0 & \cdots & 0 \\ & \ddots & & \ddots & & \vdots \\ & & \ddots & & \ddots & 0 \\ & & & \ddots & & * \\ & & & & \ddots & \vdots \\ & * & & & & * \end{pmatrix}$$

(μ) is the number of the first non-vanishing element of the first column of $Q_{-\mu}$, $(n - \mu)$ is the number of the last non-vanishing element of the first row of $Q_{-\mu+n}$. Indeed, $Q = ZQ^{(1)}Z^{-1}$, and the expansion into series of the matrix $Q^{(1)}$ starts from z^0. Therefore the expansion of Q_{ij} starts from z^{i-j}. Only those $Q_{ij,-\mu}$ do not vanish for which $\mu \leq i-j$ and those $Q_{ij,-\mu+n}$ for which $\mu - n \leq i - j$.

Let us write $\tilde{Q}$ as $\tilde{Q} = z^m Q^* + q$, where $q = 0(z^{-n})$. We have $\operatorname{tr}\tilde{Q}^k = \operatorname{tr}(z^m Q^*)^k + m\operatorname{tr}(z^m Q^*)^{k-1}q + \ldots$. The first term is constant. From (15.2.5) we find

$$(z^m Q^*)^{k-1} = n^{-(k-1)} z^{m(k-1)} \sum_{\alpha=0}^{n-1} \varepsilon^{-\alpha\mu(k-1)} Q^\alpha$$

(since $Q^\alpha Q^\beta = \delta_{\alpha\beta} Q^\alpha$) and

$$(z^m Q^*)^{k-1} = n^{-(k-1)} z^{m(k-1)} (Q_{-\rho} z^\rho + Q_{-\rho+n} z^{\rho-n} + \ldots)$$

where $0 \leq \rho \leq n - 1$ and $\rho \equiv \mu(k - 1) \pmod{n}$ i.e. $\mu(k-1) = [\mu(k-1)/n]n + \rho$. The highest power of z is $z^{m(k-1)+\mu(k-1)-[\mu(k-1)/n]n} = z^{n(S_1(k-1)-[\mu(k-1)/n])}$. The term $Q_{-\rho}$ is a lower triangular matrix; it should be multiplied by q and the trace taken. The highest term in the expansion of q is $Q_{m+n}z^{-n}$. In the proof of 15.1.9 we have seen that Q_{m+n} is a strictly triangular matrix, thus $\operatorname{tr} Q_{-\rho}Q_{m+n} = 0$. The first non-vanishing term is just the next one: $z^{n(S_1(k-1)-[\mu(k-1)/n]-2)}$.

Thus, the full number of the first integrals is $\sum_{k=2}^{n}(S_1(k-1)-[\mu(k-1)/n]-1)$. We calculate: $\sum_{k=2}^{n}[\mu(k-1)/n] = \sum_{k=2}^{n}\mu(k-1)/n - \sum_{k=2}^{n}\{\mu(k-1)/n\}$. The numbers μ and n are mutually simple, hence $\sum_{k=2}^{n}\{\mu(k-1)/n\} = \sum_{k=2}^{n}(k-1)/n = (n-1)/2$. Now $\sum_{k=2}^{n}[\mu(k-1)/n] = [(\mu-1)(n-1)/2$. In all we have $S_{1n}(n-1)/2 - (\mu-1)(n-1)/2 - (n-1) = (m-1)(n-1)$ first integrals. □

We see that the number of first integrals of the second kind is equal to half of the dimension of the phase space.

15.2.8. Instead of $\operatorname{tr}\tilde{Q}^k$, are coefficients of the characteristic polynomials of the matrix $\tilde{Q}$,

$$f(w, \varsigma) = \det(\tilde{Q} - wI) = \sum_{l=0}^{n} J_l w^l = \sum_{l=0}^{n} \sum_{k=0}^{k_0(l)} J_{lk} w^l \varsigma^k$$

can be considered. Here $k_0(l) = (n-l)S_1 - [\mu(n-l)/n]$. The highest non-trivial coefficient J_{lk} for given l is for $k = (n-l-1)S_1 - [\mu(n-l-1)/n] - 2$. The equation $f(w,\varsigma) = 0$ determines an algebraic function $w(\varsigma)$. It Riemann surface is n-sheeted.

15.2.9. Proposition. The behaviour of $w(\varsigma)$ at infinity is

$$w = n^{-1}\varsigma^{m/n} + 0(\varsigma^{-1})$$

i.e. at infinity there is a branch point of multiplicity $n-1$.

Proof: We have

$$\tilde{Q} = n^{-1} z^m \sum_{\alpha=0}^{n-1} \varepsilon^{-\alpha\mu} Q^\alpha + 0(z^{-n}).$$

The eigenvalues of the first term are $n^{-1}z^m\varepsilon^{-\alpha\mu} = n^{-1}z^m\varepsilon^{\alpha m} = n^{-1}(\varepsilon^\alpha z)^m$ which are branches of the multivalued function $n^{-1}\varsigma^{m/n}$. □

15.2.10. Proposition. The sum of the multiplicities of all the branch points in the finite part of the Riemann surface is $m(n-1)$. The genus of the Riemann surface is $g = (m-1)(n-1)/2$.

The proof is the same as in 13.1.18 and 13.1.19. Calculating the genus one must take into account the branch point of the multiplicity $n-1$ at infinity. □

As in Chap. 13 a spectral projector

$$g(P) = f_w^{-1} \sum_{l=1}^{n} J_l(\varsigma) \sum_{k=0}^{l-1} w^k \tilde{Q}^{l-1-k}$$

is introduced.

15.2.11. Proposition. If $\varsigma \to \infty$ then

$$g(P) = Q(\varsigma) + 0(\varsigma^{-\infty})$$

where $Q(\varsigma)$ is a series in $\varsigma^{-1/n}$ whose branches are Q^α (see 10.2.9). The proof is the same as in 13.1.16. □

15.2.12. Proposition. The divisor of the zeros of the function $g_{ij}(P)$ is the sum of two divisors, $d_i^{(1)} + d_j^{(2)}$. The number of zeros in the finite part of the Riemann surface is

$$\deg(d_i^{(1)}) + \deg(d_j^{(2)}) = m(n-1) + i - j.$$

Proof: The first assertion can be proved as in 13.1.17. Further, at infinity $g_{ij}(P)$ behaves like $Q_{ij}^\alpha \sim z^{i-j}$. The complete divisor has degree zero. Taking into account that the degree of the branch points divisor is $m(n-1)$ (which is the divisor of poles in the finite part of the Riemann surface) and that z^{-1} is the local parameter at infinity we obtain

$$\deg(d_i^{(1)}) + \deg(d_j^{(2)}) - m(n-1) - i + j = 0 \,.\square$$

15.2.13. Proposition.

$$\deg(d_i^{(1)}) = g + i, \deg(d_j^{(2)}) = g + n - 1 - j$$

Proof: The situation here is distinct from that in Chap. 13: rows and columns are not equivalent. This is caused by the fact that one row of the matrix U is distinguished. Temporarily the theory will be extended. Instead of the matrix U we take the following:

$$U^e = \begin{pmatrix} & & & 0 \\ & -U_{n-2} & & \\ & \vdots & & 0 \\ & -U_1 & & \\ -U_0 & -U_1 & \cdots & -U_{n-2} & 0 \end{pmatrix}$$

The algebra $\mathcal{A}$ must be extended by adding the generators $\{v_i\}$. All the constructions including the projectors $g^e(P)$ (the subscript e means "extended") can be repeated. The old g_{ij} can be obtained from the new ones if we put $\{v_i = 0\}$. Proposition 15.2.12 remains valid. If a meromorphic function continuously depends on parameters, then on a submanifold of parameters some zeros can disappear, but not appear. Therefore $\deg(d_i^{(1)}) \leq \deg(d_i^{(1)e})$, $\deg(d_j^{(2)}) \leq \deg(d_j^{(2)e})$. From $\deg(d_i^{(1)e}) + \deg(d_j^{(2)e}) = \deg(d_i^{(1)}) + \deg(d_j^{(2)})$ it follows that $\deg(d_i^{(1)e}) = \deg(d_i^{(1)})$, $\deg(d_j^{(2)e}) = \deg(d_j^{(2)})$. In the extended theory there is a symmetry with respect to the collateral diagonal; $a_{ij}^* = a_{n-1-j,n-1-i}$. This operation has the usual property $(AB)^* = B^*A^*$, and if $Q(U(x))$ is a resolvent then so is $Q^*(U^*(-x))$. Hence g_{ij}^e is equal to $g_{n-1-j,n-1-i}^e$ at another point of the phase space and $\deg(d_i^{(1)}) = \deg(d_{n-1-i}^{(2)})$, whence $\deg(d_i^{(1)}) = \deg(d_{n-1-i}^{(2)}) = [m(n-1) + i - n + 1 + i]/2 = (m-1)(n-1)/2 + i = g + i$.
$\square$

15.2.14. Proposition. The relation between first integrals of the first and of the second kind is (if $Q^{(1)} = \sum b_\alpha Q^\alpha$)

$$F_k^{(1)} = \sum_{\alpha=0}^{n-1} b_\alpha w_{\alpha,k}$$

where $w_{\alpha,k}$ is the coefficient in z^{-k} of the asymptotic expansion of $w(P)$ if $P \to \{\alpha\}$.

Proof: Eq. (15.2.3) yields

$$F_k^{(1)} = \operatorname{tr} \tilde{Q} Q^{(1)}|_k = \operatorname{tr} \tilde{Q} \sum_0^{n-1} b_\alpha Q^\alpha|_k = \operatorname{tr} \sum_{\alpha=0}^{n-1} b_\alpha g(P)\tilde{Q}|_{k, p \to \{\alpha\}}$$

$$= \sum_{\alpha=0}^{n-1} b_\alpha w_{\alpha,k} \,.\square$$

It is not difficult to express $w_{\alpha,k}$ in terms of the coefficients J_{lk} of the characteristic polynomial.

15.3. Hamilton structure. Integration.

15.3.1. We consider a resolvent (see (1.7.2))

$$X = \sum_{(\varepsilon)} \varepsilon T_{(\varepsilon)} = \sum_{k=-\infty}^{\infty} z^{-n(k+1)+1} L_-^{k+1/n} = X_{-1}z + X_{-1+n}z^{-n+1}$$
$$+ X_{-1+2n}z^{-2n+1} + \ldots \in R_-/R_{(-\infty,-n-1)}$$

Here $X_{-1} = L_-^{-1+1/n} = \partial^{-n+1}$ (since the terms with $\partial^{-n-1}, \partial^{-n-2}, \ldots$ are insignificant in $R_-/R_{(-\infty,-n-1)}$). The matrix resolvent related to this resolvent is $Q_X = \sum\limits_{\alpha=0}^{n-1} \varepsilon^\alpha Q^\alpha$ (see 10.2.12). The corresponding first integrals are $F_k^{(1)} = \sum\limits_{\alpha=0}^{n-1} \varepsilon^\alpha F_k^\alpha$. We have for $k = n-1$:

$$\partial F_{n-1}^{(1)} = \operatorname{tr}[A, Q_X] Q_{m+n}|_{n-1} = -\operatorname{tr}[-\partial + U + J, Q_X]Q_{m+n}|_{-1}$$
$$= -\operatorname{tr} U_{H^{(0)}(X_1)} \cdot Q_m$$

(see 10.1.4). Further, $H^{(0)}(X_{-1}) = (L\partial^{-n+1})_+ L - L(\partial^{-n+1}L)_+ = L'$; $U_{H^0(X_1)} = U'$. Thus, $\partial F_{n-1}^{(1)} = \operatorname{tr} U' \delta\Lambda/\delta U$. This implies that the characteristic of the first integral $F_{n-1}^{(1)}$ is U'; then the proposition 11.3.10 yields the following.

15.3.2. Proposition. The Hamiltonian of the system of equations (15.1.5) is

$$\mathcal{H} = F_{n-1}^{(1)} = \sum_{\alpha=0}^{n-1} \varepsilon^\alpha w_{\alpha, n-1}.$$

The Hamiltonian could be expressed in terms of coefficients $\{J_{lk}\}$; we shall not do it.

15.3.3. Proposition. The form $\Omega^{(1)}$ is

$$\Omega^{(1)} = \operatorname{res}_{\varsigma = \infty} nw(P)(\delta g \cdot g)_{jl} \cdot (g_{jl})^{-1} d\varsigma$$

where j and l are arbitrary.

Proof. We start from 10.2.10 whence

$$\Omega^{(1)} = (n+m)^{-1} \operatorname{tr}(\delta\phi B\psi_\varsigma - \phi_\varsigma B\delta\psi)|_{m+2n}$$
$$= -(n+m)^{-1} \operatorname{tr}(\phi B\delta\psi)_\varsigma|_{m+2n} + \delta(\ ).$$

The second term is a complete differential and can be dropped. Then

$$\Omega^{(1)} = \operatorname{tr}(\phi B\delta\psi)|_{m+n} = \sum_{\alpha=0}^{n-1} b_\alpha \sum_{i=0}^{n-1} \phi_{i\alpha} \delta\psi_{\alpha i}|_{m+n} = \sum_{\alpha=0}^{n-1} b_\alpha \sum_{i=0}^{n-1} \frac{\delta(\phi_{j\alpha}\psi_{\alpha i})}{\phi_{j\alpha}} \cdot$$

$$\frac{\phi_{i\alpha}\psi_{\alpha l}}{\psi_{\alpha l}}|_{m+n} + \delta(\ ) = \sum_{\alpha=0}^{n-1} b_\alpha \frac{(\delta Q^\alpha \cdot Q^\alpha)_{jl}}{Q_{jl}^\alpha}|_{m+n}$$

We have agreed to take $b_0 = 1, b_1 = \ldots = b_{n-1} = 0$, then $\Omega^{(1)} = (\delta Q^0 \cdot Q^0)_{jl}/Q_{jl}^0|_{m+n}$. Recall the asymptotics 15.2.9 and 15/2.11: $\Omega^{(1)} = nw(P)(\delta y \cdot y)_{jl}/g_{jl}|_1$ where the subscript 1 denote the coefficient in ς^{-1}. □

There will be an essential distinction from Chap. 13. Now it is significant which the divisors $d_i^{(1)}$ and $d_j^{(2)}$ are taken as coordinates on a level manifold ($J = c$). These can be only $d_0^{(1)}$ or $d_{n-1}^{(2)}$ whose orders are equal to g, the genus of the Riemann surface, i.e. the dimension of the manifold ($J = c$).

15.3.4. Proposition. The form $\Omega^{(1)}$ can also be written as

$$\Omega^{(1)} = n \sum_{P^* \in d_0^{(1)}} w(P^*) \delta\varsigma_{P^*} + \sum a_{lk}(J)\delta J_{lk}.$$

The proof is similar to that in 13.3.3. □

15.3.5. Proposition. The angle variables θ_{lk} related to the action variables J_{lk} are given by the Abel mapping

$$\theta_{lk} = n \sum_{P^* \in d_0^{(1)}} \int_{P_0}^{P^*} w^l \varsigma^k / f_w \, d\varsigma.$$

The proof is as in 13.3.6 (unessential constants a_{lk} are dropped). □

Further, everything from Chaps. 13 and 14 can be literally repeated. Having the Hamiltonian, it is easy to obtain the linear dependence of θ_{lk} on x (and on t for any non-stationary equation of the KdV hierarchy). The return to the original variables $\{u\}$ does not differ in principle from that in Chap. 14.

CHAPTER 16. Matrix Differential Operators Polynomially Depending on a Parameter.

16.1. Resolvent.

16.1.1. In this chapter the results of Chap. 9 will be extended to operators of a more general form:[a]

$$L = -\partial + U = -\partial + U_0 + U_1\varsigma + \ldots + U_m\varsigma^m + A\varsigma^{m+1}$$

where U_i and A are matrices, $A = \text{diag}(a_1, \ldots, a_n)$, $a_i \neq a_j$ if $i \neq j$. The differential algebra $\mathcal{A}$ is generated by elements of matrices $\{U_i\}$.

A resolvent is a series $R = \sum\limits_{i_0}^{\infty} R_i \varsigma^{-i}$ which satisfies the equation

(16.1.2) $$[L, R] = 0.$$

The resolvents form a commutative algebra.

16.1.3. *Proposition.* Resolvents exist that have any set of constants in the diagonals of matrices R_i; these constants uniquely determine a resolvent. The non-diagonal terms do not contain constants.

Proof: Let us write the equation (16.1.2) in detail. Without loss of generality it can be agreed that R_0 is the first non-vanishing coefficient. Equate

---

[a] We follow the articles [76], [77].

to zero terms of all the powers of ς:

$$\varsigma^{m+1} : [A, R_0] = 0$$
$$\varsigma^m : [A, R_1] + [U_m, R_0] = 0$$
$$\dots\dots\dots\dots\dots\dots\dots\dots\dots\dots$$

(16.1.4)
$$\varsigma^1 : [A, R_m] + \dots + [U_1, R_0] = 0$$
$$\varsigma^0 : -R'_0 + [A, R_{m+1}] + \dots + [U_0, R_0] = 0$$
$$\dots\dots\dots\dots\dots\dots\dots\dots\dots\dots$$
$$\varsigma^{-k} : -R'_k + [A, R_{m+k+1}] + \dots + [U_0, R_k] = 0$$
$$\dots\dots\dots\dots\dots\dots\dots\dots\dots\dots$$

The first equation implies that R_0 is a diagonal matrix. The first $m+1$ of the equations determine the non-diagonal parts of the matrices $R_0, R_1, \dots, R_m$ from the non-diagonal parts of these equations. The diagonal parts of these equations automatically vanish. This follows from the lemma below.

16.1.5. Lemma. Let $\tilde{U} = U\varsigma^{-m-1} = A + U_m\varsigma^{-1} + \dots + U_0\varsigma^{-m-1}$, $V = V_0 + V_1\varsigma^{-1} + V_1\varsigma^{-2} + \dots$ and $[U, V] = \sum_l^\infty C_k\varsigma^{-k}, C_l \neq 0$. Then C_l cannot be a diagonal matrix.

Proof: We have $0 = \operatorname{tr}[U^p, V] = p\operatorname{tr} U^{p-1}[U, V]$. The expansion of the right-hand side of this equality in ς^{-1} starts from $p\operatorname{tr} A^{p-1}C_l$ (the coefficient in ς^{-l}). Thus $\operatorname{tr} A^{p-1}C_l = 0$ for any p. If C_l is diagonal this would imply that $C_l = 0$. □

Further, the non-diagonal part of the equation in ς^0 determines the non-diagonal elements of R_{m+1}. The matrix R'_0 is not involved in the non-diagonal part, hence the non-diagonal part of the matrix $[A, R_{m+1}] + \dots + [U_0, R_0]$ vanishes; then the lemma is applicable, and the diagonal part vanishes, too: $R'_0 = 0, R_0 = \text{const}$. We continue further. For any k the equation in ς^{-k} determines the non-diagonal part of R_{m+k+1} and the diagonal part of R_k. The latter requires an integration. Beforehand it is not clear whether this integration leads beyond the algebra $\mathcal{A}$, then we must extend the algebra (see 9A).

It remains to prove that the elements of all the R_k belong to $\mathcal{A}$. Let this be already proved for the non-diagonal parts of all $R_i, i < m + k + 1$ and the diagonal parts of $R_i, i < k$. Then this assertion is true for the non-diagonal part of R_{m+k+1}. Further, R^p is a resolvent, for any p. Therefore $\operatorname{tr} R^p = \text{const}$. The terms with ς^{-k} have the form $p\operatorname{tr} R_0^{p-1}R_k + F_p(U, R)$

where F_p depend only on matrices $\{U_i\}$ and $\{R_i\}$ with $i < k$. Thus, all the diagonal elements of R_k belong to $\mathcal{A}$ (properly speaking this is proved when all the diagonal elements of R_0 are distinct; however this restriction is unimportant because all R_k linearly depend on R_0).

Each diagonal is obtained by an integration which introduces arbitrary constants. It is easy to see that the non-diagonal elements cannot contain constants. □

16.1.6. Remark. We have seen that in the recurrence process of determining R_k the diagonals are determined with a delay on $m+1$ numbers. If a truncated system is considered up to the equation in ς^{-k}, then this determines the diagonal elements of all the matrices up to R_{m+k+1}, and the diagonal ones only up to R_k.

The above recurrence relations connect many coefficients together, instead of two as it was the case in Chap. 9. The situation can be formally improved by introducing the quantities

$$S_p = \sum_{k=0}^{m} R_{p+k} \varsigma^{-k-1}.$$

16.1.7. Proposition. Eq. (16.1.2) is equivalent to a set of recurrence relations

$$([L, S_p]\varsigma^{m+1})_{(0,m)} + ([L, S_{p+m+1}])_{(0,m)} = 0$$

where the subscript $(0, m)$ means cutting out a segment of the series from ς^0 to ς^m. We have $m+1$ chains of recurrence relations connecting S_p with S_{p+m+1}.

Proof: For a number p we can write R as $\sum_{l=-\infty}^{\infty} S_{p+(m+1)l} \varsigma^{-p-(m+1)l+1}$, the series being one-way infinite. Then

$$\sum_{l=-\infty}^{\infty} [L, S_{p+(m+1)l}] \varsigma^{-(m+1)l} = 0.$$

Every commutator contains terms from ς^{-m-1} to ς^m. Therefore any power ς^k is involved in two neighbouring terms, the sum of which yields the required relation. □

16.2. Another hierarchy of equations.

16.2.1. We construct the equations

(16.2.2) (a) $\dot{U} = ([L, S_p]\varsigma^{m+1})_{(0,m)}$ and (b) $\dot{U} = ([L, S_{p+m+1}])_{(0,m)}$

(owing to 16.1.7 they are equivalent). In more detail,

(16.2.3)
(a) $\dot{U}_j = \sum_{s=m-j}^{m} [U_{j+s-m}, R_{p+s}] - R'_{p+m-j}$,

(b) $\dot{U}_j = \sum_{s=m+1}^{2m+1-j} [U_{j+s-m}, R_{p+s}]$.

16.2.4. Let us show the relation of these equations to the so-called Zakharov-Shabat scheme ([1] p. 222). In this book the zero curvature equations

$$\dot{U} + V' = [U, V]$$

are considered. The solutions in the form

$$U = U_0 + \ldots + U_{m+1}\varsigma^{m+1}, V = V_0 + \ldots + V_{m+p}\varsigma^{m+p}$$

must satisfy the equation identically with respect to ς. This yields a system of equations consisting of $p + 2m + 2$ equations with $p + 2m + 3$ unknown matrices $\{U_i, V_i\}$. One of the matrices can be chosen arbitrarily; let $U_{m+1} = A$. Now the number of equations is equal to the number of the unknown matrices. Is this system closed? The answer is negative; we shall see that there is a possibility of choosing arbitrary $m + 1$ diagonal matrices. If the series for U and V are substituted into the zero curvature equation then the equations for higher powers, from ς^{m+1} to ς^{2m+p+1}, coincide with the first $p + m + 1$ of the equations (16.1.4), where $V_{m+p-i} = R_i$. The rest of the equations, for the powers from ς^0 to ς^m coincide with the system (16.2.3a). This system involves $U_0, \ldots, U_m; R_0, \ldots, R_{m+p}$. We have already seen (remark 16.1.6) that the first $p + m + 1$ equations (16.1.4) do not express the matrices $R_0, \ldots, R_{m+p}$ uniquely in terms of $U_0, \ldots, U_m$; it remains free to choose $m + 1$ diagonal matrices. Thus, the following proposition.

16.2.5. Proposition. *The Zakharov-Shabat system is underdetermined; there remains the freedom to choose $m + 1$ arbitrary diagonal matrices.*

For the example in Chap. 9 where $m = 0$ we could choose 1 diagonal matrix. Therefore, we put the diagonal of the matrix U to be zero. In the general case we do away with this freedom assuming that the matrices $R_0, \ldots, R_{m+p}$ on the whole, including diagonals, are determined from the equation for resolvents, i.e. we make use of additional equations not involved in the Zakharov-Shabat system.

(Another explanation: the situation is similar to the following. Let the matrix X be the solution of an equation $[B, X] = 0$. The number of equations is here equal to that of the unknown functions. However, the system is underdetermined just on a diagonal matrix. This is especially clear if we use a coordinate system where B is a diagonal matrix; then X is any diagonal matrix in a generic case otherwise there is still more freedom.)

And another remark: we put arbitrarily $U_{m+1} = A$ in the Zakharov-Shabat equation. Such a solution can be obtained from any solution by a gauge transformation

$$U \mapsto TUT^{-1} + T'T^{-1}, V \mapsto TVT^{-1} - \dot{T}T^{-1}$$

preserving the equation. After that a possibility remains to perform a gauge transformation with an arbitrary diagonal matrix T. With the aid of this transformation it is possible, e.g. to annul the diagonal of the matrix U_0. After that m diagonal matrices remain arbitrary.

16.3. Two Hamiltonian structures.

16.3.1. The recurrence relations 16.1.7 and two forms of the equations (16.2.2(a) and (b)) prompt us how to construct Hamiltonian structures.

As usual, the space of functionals Ω^0 is $\tilde{\mathcal{A}} = \{\int f dx, f \in \mathcal{A}\}$. The vector fields which form a Lie algebra $\mathfrak{E}$ acting on $\tilde{\mathcal{A}}$ to the left are $\partial_M, M = M_0 + M_1\varsigma + \ldots + M_m\varsigma^m$; $\partial_M \int f dx = \int \text{tr} \sum_{i,j} M_i^{(j)} \partial f / \partial U_i^{(j)} dx$, where $(\partial f / \partial U_i^{(j)})_{\alpha\beta} = \partial f / \partial (U_i^{(j)})_{\beta\alpha}$. It can also be written $\partial_M \int f dx = \int \text{tr res} \sum M(\partial f / \partial U^{(j)}) dx$, where $\partial f / \partial U^{(j)} = \sum_0^m \partial f / \partial U_i^{(j)} \cdot \varsigma^{-i-1}$, res denotes the coefficient in ς^{-1}.

The dual to $\mathfrak{E}$ space $\Omega^{(1)}$ comprises sums $X = \sum_0^m X_i\varsigma^{-i-1}$, elements of all the matrices belonging to $\mathcal{A}$. The coupling is defined as $\langle M, X \rangle = \int \text{tr res } MX dx$.

The family of Hamiltonian mappings is given by

(16.3.2) $$X \mapsto H^{(z)}(X) = ([L,X](\varsigma^{m+1}+z))_{(0,m)}$$

z being a parameter and the superscript (z) being usually dropped. Special cases of this mapping are

$$H^{(0)}(X) = ([L,X]\varsigma^{m+1})_{(0,m)} \quad \text{and} \quad H^{(\infty)}(X) = ([L,X])_{(0,m)}.$$

As usual, it has to be proved that the elements $\partial_{H(X)} \in \mathfrak{E}$ form a Lie subalgebra in $\mathfrak{E}$. Then we define a 2-form

(16.3.3) $$\omega(\partial_{H(X)}, \partial_{H(Y)}) = \langle H(X), Y \rangle = \int \operatorname{tr}\operatorname{res}(\varsigma^{m+1}+z)[L,X]Y\,dx$$

(depending on the parameter z) and prove that it is closed.

We denote

(16.3.4) $$X, Y \in \Omega^{(1)}, [X,Y]_z = ([X,Y](\varsigma^{m+1}+z))_{(-1,-m-1)}$$

(the subscript $(-1,-m-1)$ denotes a segment from ς^{-1} to ς^{-m-1}).

16.3.5. Lemma. Expression (16.3.4) has all the properties of a commutator: turning $\Omega^{(1)}$ for every fixed z to a Lie algebra.

Proof: The left-hand side of the Jacobi identity is

$$\begin{aligned}
[[X,Y]_z, Z]_z + \text{c.p.} =\, & z^2[[X,Y]_{(-1,-m-1)}, Z]_{(-1,-m-1)} \\
& + ([([X,Y]\varsigma^{m+1})_{(-1,-m-1)}, Z]\varsigma^{m+1})_{(-1,-m-1)} \\
& + z([[X,Y]_{(-1,-m-1)}, Z]\varsigma^{m+1})_{(-1,-m-1)} \\
& + z[([X,Y]\varsigma^{m+1})_{(-1,-m-1)}, Z]_{(-1,-m-1)} + \text{c.p.}
\end{aligned}$$

It is easy to see that in the first two terms the inner subscripts $(-1,-m-1)$ can be dropped. The remaining two terms are

$$\begin{aligned}
& z([[X,Y]_{(-1,-m-1)} \cdot \varsigma^{m+1} + ([X,Y]\varsigma^{m+1})_{(-1,-m-1)}, Z])_{(-1,-m-1)} \\
& = z([[X,Y]\varsigma^{m+1}, Z])_{(-1,-m-1)}.
\end{aligned}$$

In all we have

$$((z^2 + z\varsigma^{m+1} + \varsigma^{2m+2})[[X,Y],Z]_{(-1,-m-1)} + \text{c.p.} = 0 \,.\square$$

16.3.6. Proposition. The relation

$$[\partial_{H(X)}, \partial_{H(Y)}] = \partial_{H([X,Y]_z + \partial_{H(X)}Y - \partial_{H(Y)}X)}$$

holds.

Proof:

$$\begin{aligned}
[\partial_{H(X)}, \partial_{H(Y)}] &= \partial_{\partial_{H(X)}H(Y) - \partial_{H(Y)}H(X)} \\
&= \partial_{\partial_{H(X)}([L,Y](\varsigma^{m+1}+z)_{(0,m)}} - (X \leftrightarrow Y) \\
&= \partial_{([H(X),Y](\varsigma^{m+1}+z))_{(0,m)} + H(\partial_{H(X)}Y)} - (X \leftrightarrow Y).
\end{aligned}$$

Now

$$([H(X),Y](\varsigma^{m+1}+z))_{(0,m)} - (X \leftrightarrow Y) = ([([L,X](\varsigma^{m+1}+z))_{(0,m)}, Y]$$
$$(\varsigma^{m+1}+z))_{(0,m)} - (X \leftrightarrow Y) = z^2([[L,X],Y])_{(0,m)}$$
$$+ ([[L,X]\varsigma^{m+1},Y]\varsigma^{m+1})_{(0,m)} + z([[L,X]_{(0,m)},Y]\varsigma^{m+1}$$
$$+ [([L,X]\varsigma^{m+1})_{(0,m)}, Y])_{(0,m)}.$$

The coefficient of z:

$$([[L,X]_{(0,m)}\varsigma^{m+1}, Y] + [([L,X]\varsigma^{m+1})_{(0,m)}, Y])_{(0,m)}$$
$$= [[L,X]\varsigma^{m+1}, Y]_{(0,m)}.$$

In all:

$$([H(X),Y](\varsigma^{m+1}+z))_{(0,m)} - (X \leftrightarrow Y)$$
$$= ((z^2 + z\varsigma^{m+1} + \varsigma^{2m+2})[[L,X],Y])_{(0,m)} - (X \leftrightarrow Y)$$
$$= ((z^2 + z\varsigma^{m+1} + \varsigma^{2m+2})[L,[X,Y]])_{(0,m)}.$$

On the other hand

$$\begin{aligned}
H([X,Y]_z) &= ([L,[X,Y]_z](\varsigma^{m+1}+z))_{(0,m)} \\
&= ([L,([X,Y](\varsigma^{m+1}+z))_{(-1,-m-1)}](\varsigma^{m+1}+z))_{(0,m)} \\
&= ([L,[X,Y]]\varsigma^{2m+2} + z^2[L,[X,Y]] + z([L,[X,Y]_{(-1,-m-1)} \\
&\quad \varsigma^{m+1} + ([X,Y]\varsigma^{m+1})_{(-1,-m-1)}])_{(0,m)} \\
&= ([L,[X,Y]] \cdot (\varsigma^{2m+2} + z^2 + z\varsigma^{m+1}))_{(0,m)}
\end{aligned}$$

which is just the same expression. □

16.3.7. Proposition. The form (16.3.3) is closed.

Proof: For arbitrary X, Y and Z we have

$$d\omega(\partial_{H(X)}, \partial_{H(Y)}, \partial_{H(Z)}) = \partial_{H(X)}\omega(\partial_{H(Y)}, \partial_{H(Z)})$$
$$- \omega([\partial_{H(X)}, \partial_{H(Y)}], \partial_{H(Z)}) + \text{c.p.} = \text{tr res} \int \{(\varsigma^{m+1} + z)$$
$$([H(X), Y]Z + [L, \partial_{H(X)}Y]Z + [L, Y]\partial_{H(X)}Z)$$
$$+ (\varsigma^{m+1} + z)H(Z)[X, Y] + H(Z)(\partial_{H(X)}Y - \partial_{H(Y)}X)\}$$
$$dx + \text{c.p.} = 2\,\text{tr res} \int (\varsigma^{m+1} + z)[H(Z), X]Y\,dx + \text{c.p.}$$
$$= 2\text{tr res} \int (\varsigma^{m+1} + z)[((\varsigma^{m+1} + z)[L, Z])_{(0,m)}, X]Y\,dx + \text{c.p.}$$
$$= 2\,\text{tr res} \int (\varsigma^{2m+2} + z^2 + z\varsigma^{m+1})[[L, Z], X]Y\,dx + \text{c.p.}$$
$$= 2\,\text{tr res} \int (\varsigma^{2m+2} + z^2 + z\varsigma^{m+1})([-Z', X]Y + [[U, Z], X]Y)dx$$
$$+ \text{c.p.} = 2\,\text{tr res} \int (\varsigma^{2m+2} + z^2 + z\varsigma^{m+1})(-([X, Y]Z)'$$
$$- U([[X, Y], Z] + [[Y, Z], X] + [Z, X], Y]))dx = 0. \square$$

Thus, the mapping $X \mapsto \partial_{H(X)}$ turned out to be Hamiltonian. As usual, a relation between the functionals $\tilde{f} \in \tilde{\mathcal{A}}$ and the vector fields is established: $\tilde{f} \mapsto \partial_{H(d\tilde{f})}$. It is easy to verify that $d\tilde{f} = \sum_0^m \delta f/\delta U_i \varsigma^{-i-1} = \delta f/\delta U$. Therefore this relation is given by the formula

(16.3.8) $$\tilde{f} \mapsto \partial_{H(\delta f/\delta U)}$$

and the Poisson bracket is

$$\{\tilde{f}, \tilde{g}\} = \partial_{H(\delta f/\delta U)}\tilde{g} = \int \text{tr res}(\varsigma^{m+1} + \varsigma)[L, \delta f/\delta U]\delta g/\delta U\,dx.$$

If a functional $\tilde{f}$ is taken as a Hamiltonian then the corresponding Hamilton equation is

$$\dot{U} = H(\delta f/\delta U)$$

In detail,

(16.3.9)
$$\dot{U}_j = \sum_{s=m-j}^{m} [U_{j+s-m}, \delta f/\delta U_s] - (\delta f/\delta U_{m-j})' + z \sum_{s=0}^{m-j} [U_{j+s+1}, \delta f/\delta U_s], j = 0, \ldots, m$$

In the two limiting cases where $z = 0$ and $z = \infty$, we have

(16.3.10)
$$\text{a)} \quad \dot{U}_j = \sum_{s=m-j}^{m} [U_{j+s-m}, \delta f/\delta U_s] - (\delta f/\delta U_{m-j})'$$
$$\text{b)} \quad \dot{U}_j = \sum_{s=0}^{m-j} [U_{j+s+1}, \delta f/\delta U_s].$$

16.4. Poisson-Lie-Berezin-Kirillov-Costant bracket, central extension, and Kac-Moody algebras.

16.4.1. This section does not contain new results but only an interpretation of those obtained earlier.

First we consider the limiting case $z = \infty$ and show that the symplectic form (16.3.3) is the natural form on the orbit of the coadjoint representation (Sec. 2.4). The Lie algebra here is $\Omega^{(1)}$ with commutator $[X, Y]_\infty = [X, Y]_{(-1, -m-1)}$. The coalgebra is the space R_+ of the series $M = \sum_0^m M_i \varsigma^i$, with the coupling as above. The coadjoint representation is given by the formula

$$\text{ad}^*(X)M = [M, X]_{(0, m)}.$$

In particular if $M = U$ then $\text{ad}^*(X)U = [U, X]_{(0, m)}$. Instead of this, the equivalent formula $\text{ad}^*(X)U = [-\partial + U, X]_{(0, m)} = [L, X]_{(0, m)}$ fits, which is exactly $H^{(\infty)}(X)$. Thus, $\omega(\partial_{H(X)}, \partial_{H(Y)}) = \langle \text{ad}^*(X)U, Y \rangle$ i.e. (2.4.5) is obtained.

16.4.2. Now we turn to another limiting case $z = 0$. It is natural to consider $\Omega^{(1)}$ as a Lie algebra with respect to the commutator

$$[X, Y]_0 = ([X, Y]\varsigma^{m+1})_{(-1, -m-1)}.$$

The coalgebra is R_+ with the same coupling. The coadjoint representation is

$$\mathrm{ad}^*(X)M = ([M,X]\varsigma^{m+1})_{(0,m)}.$$

If $M = U$ this will be $\mathrm{ad}^*(X)U = ([U,X]\varsigma^{m+1})_{(0,m)}$. However, this expression does not coincide with $([L,X]\varsigma^{m+1})_{(0,m)} = -X'\varsigma^{m+1} + ([U,X]\varsigma^{m+1})_{(0,m)}$ i.e. with $H(X)$ for $z = 0$. The difference is in the one term $X'\varsigma^{m+1}$. If we want to maintain the group interpretation of the symplectic form and of the equation we have to modify the definition of the algebra $\Omega^{(1)}$. Namely, we define it as the central extension with the aid of the cocycle $F(X,Y) = \int \mathrm{tr}\,\mathrm{res}\,\varsigma^{m+1}XY'dx$ (cf. 9.2.8). The fact that this is a cocycle i.e. $F([X,Y],Z) + \mathrm{c.p.} = 0$ can be easily checked. The elements of the extended algebra are pairs $(\tilde{f},X)$ where $\tilde{f} \in \tilde{A}$. The commutator is $[(\tilde{f},X),(\tilde{g},Y)] = (F(X,Y),[X,Y]_0)$. The dual space is the set of pairs (λ, M), where λ is a number and $M \in R_+$. The coupling is

$$\langle(\lambda,M),(\tilde{f},x)\rangle = \lambda\tilde{f} + \mathrm{tr}\,\mathrm{res}\int MXdx \in \tilde{A}.$$

The coadjoint representation is

$$\mathrm{ad}^*(\tilde{f},X)(\lambda,M) = (0,((-\lambda X' + [M,X])\varsigma^{m+1})_{(0,m)})$$

In particular, for $(\lambda, M) = (1, U)$ we get $(0,((-X' + [U,X])\varsigma^{m+1})_{(0,m)})$. The symplectic form defined on these vectors has the required form $\int \mathrm{tr}\,\mathrm{res}(-X' + [U,X])Y\varsigma^{m+1}dx$.

16.4.3. The construction admits of the following generalization. As far as the Lie algebra of all the matrices $n \times n$ is considered, we can restrict ourself with a matrix subalgebra $\mathfrak{G}$. The series $\sum X_i \varsigma^{-i}$ form a new algebra which is infinite-dimensional and is called the loop algebra. The central extension of this algebra with a cocycle is a Kac-Moody algebra. The whole theory can be built on these algebras. For more detail we refer to the works by Reymann and Semenov-Tjan-Shansky [78], Adler and van Moerbeke [79] Lesnov, Saveljev and Smirnov [107], Kac [80], Drinfeld and Sokolov [60,61], Flashka, Newell and Ratiu [81].

16.5. Hamiltonians in involution.

16.5.1. The recurrence relation 16.1.7 can be written in a simpler way if a generator for $S_p \in \Omega^{(1)}$ is introduced:

$$S = \sum_{k=0}^{\infty} z^{-k} S_{p_0+k(m+1)}$$

The Hamilton mapping $H^{(z)}(X)$ can be extended to the series $\sum z^{-k} X_k$, $X_k \in \Omega^{(1)}$. Then 16.1.7 is equivalent to the fact that S belongs to the kernel of the mapping $H^{(z)}$.

If we put $z = \varsigma^{m+1}$ then the expression S reduces to the resolvent R. In S there are two parameters instead of one in R; we call S the polarization of the resolvent.

Let S be a polarization of a resolvent and

(16.5.2) $$\mathcal{H}_p = \int \operatorname{tr} \operatorname{res} U_\varsigma S_p \, dx / (-p+1), U_\varsigma = \partial U / \partial \varsigma$$

be taken as the Hamiltonian.

16.5.3. Proposition. If the resolvent has constants only in R_0 then $\delta \mathcal{H}_p / \delta U = S_{p-1}$.

Proof: Similarly to that in Chap. 9, let us obtain the resolvent by dressing the constant matrix R_0.

We look for solutions ϕ and Λ of the equation

(16.5.4) $$-\phi' + U\phi = \phi \Lambda$$

in the form of series $\phi = \sum_0^\infty \phi_i \varsigma^{-i}, \Lambda = \sum_{-m-1}^\infty \Lambda_i \varsigma^{-i}; (\phi_i)_{\alpha,\beta}, (\Lambda_i)_{\alpha\beta} \in \mathcal{A}$, the matrices Λ_i being diagonal. The equation is equivalent to the recurrence relations

$$A\phi_r - \phi_r \Lambda_{-m-1} + \phi_0 \Lambda_{r-m-1} + F_r(\phi_0, \ldots, \phi_{r-1}, \Lambda_{-m-1}, \ldots, \Lambda_{r-m-2}) = 0$$

where F_r are some functions of the written arguments. The first of these formulas is $A\phi_0 - \phi_0 \Lambda_{-m-1} = 0$. Let $\Lambda_{-m-1} = A, \phi_0 = 1$ then

$$[A, \phi_r] + \Lambda_{r-m+1} + F_r = 0.$$

The non-diagonal part of this equality determines the non-diagonal elements of ϕ_r, the diagonal ones being put to be zero. The diagonal part of this equality determines Λ_{r-m+1}. The inverse matrix $\psi = \phi^{-1}$ satisfies the equation

(16.5.5) $$\psi' + \psi U = \Lambda \psi$$

16.5.6. Lemma. The equality

$$-\partial + U = \phi(-\partial + \Lambda)\phi^{-1}$$

holds.

Proof: This equality is evidently equivalent to (16.5.4), cd. 9.3.1. □

16.5.7. Lemma. A resolvent with constants only in R_0 is

$$R = \phi R_0 \phi^{-1}$$

Proof: See 9.3.4. □

Now the proof of the proposition 16.5.3 can be obtained as in 9.3.7. More exactly, the relation

(16.5.8) $\qquad \delta \operatorname{tr} U_\varsigma R = \operatorname{tr}(\delta U \cdot R)_\varsigma + \partial \operatorname{tr}(-\delta\phi \cdot R_0 \cdot \psi_\varsigma + \phi_\varsigma R_0 \delta\psi)$

holds. The coefficient in ς^{-p} yields

$$\delta \operatorname{tr} \sum_{k=0}^{m+1} (U_\varsigma)_k R_{p+k} = (-p+1) \sum_{k=0}^{m} \operatorname{tr} \delta U_k R_{p+k-1} + \partial(\ )$$

i.e.

$$\delta \operatorname{tr} \operatorname{res}(U_\varsigma S_p) = (-p+1) \cdot \operatorname{tr} \operatorname{res}(\delta U \cdot S_{p-1}) + \partial(\ ).$$

which yields the required assertion. □

16.5.9. Proposition. Any two Hamiltonians $\mathcal{H}_p^{(1)}$ and $\mathcal{H}_q^{(2)}$ are in involution (the superscripts (1) and (2) mean that the Hamiltonians can be constructed with the aid of distinct resolvents i.e. starting with distinct diagonal matrices).

The proof is similar to that in 9.3.13 using 16.5.3 and 16.1.7. □

16.5.10. Proposition. The equation (16.2.3a) is related to the Hamiltonian $\mathcal{H}_{p+1}$ with respect to the form $\omega^{(0)}$, and the equation (16.2.3b) is related to the Hamiltonian $\mathcal{H}_{p+m+1}$ with respect to the form $\omega^{(\infty)}$.

Proof. See (16.3.10) and 16.5.3. □

Thus, a Hamiltonian system (16.2.2) and an involutive system of first integrals are constructed.

16.6. Still more general equation.

16.6.1. Take now the operator

$$L = -\varsigma^r \cdot \partial + U_0 + U_1 \varsigma + \ldots + U_m \varsigma^m + A \varsigma^{m+1}$$

where r is an integer, $0 \leq r \leq m$. The operator L from 16.1.1 is a special case for $r = 0$. The resolvent can be defined as above, and it maintains all its properties. In the same manner S_p is introduced, and the proposition 16.1.7 remains valid. However, instead of (16.2.3) we obtain from (16.2.2) the other equations:

(16.6.2)

(a) $\dot{U}_j = \sum_{s=m-j}^{m} [U_{j+s-m}, R_{p+s}] - R'_{p+m-j+r}$,

(b) $\dot{U}_j = \sum_{s=m+1}^{2m+1-j} [U_{j+s-m}, R_{p+s}] - R'_{p+m-j+r} \cdot \eta_{-j+r-1}$,

$j = 0, \ldots, m$

where $\eta_k = 0$ if $k < 0$ and $\eta_k = 1$ if $k \geq 0$.

The same formula (16.3.2) remains valid for the Hamiltonian mapping $H^{(z)}$, and the formulas for the symplectic form and Poisson bracket also remain unchanged. The Hamiltonians are (16.5.2). If f is a Hamiltonian then the corresponding Hamilton equations are

(16.6.3)

(a) $\dot{U}_j = \sum_{s=m-j}^{m} [U_{j+s-m}, \delta f/\delta U_s] - (\delta f/\delta U_{m-j+r})'$

(b) $\dot{U}_j = \sum_{s=0}^{m-j} [U_{j+s+1}, \delta f/\delta U_s] - (\delta f/\delta U_{-j-1+r})'$

We shall see that this generalization is necessary if we want to include in the consideration the so-called chiral field equation.

16.6.4. Proposition. The equations (16.6.2) are of the Hamilton type with Hamiltonians $\mathcal{H}_{p+1}$ and $\mathcal{H}_{p+m+2}$, respectively.

The proof is an obvious corollary of the proposition 16.5.3. □

16.7.   Principal chiral field equation.

16.7.1. Let us consider the zero curvature equation

$$\dot{U} + V' = [U, V]$$

where

$$U = \varsigma^{-1} U_0 + U_1 + \varsigma A, \quad V = -\varsigma^{-1} U_0 + U_1 + \varsigma A.$$

The matrices U_0, U_1 and A have the same meanings as above. This equation is equivalent to the system

(16.7.2) $\quad \dot{U}_0 = U'_0 + 2[U_0, U_1], \dot{U}_1 = -U'_1 - 2[A, U_0]$.

Let us show a relation to the so-called chiral field equation

(16.7.3) $$M_\eta + M_\xi = 0, M_\eta - N_\xi = [N, M]$$

At first we pass, in (16.7.2), to the cone variables $\xi = -x + t, \eta = -x - t$:

(16.7.4) $$U_{0\xi} = [U_0, U_1], U_{1\eta} = [A, U_0].$$

If a solution of (16.7.4) is known then we find h from

$$U_1 + A = h^{-1} h_\xi, U_0 = h^{-1} h_\eta$$

(Eq. (16.7.4) just guarantees the compatibility of these equations.) Then we put $M = 2hAh^{-1}$ and $N = 2hU_0 h^{-1}$. The matrices M and N satisfy Eq. (16.7.3). Thus, a solution of (16.7.2) generates a solution of (16.7.3) having the property that the matrix M has a constant spectrum. It is easy to see that the converse is also true.

The systems turned out to be not exactly equivalent: the fact that A is constant is equivalent to that M has a constant spectrum. The equivalence will be complete if we consider A as depending on $\xi = -x + t$ i.e. $\dot{A} + A' = 0$. This will not violate the equations. We will not go deeper into these subtleties.

16.7.5. Remark. The system can be reduced assuming that the matrices U_0 and U_1 belong to a subalgebra of the Lie algebra $\mathfrak{Gl}(n)$.

16.7.6. Now return to the Hamilton equations of Sec. 16.6 and let $m = r = 1$. Equations (16.6.2b) with $p = 0$ become

$$\dot{U}_0 = [U_1, R_2] + [A, R_3] - R_2', \dot{U}_1 = [A, R_2]$$

Let R be a resolvent $R^{(-2A)}$ which starts with $R_0 = -2A$. The equation $[L, R^{(-2A)}] = 0$ yields the recurrence equation

$$[A, R_{i+1}] + [U_1, R_i] - R_i' + [U_0, R_{i-1}] = 0.$$

We find in succession:

$$R_0 = -2A, R_1 = -2U_1, [A, R_2] = -2U_1' - 2[A, U_0],$$
$$[A, R_3] + [U_1, R_2] - R_2' - 2[U_0, U_1] = 0.$$

Substituting this into the above system we have

(16.7.7) $\quad \dot{U}_0 = 2[U_0, U_1], \dot{U}_1 = -2U_1' - 2[A, U_0].$

This system is related to the Hamiltonian $\mathcal{H}_3^{(-2A)}$.

Let us compare this system with (16.7.2). They do not coincide completely: the right-hand sides differs by (U_0', U_1') which corresponds to an additional vector field ∂. This means that one system can be obtained from another by transition to a moving frame: $x \mapsto x + t$. However, we shall not restrict ourself by this remark, but write the Hamiltonian of this additional vector field ∂ and hence the Hamiltonian of the equation (16.7.2).

First we note that using the Hamiltonian mapping $H^{(\infty)}$, the system can be reduced to the submanifold $\text{diag}U_1 = 0$ since $\text{diag}\dot{U}_1 = \text{diag}[A, R_2] = 0$. Now let $U_1 = [A, \varphi]$.

16.7.8. Proposition. The Hamiltonian of the vector field ∂ with respect to the symplectic form $\omega^{(\infty)}$ is

$$\hat{\mathcal{H}} = \text{tr} \int \left\{ \frac{1}{3}[[A, \varphi], \varphi']\varphi + \varphi' U_0 + \frac{1}{2}\varphi'^2 \right\} dx$$

Proof: Let us find the connection between the variational derivatives $\delta f/\delta U_1$ and $\delta f/\delta \varphi$ for any $\tilde{f} = \int f dx$. We have

$$\delta \tilde{f} = \text{tr} \int \left(\frac{\delta f}{\delta U_0} \delta U_0 + \frac{\delta f}{\delta U_1} \delta U_1 \right) dx = \text{tr} \int \left(\frac{\delta f}{\delta U_0} \delta U_0 + \frac{\delta f}{\delta U_1}[A, \delta \varphi] \right) dx$$

$$= \text{tr} \int \left(\frac{\delta f}{\delta U_0} \delta U_0 + \left[\frac{\delta f}{\delta U_1}, A \right] \delta \varphi \right) dx$$

whence $\delta f/\delta \varphi = -[A, \delta f/\delta U_1]$. Now

$$\delta \hat{\mathcal{H}} = \text{tr} \int \left\{ \frac{1}{3}[[A, \delta\varphi], \varphi']\varphi + \frac{1}{3}[[A, \varphi], \delta\varphi']\varphi \right.$$
$$\left. + \frac{1}{3}[[A, \varphi], \varphi']\delta\varphi + \delta\varphi' U_0 + \varphi' \delta U_0 + \varphi' \delta\varphi' \right\} dx$$
$$= \text{tr} \int \left\{ [[A, \varphi], \varphi']\delta\varphi - U_0'\delta\varphi - \varphi'' \delta\varphi + \varphi' \delta U_0 \right\} dx$$

whence $\delta \hat{\mathcal{H}}/\delta \varphi = [[A, \varphi], \varphi'] - U_0' - \varphi''$ and $\delta \hat{\mathcal{H}}/\delta U_0 = \varphi'$ for $r = m = 1$. Equation (16.6.3b) becomes

$$\dot{U}_0 = [U_1, \delta \hat{\mathcal{H}}/\delta U_0] + [A, \delta \mathcal{H}/\delta U_1] - (\delta \hat{\mathcal{H}}/\delta U_0)', \dot{U}_1 = [A, \delta \hat{\mathcal{H}}/\delta U_0]$$

taking into account that $[A, \delta\mathcal{H}/\delta U_1] = -\delta\hat{\mathcal{H}}/\delta\varphi$ and the above relations for the variational derivatives, we get

$$\dot{U}_0 = [U_1, \varphi'] - [[A, \varphi], \varphi'] + U_0' + \varphi'' - \varphi'' = U_0'$$
$$\dot{U}_1 = [A, \varphi'] = U_1'$$

as required. □

16.7.9. Corollary. For any $\tilde{f} \in \tilde{\mathcal{A}}$

$$\{\hat{\mathcal{H}}, \tilde{f}\}^{(\infty)} = 0.$$

Proof:

$$\{\hat{\mathcal{H}}, \tilde{f}\}^{(\infty)} = \partial \tilde{f} = \int \partial f \, dx = \int f' \, dx = 0. \square$$

16.7.10. Corollary. All the $\mathcal{H}_p^{(B)}$, for any p and matrices B are first integrals in involution of Eq. (16.7.2).

Proof. This follows from 16.5.9, 16.7.9 and the fact that the Hamiltonian of the system is $\mathcal{H}_3^{(-2A)} + \hat{\mathcal{H}}$. □

16.7.11. Proposition. The Hamiltonian can also be written as

$$\mathcal{H} = -\mathrm{tr} \int U_0^2 \, dx - \hat{\mathcal{H}}$$

Proof:

$$\delta \, \mathrm{tr} \int U_0^2 \, dx = 2\mathrm{tr} \int U_0 \delta U_0 \, dx \,, \quad \delta \, \mathrm{tr} \int U_0^2 \, dx / \delta U_0 = 2U_0 \,.$$

Therefore the system related to the above Hamiltonian is

$$\dot{U}_0 = -[U_1, 2U_0] + 2U_0' - U_0'$$
$$\dot{U}_1 = -[A, 2U_0] - U_1'$$

i.e. (16.7.2). □

CHAPTER 17. Multi-time Lagrangian and Hamiltonian Formalism.

17.1. Introduction.

17.1.1. In the above examples the space coordinate and time played quite distinct roles: the space coordinate was merely an index numbering freedom degrees, and the time coordinate was the usual physical time in which the system evolves. Such a theory is satisfactory unless we turn our attention to relativistic invariant equations, e.g. chiral fields, sine-Gordon, and others. Also, considering the KP-hierarchy (5.2.4) for arbitrary m and n, the variables x_m and x_n are quite equal in rights and there is no reason to prefer one to the other by choosing it as time. In such cases a new field theory is useful which involves many time variables. This will be done in this chapter. The above symplectic form and the Hamiltonian (if one of the variables is fixed as time) will now be merely one of the components of a vector or tensor quantity (e.g. the old Hamiltonian is one of the components of the energy-momentum tensor; however a new Hamiltonian which is a scalar will be introduced called the field Hamiltonian). One subtle problem is the relation between first integrals and vector fields which will enable us to construct symmetries if a first integral is known, and another is the problem of Poisson brackets. Here the situation is more complicated than in the single-time theory.

The multi-time formalism is of interest even for our old examples (e.g.

KdV) where the variables are involved in a distinctly asymmetric way. The first component of the compound symplectic form appearing in this theory coincides with the form introduced earlier and the second one is in a direct relation to the stationary KdV equation. Thus, the theories of stationary and of non-stationary equations turn out to be more closely connected with each other. In particular a new interpretation can be given to the result by Bogoyavlenskii and Novikov [67] and Gelfand and Dickey [25] about the relation betwen first integrals of stationary and of non-stationary equations and the associated vector field.

17.1.2. The multi-time canonical equations are written in a book by de Donder [84]. Let, for simplicity, a Lagrangian Λ depend on independent variables $t_1, \ldots, t_n$, and some other variables $q_1, \ldots, q_m$ and their first derivatives $q_{i,j} = \partial q_i / \partial t_j$. Then the variational equation $\{\delta\Lambda/\delta q_i = 0\}$ can be reduced to the canonical Hamiltonian form

$$(17.1.3) \qquad \partial q_k / \partial t_j = \partial \mathcal{H} / \partial p_{kj}, \sum_{j=1}^{n} \partial p_{kj}/\partial t_j = -\partial \mathcal{H}/\partial q_k$$

where $p_{kj} = \partial \Lambda / \partial q_{k,j}$ are "momenta", and $\mathcal{H}$ a Hamiltonian. If the Lagrangian depends on higher derivatives the canonical form of the equations is more complicated, see below (17.5.18). We shall write canonical equations in an invariant form which does not require finding the canonical variables "coordinate-momentum". A formal framework for such study is represented by the so-called "variational bi-complex".

17.1.4. The multi-time calculus of variations is of course not new. The earlier works were by Caratheodory [82], Weyl [83], de Donder [84] and the more recent by Dedecker and Tulczyjev [85–87], Vinogradov (see survey [88]), Vinogradov and Kupershmidt [89], Tsujishita [90], Goldshmidt and Sternberg [91]. For the connection with the completely integrable systems see also Manin [108]. The latter author noted that in this study essential different languages can be used, namely those of analysis, algebra and geometry. We are close to the algebraic concepts of Dedecker and Tulczyjev's works. The so-called variational bi-complex is used i.e. a complex with two differentials d and δ. A similar complex was introduced by Gabrielov, Gelfand and Losik [92]. We use the results of Dedecker and Tulczyjev about the exactness of this bi-complex. Some of our results appear newer: the construction of the symplectic form and of the energy-momentum tensor as well as the relation between first integrals and vector fields (theory of characteristic of first integrals similar to that in Chap. 11). Martinez Alonso [93] has used a similar notion (integrating multipliers). However, he did not use symplectic forms and he therefore did not have a theorem like the one we call "main proposition about first integrals". Our point of view is close to that in the works of Shadwick [94–98] who used the multi-time formalism in the form of Goldsmith and Sternberg.

17.2. Variational bi-complex.

17.2.1. Let $\mathcal{K}$ be a differential algebra ("algebra of coefficients") with commuting differentiations $\partial_j, j = 1, \ldots, n$. An example: smooth functions of variables $x_j, \partial_j = \partial/\partial x_j$. A second example: contants $\mathcal{K} = \mathbb{R}(\mathbb{C}), \partial_j \mathcal{K} = 0$. This example will play the main role.

17.2.2. Let us extend this algebra adding independent generators $u_k^{(i)}, k = 1, \ldots, m$, (i) being a multiindex $(i) = (i_1, \ldots, i_n)$. It is assumed $\partial_j u_k^{(i)} = u_k^{(i)+e_j}$ where $(i)+e_j = (i_1, \ldots, i_j+1, \ldots, i_n)$. We denote $\partial^{(i)} = \partial_1^{i_1} \ldots \partial_n^{i_n}$. This differential algebra is denoted as $\mathcal{A}$. Let $\overset{\circ}{\mathcal{A}}$ be the subalgebra comprising the differential polynomials in $\{u_k\}$ having no terms free of the generators $\{u_k\}$, i.e. "polynomials without free terms".

17.2.3. The space $\mathcal{A}^{(p,q)}$ comprises formal sums

$$\omega^{p,q} = \sum f_{(k),(j)}^{(i)} \delta u_{k_1}^{(i_1)} \wedge \ldots \wedge \delta u_{k_p}^{(i_p)} \wedge dx_{j_1} \wedge \ldots \wedge dx_{i_q}$$

$(i) = ((i_1), \ldots, (i_p)), (i_1) = (i_{11}, \ldots, i_{1n}), \ldots, (k) = (k_1, \ldots, k_p), (j) = (j_1, \ldots, j_q)$ which are called p, q-forms. All the differentials $\{\delta u_k^{(i)}, dx_j\}$ are anticommuting. The external product is as usual. Now we have a Grassman algebra with the following generators: (i) elements of $\mathcal{K}$, (ii) $u_k^{(i)}$, (iii) $\delta u_k^{(i)}$, (iv) dx_j. The differentials $\delta u_k^{(i)}$ will be called variations.

Let $\mathcal{A}^{**} = \{\mathcal{A}^{(p,q)}\}$. Now we define operations $d : \mathcal{A}^{(p,q)} \to \mathcal{A}^{(p,q+1)}$ and $\delta : \mathcal{A}^{(p,q)} \to \mathcal{A}^{(p+1,q)}$. (i) They are derivations

$$d(\omega_1^{p_1,q_1} \wedge \omega_2^{p_2,q_2}) = d\omega_1^{p_1,q_1} \wedge \omega_2^{p_2,q_2} + (-1)^{p_1+q_1} \omega_1^{p_1,q_1} \wedge d\omega_2^{p_2,q_2},$$
$$\delta(\omega_1^{p_1,q_1} \wedge \omega_2^{p_2,q_2}) = \delta\omega_1^{p_1,q_1} \wedge \omega_2^{p_2,q_2} + (-1)^{p_1+q_1} \omega_1^{p_1,q_1} \wedge \delta\omega_2^{p_2,q_2}.$$

(ii) On the generators:

$$f \in \mathcal{A}, df = \sum \partial_j f dx_j = \sum (\partial f/\partial x_j + \partial f/\partial u_k^{(i)} \cdot u_k^{(i)+e_j}) dx_j,$$
$$\delta f = \sum \partial f/\partial u_k^{(i)} \cdot \delta u_k^{(i)}, d(\delta u_k^{(i)}) = -\delta du_k^{(i)}$$

$$= -\sum \delta u_k^{(i)+e_j} \wedge dx_j \, , d(dx_j) = 0 \, , \delta(dx_j) = \delta(\delta u_k^{(i)}) = 0 \, .$$

This determines the action of d and δ on any form.

17.2.4. Lemma.

$$d^2 = \delta^2 = 0 \, , d\delta = -\delta d \, .$$

Proof: Owing to (i) this must be verified on the generators which can be easily done.

Here is the mapping diagram

$$\begin{array}{ccccccc}
\delta \uparrow & & \delta \uparrow & & & & \delta \uparrow \\
0 \to \mathcal{A}^{(2,0)} & \xrightarrow{d} & \mathcal{A}^{(2,1)} & \xrightarrow{d} & \ldots & \xrightarrow{d} & \mathcal{A}^{(2,n)} \\
\delta \uparrow & & \delta \uparrow & & & & \delta \uparrow \\
0 \to \mathcal{A}^{(1,0)} & \xrightarrow{d} & \mathcal{A}^{(1,1)} & \xrightarrow{d} & \ldots & \xrightarrow{d} & \mathcal{A}^{(1,n)} \\
\delta \uparrow & & \delta \uparrow & & & & \delta \uparrow \\
0 \to \overset{\circ}{\mathcal{A}}{}^{(0,0)} & \xrightarrow{d} & \overset{\circ}{\mathcal{A}}{}^{(0,1)} & \xrightarrow{d} & \ldots & \xrightarrow{d} & \overset{\circ}{\mathcal{A}}{}^{(0,n)} \\
\uparrow & & \uparrow & & & & \uparrow \\
0 & & 0 & & & & 0
\end{array}$$

where $\overset{\circ}{\mathcal{A}}{}^{(0,q)}$ consists of forms whose coefficients belong to $\overset{\circ}{\mathcal{A}}$.

17.2.5. The bi-complex $\{\mathcal{A}^{(p,q)}\}$ generates an associated complex with elements $\mathcal{A}^{(r)} = \underset{p+q=r}{\oplus} \mathcal{A}^{(p,q)}$ and a derivation $d + \delta$ since $(d + \delta)^2 = d^2 + d\delta + \delta d + \delta^2 = 0$.

17.2.6. The dual to the space of 1-form $\mathcal{A}^{(1)}$ is the space $T\mathcal{A}$ of "vector fields" which are the formal sums

$$\xi = \sum_{k,(i)} \xi_{k,(i)} \partial/\partial u_k^{(i)} + \sum_j \xi_j^* \partial_j \, .$$

The coupling of vector fields and forms of the substitution of vector fields into forms is the following. Only $i(\partial/\partial u_k^{(i)})\delta u_k^{(i)} = 1$ and $i(\partial_j)dx_j = 1$ do not vanish. The substitution of a vector field into a form of any rank is determined as usual: in succession into all the differentials $\delta u_k^{(i)}, dx_j$, changing the sign if the differential stands on even place, e.g.

$$i(\partial_j)\delta u_k^{(i)} \wedge dx_j \wedge dx_l = -\delta u_k^{(i)} \wedge dx_l \, , i(\partial/\partial u_k^{(i)})\delta u_l^{(m)} \wedge \delta u_k^{(i)} \wedge dx_j$$
$$= -\delta u_l^{(m)} \wedge dx_j \, .$$

The operation of the substitution of the vector field is a derivation

$$i(\xi)(\omega_1^{p_1,q_1} \wedge \omega_2^{p_2,q_2}) = (i(\xi)\omega_1^{p_1,q_1}) \wedge \omega^{p_2,q_2} + (-1)^{p_1+q_1}\omega_1^{p_1,q_1} \wedge i(\xi)\omega_2^{p_2,q_2}$$

Apart from the operation $i(\xi)$ there is another operation between vector fields and forms: the Lie derivative

$$L_\xi = (\delta + d)i(\xi) + i(\xi)(\delta + d).$$

17.2.7. The special vector fields

$$\tilde{\partial}_j = \sum u_k^{(i)+e_j} \partial/\partial u_k^{(i)}$$

play an important part. Let us calculate the action of the vector fields ∂_j and $\tilde{\partial}_j$ as the Lie derivatives on the functions, i.e. on the elements of $\mathcal{A}^{(0,0)} = \mathcal{A}$:

$$L_{\partial_j}f = i(\partial_j)(\delta+d)f = i(\partial_j)\left(\sum \partial f/\partial u_k^{(i)} \cdot \delta u_k^{(i)} + \sum \partial_j f dx_j\right) = \partial_j f,$$

$$L_{\tilde{\partial}_j}f = i(\tilde{\partial}_j)\left(\sum \partial f/\partial u_k^{(i)} \cdot \delta u_k^{(i)} + \sum \partial_j f \cdot dx_j\right) = \sum \partial f/\partial u_k^{(j)} \cdot u_k^{(i)+e_j}.$$

Thus, $L_{\partial_j}f$ is a total derivative of the function f with respect to x_j, $L_{\tilde{\partial}_j}$ is the total derivative minus the partial one. (The latter takes into account only the explicit dependence of f on x_j.) In other words, in calculating $L_{\tilde{\partial}_j}$ one must not differentiate the coefficients belonging to $\mathcal{K}$. We further use the simpler notations $L_{\partial_j}f = \partial_j f$, $L_{\tilde{\partial}_j}f = \tilde{\partial}_j f$ and, in general, $L_\xi \omega = \xi \omega$. Partial derivatives will be denoted as $\partial'_j = \partial_j - \tilde{\partial}_j$.

17.2.8. *Proposition.* The action of L_{∂_j} on a form ω consists in differentiating with respect to x_j all the coefficients and all $u_k^{(i)}$ which appear as factors $\delta u_k^{(i)}$ (e.g. $\partial_j(f\delta u_l^{(m)} \wedge dx_r) = (\delta_j f)\delta u_l^{(m)} \wedge dx_r + f\delta u_l^{(m)+e_j} \wedge dx_r$). The action of $L_{\tilde{\partial}_j}$ is the same but the coefficients from $\mathcal{K}$ must not be differentiated.

Proof: It is sufficient to verify this on the generators, which we leave to the reader. □

17.2.9. *Proposition.* Vector fields form a Lie algebra with respect to the commutator defined by the rules: if $\xi, \eta \in T\mathcal{A}; f, g \in \mathcal{A}$ then

$$[f\xi, g\eta] = fg[\xi,\eta] + f(\xi g)\eta - g(\eta f)\xi,$$

$$[\partial/\partial u_k^{(i)}, \partial/\partial u_l^{(m)}] = [\partial_j, \partial_k] = 0$$

$$[\partial/\partial u_k^{(i)}, \partial_j] = \begin{cases} 0, \text{if } i_j = 0 \\ \partial/\partial u_k^{(i)-e_j}, \text{otherwise} \end{cases}$$

(the last rule can be more simply formulated: $[\partial/\partial u_k^{(i)}, \partial_j]$ is always $\partial/\partial u_k^{(i)-e_j}$, this expression being zero if the multi-index $(i) - e_j$ contains negative components).

Proof: It is easy to verify that the commutator thus defined has the property

$$[\xi,\eta]f = \xi(\eta f) - \eta(\xi f)$$

for any $f \in \mathcal{A}$. Then $([[\xi,\eta],\varsigma]+\text{c.p.})f = 0$. This implies the Jacobi identity since f is arbitrary. □

17.2.10. Lemma.

$$\delta i(\partial_j) = -i(\partial_j)\delta \,, di(\tilde{\partial}_j) = -i(\tilde{\partial}_j)d\,.$$

Proof: It suffices to check this on the generators which we leave to the reader. □

17.2.11. Corollary.

$$\partial_j \omega = (di(\partial_j) + i(\partial_j)d)\omega\,, \tilde{\partial}_j\omega = (\delta_i(\tilde{\partial}_j) + i(\tilde{\partial}_j)\delta)\omega\,.$$

17.2.12. Lemma.

$$d\omega = \sum dx_i \wedge \partial_j \omega$$

Proof: Check on the generators. □

17.2.13. Lemma. The operation

$$\delta i(\partial/\partial u_k^{(i)}) + i(\partial/\partial u_k^{(i)})\delta$$

acts as $\partial/\partial u_k^{(i)}$ on the coefficients of a form.

Proof: Check this on the generators. □

17.2.14. Definition. The vector field ∂_h related to the set $h = (h_1,\ldots,h_m) \in \mathcal{A}^m$ is

$$\partial_h = \sum h_k^{(i)} \partial/\partial u_k^{(i)}\,, h_k^{(i)} = \partial^{(i)} h_k = \partial_1^{i_1}\ldots\partial_n^{i_n} h_k\,.$$

17.2.15. Lemma.

$$[\partial_h,\partial_j] = 0\,, j = 1,\ldots,n\,.$$

Proof:

$$[\sum h_k^{(i)}\partial/\partial u_k^{(i)}, \partial_j] = -\sum h_k^{(i)+e_j}\partial/\partial u_k^{(i)} + \sum h_k^{(i)}\partial/\partial u_k^{(i)-e_j} = 0\,.\square$$

17.2.16. Lemma.
$$di(\partial_h) + i(\partial_h)d = 0,\, \partial_h d = d\partial_h\,.$$

Proof: According to 17.2.12, $d = \sum dx_j \wedge \partial_j$ whence $(\partial_h d - d\partial_h)\omega = \partial_h \sum dx_j \wedge \partial_j \omega - \sum dx_j \wedge \partial_j(\partial_h \omega)$. From $\partial_h dx_j = 0$, taking 17.2.15 into account we obtain the second equality. The first one can be checked on the generators:

$$di(\partial_h)\delta u_k^{(i)} + i(\partial_h)d\delta u_k^{(i)} = dh_k^{(i)} - i(\partial_h)\sum \delta u_k^{(i)+e_r} \wedge dx_r$$
$$= \sum h_k^{(i)+e_r} dx_r - \sum h_k^{(i)+e_r} dx_r = 0\,.\square$$

17.2.17. Proposition.

$$\partial_h i(\partial_g) - i(\partial_g)\partial_h = i([\partial_h, \partial_g])\,.$$

Proof: It is sufficient to verify this on the generators $\delta u_k^{(i)}$. The left-hand side is

$$\partial_h g_k^{(i)} - \partial_g h_k^{(i)} = \sum h_l^{(m)} \partial g_k^{(i)}/\partial u_l^{(m)} - (h \leftrightarrow g)\,.$$

It is easy to see that the right-hand side is the same. $\square$

17.3. Exactness of the bi-complex.

17.3.1. The results of this section belong mainly to Tulczyjev and Dedecker.

17.3.2. Proposition. All the vertical sequences in the bi-complex are exact.
Proof. All the $u_k^{(i)}$ can be considered as independent variables with respect to the operator δ. Therefore the assertion is none other than the Poincaré lemma (see e.g. [99]). Recall the proof in conformity with this case. Thus, let $\omega^{p,q}$ be a form and $\delta\omega^{p,q} = 0$. The form $\omega^{p-1,q}$ must be constructed such that $\delta\omega^{p-1,q} = \omega^{p,q}$. Let us write $\omega^{p,q}$ as in 17.2.3. The coefficient $f_{(k),(j)}^{(i)}$ is a polynomial in $\{u_k^{(i)}\}$ with coefficients in $\mathcal{K}$. Without loss of generality it can be assumed to be homogeneous of a degree r, since the operator δ transforms homogeneous forms to homogeneous forms. Let

$$I = \frac{1}{p+r}\sum u_k^{(i)} i(\partial/\partial u_k^{(i)})$$

be an operator acting on forms with the total degree of homogeneity $p+r$ with respect to $\{u_k^{(i)}\}$ and to $\{\delta u_k^{(i)}\}$. We shall show that $I\delta + \delta I = 1$.

Indeed,

$$\delta I\omega = \frac{1}{p+r}\sum \delta u_r^{(i)} \wedge i(\partial/\partial u_k^{(i)})\omega + \frac{1}{p+r}\sum u_k^{(i)}(\delta i(\partial/\partial u_k^{(i)})$$
$$+ i(\partial/\partial u_k^{(i)})\delta)\omega - \frac{1}{p+r}\sum u_k^{(i)} i(\partial/\partial u_k^{(i)})\delta\omega.$$

The form $\delta\omega$ is also homogeneous of the total degree $p+r$, hence the last term is $-I\delta\omega$. The first term is $(p/(p+r))\omega$. The second one in accordance with 17.2.13 is the operator $(p+r)^{-1}\sum u_k^{(i)}\partial/\partial u_k^{(i)}$ applied to the coefficients of the form ω. All of them are homogeneous of degree r; the Euler theorem about homogeneous functions yields: the second term is $r(p+r)^{-1}\omega$ and in all $\delta I\omega = \omega - I\delta\omega$ as required. Let now $\delta\omega^{p,q} = 0$. Then $\omega^{p,q} = \delta I\omega^{p,q}$ and $\omega^{p-1,q} = I\omega^{p,q}$. $\square$

7.3.3. *Remark.* We have separated homogeneous parts of a form because each of them must be divided by its own degree $p+r$. This can be done in another way: substitute $tu_k^{(i)}$ for $u_k^{(i)}$ (everywhere including differentials $\delta u_k^{(i)}$), t being a parameter. Then

$$\omega^{p-1,q} = \int_0^1 \sum u_k^{(i)} i(\partial/\partial u_k^{(i)})\omega(tu) t^{-1} dt.$$

Each term will automatically be divided by its degree.

17.3.4. *Proposition.* All the horizontal sequences in the bi-complex are exact for $p > 0$.

Proof: An operator D ("Tulczyjev's operator") will be constucted having the property
$$Dd + dD = p \cdot 1$$
for the p-th row of the diagram. This will yield the required assertion. The construction is however not very simple.

Let (m) be a multi-index. We construct an operation $\theta_{(m)}$ having properties:

(i) it is a derivation
$$\theta_{(m)}(\omega_1^{p_1,q_1} \wedge \omega_2^{p_2,q_2}) = (\theta_{(m)}\omega_1^{p_1,q_1}) \wedge \omega_2^{p_2,q_2} + \omega_1^{p_1,q_1} \wedge \theta_{(m)}\omega_2^{p_2,q_2},$$

(ii) on the generators it is defined as: for $f \in \mathcal{A}, \theta_{(m)}f = 0, \theta_{(m)}dx_j = 0, \theta_{(m)}\delta u_k^{(i)} = \binom{(i)}{(m)}\delta u_k^{(i)-(m)}$ where $\binom{(i)}{(m)} = \binom{i_1}{m_1}\cdots\binom{i_n}{m_n}$. This expression is zero if $m_r > i_r$ for any $r = 1,\ldots,n$; then $\theta_{(m)}\delta u_k^{(i)} = 0$.

A derivation is uniquely determined by its values on the generators. It is obvious that for $(m) = 0$ we have $\theta_0 \omega^{p,q} = p \omega^{p,q}$. The explicit expression for the operator $\theta_{(m)}$ is $\sum \binom{(i)}{(m)} \delta u_k^{(i)-(m)} i(\partial/\partial u_k^{(i)})$.

17.3.5. Lemma.

$$[\theta_{(m)}, \partial_\alpha] = \theta_{(m)-e_\alpha}.$$

Proof: If $\theta_{(m)}$ is a derivation then $[\theta_{(m)}, \partial_\alpha]$ is a derivation also (check it). An equality of two derivations can be verified on the generators. Less obvious case is that one:

$$[\theta_{(m)}, \partial_\alpha] \delta u_k^{(i)} = \theta_{(m)} \delta u_k^{(i)+e_\alpha} - \partial_\alpha \binom{(i)}{(m)} \delta u_k^{(i)-(m)}$$

$$= \left(\binom{(i)+e_\alpha}{(m)} - \binom{(i)}{(m)} \right) \delta u_k^{(i)-(m)+e_\alpha}$$

$$= \binom{(i)}{(m)-e_\alpha} \delta u_k^{(i)-(m)+e_\alpha} = \theta_{(m)-e_\alpha} \delta u_k^{(i)}.$$

Now let $I_\alpha, \alpha = 1, \ldots, n$ be a set of multi-indices (m) such that $m_\alpha > 0$ and $m_\beta = 0$ if $\beta > \alpha$. Let

$$\sigma^\alpha = \sum_{(m) \in I_\alpha} (-1)^{|m|} \partial^{(m)-e_\alpha} \theta_{(m)}, |m| = m_1 + \ldots + m_n.$$

17.3.6. Lemma.

$$\sigma^\alpha \partial_\beta = \begin{cases} \partial_\beta \sigma^\alpha, & \alpha < \beta \\ \theta_0 - \sum_{\gamma < \alpha} \partial_\gamma \sigma^\gamma, & \alpha = \beta \\ 0, & \alpha > \beta \end{cases}.$$

Proof:

$$\sigma^\alpha \partial_\beta = - \sum_{(m) \in I_\alpha} (-1)^{|m|} \partial^{(m)-e_\alpha} \theta_{(m)} \partial_\beta$$

$$= \partial_\beta \sigma^\alpha - \sum_{(m) \in I_\alpha} (-1)^{|m|} \partial^{(m)-e_\alpha} \theta_{(m)-e_\beta};$$

1. if $\beta > \alpha, (m) \in I_\alpha$, then $\theta_{(m)-e_\beta} = 0, \sigma^\alpha \partial_\beta = \partial_\beta \sigma^\alpha$,

2. if $\beta < \alpha$, then $\sigma^\alpha \partial_\beta = \partial_\beta \sigma^\alpha - \sum (-1)^{|m|} \partial_\beta \partial^{(m)-e_\alpha-e_\beta} \theta_{(m)-e_\beta}$
   $= \partial_\beta \sigma^\alpha - \partial_\beta \sigma^\alpha = 0$,
3. if $\beta = \alpha$, then $\sigma^\alpha \partial_\alpha = \partial_\alpha \sigma^\alpha - \sum (-1)^{|m|} \partial^{(m)-e_\alpha} \theta_{(m)-e_\alpha} = \partial_\alpha \sigma^\alpha - \sum_{\gamma \leq \alpha} \partial^\gamma \sigma^\gamma + \theta_0 = \theta_0 - \sum_{\gamma < \alpha} \partial_\gamma \sigma^\gamma$. □

An arbitrary p,q-form $\omega = \omega^{p,q}$ can be written as

$$\omega = \sum_{\alpha_1 < \ldots < \alpha_{n-q}} i(\partial_{\alpha_1}) \ldots i(\partial_{\alpha_{n-q}}) dx_1 \wedge \ldots \wedge dx_n \wedge a_{\alpha_1 \ldots \alpha_{n-q}}$$

where $a_{\alpha_1 \ldots \alpha_{n-q}} \in \mathcal{A}^{p,0}$. Let

$$D\omega = \sum_{\alpha_0 < \alpha_1 < \ldots < \alpha_{n-q}} i(\partial_{\alpha_0}) i(\partial_{\alpha_1}) \ldots i(\partial_{\alpha_{n-q}}) dx_1 \wedge \ldots \wedge dx_n \wedge \sigma^{\alpha_0} a_{\alpha_1 \ldots \alpha_{n-q}}$$

This is Tulczyjev's operator. If $q = 0$ then $D = 0$.

17.3.7. Lemma. The relation

$$dD + Dd = \theta_0$$

holds.

Proof:

$$dD\omega = \sum_{\beta=1}^{n} dx_\beta \wedge \partial_\beta D\omega = \sum dx_\beta \wedge i(\partial_{\alpha_0}) \ldots i(\partial_{\alpha_{n-q}}) dx_1 \wedge \ldots \wedge dx_n$$

$$\wedge \partial_\beta \sigma^{\alpha_0} a_{\alpha_1 \ldots \alpha_{n-q}} = \sum_{r=0}^{n-q} (-1)^r \sum_{\alpha_0 < \ldots < \alpha_{n-q}} i(\partial_{\alpha_0}) \ldots \widehat{i(\partial_{\alpha_r})} \ldots i(\partial_{\alpha_{n-q}})$$

$$\cdot dx_1 \wedge \ldots \wedge dx_n \wedge \partial_{\alpha_r} \sigma^{\alpha_0} a_{\alpha_1 \ldots \alpha_{n-q}};$$

$$d\omega = \sum_\beta \sum_{\alpha_1 < \ldots < \alpha_{n-q}} dx_\beta \wedge i(\partial_{\alpha_1}) \ldots i(\partial_{\alpha_{n-q}}) dx_1 \wedge \ldots \wedge dx_n \partial_\beta a_{\alpha_1 \ldots \alpha_{n-q}}$$

$$= \sum_{r=1}^{n-q} (-1)^{r-1} \sum_{\alpha_1 < \ldots < \alpha_{n-q}} i(\partial_{\alpha_1}) \ldots \widehat{i(\partial_{\alpha_r})} \ldots i(\partial_{\alpha_{n-q}}) dx_1 \wedge \ldots \wedge dx_n$$

$$\wedge \partial_{\alpha_r} a_{\alpha_1 \ldots \alpha_{n-q}};$$

$$Dd\omega = \sum_{r=1}^{n-q} (-1)^{r-1} \sum_{\alpha_0 < \ldots < \alpha_{n-q}} i(\partial_{\alpha_0}) i(\partial_{\alpha_1}) \ldots \widehat{i(\partial_{\alpha_r})} \ldots i(\partial_{\alpha_{n-q}})$$

$$dx_1 \wedge \ldots \wedge dx_n \wedge \sigma^{\alpha_0} \partial_{\alpha_r} a_{\alpha_1 \ldots \alpha_{n-q}} + \sum_{\alpha_1 < \ldots < \alpha_{n-q}} i(\partial_{\alpha_1}) \ldots i(\partial_{\alpha_{n-q}})$$

$$dx_1 \wedge \ldots \wedge dx_n \wedge \sigma^{\alpha_1} \partial_{\alpha_1} a_{\alpha_1 \ldots \alpha_{n-q}}$$

(the reason for the appearance of the last term is the following). If $r = 1$ then the number α_0 changes from 0 to $\alpha_2 - 1$. When $\alpha_0 > \alpha_1$ then $\sigma^{\alpha_0}\partial_{\alpha_1} = 0$ (see 17.3.6). When $\alpha_0 = \alpha_1$ then this additional term appears; the rest, for $\alpha_0 < \alpha_1$ is involved in the first term). Now

$$dD\omega + Dd\omega = \sum_{\alpha_0 < \alpha_1 < \ldots < \alpha_{n-q}} i(\partial_{\alpha_1})\ldots i(\partial_{\alpha_{n-q}})dx_1 \wedge \ldots \wedge dx_n$$
$$\wedge \partial_{\alpha_0}\sigma^{\alpha_0} a_{\alpha_1\ldots\alpha_{n-q}} + \sum_{\alpha_1 < \ldots < \alpha_{n-q}} i(\partial_{\alpha_1})\ldots i(\partial_{\alpha_{n-q}})$$
$$dx_1 \wedge \ldots \wedge dx_n \wedge \sigma^{\alpha_1}\partial_{\alpha_1} a_{\alpha_1\ldots\alpha_{n-q}}.$$

Taking into account 17.3.6 $\sigma^{\alpha_1}\partial_{\alpha_1} = \theta_0 - \sum_{\alpha_0 < \alpha_1} \partial_{\alpha_0}\sigma^{\alpha_0}$ we obtain the required assertion. □

Now we finish the proof of the proposition 17.3.4. If $d\omega^{p,q} = 0$ then $dD\omega^{p,q} = \theta_0 \omega^{p,q} = p\omega^{p,q}$ i.e. $\omega^{p,q} = d(p^{-1}D\omega^{p,q})$. The closed form is exact. □

17.3.8. *Remark.* It is evident that $D\omega = 0$ if the form ω contains only δu_k with zero multi-index.

We have not proved yet that the bottom row is exact. Here we use the method of diagram chasing.

17.3.9. *Proposition.* If for a form $\omega^{p,q} \in \mathcal{A}^{(p,q)}$ another form $\omega^{p+1,q-1} \in \mathcal{A}^{(p+1,q-1)}$ exists for which $\delta\omega^{p,q} = -d\omega^{p+1,q-1}$, then forms $\varphi^{p-1,q} \in \mathcal{A}^{(p-1,q)}$ and $\varphi^{p,q-1} \in \mathcal{A}^{(p,q-1)}$ exist such that

$$\omega^{p,q} = \delta\varphi^{p-1,q} + d\varphi^{p,q-1}$$

(a form $\varphi^{p,q}$ is zero if p or q are negative). All the forms from $\mathcal{A}^{(0,q)}$ are assumed being in $\overset{\circ}{\mathcal{A}}{}^{(0,q)}$.

Proof: We construct a sequence of forms $\omega^{p+r,q-r}, r = 0, 1, \ldots, q$ with the property

$$\delta\omega^{p+r,q-r} = -d\omega^{p+r+1,q-r-1}.$$

We have already two first terms of this sequence, for $r = 0$ and 1. Let $\omega^{p+r,q-r}$ be already built: $d\omega^{p+r,q-r} = -\delta\omega^{p+r-1,q-r+1}$, then $\delta d\omega^{p+r,q-r} = 0, d\delta\omega^{p+r,q-r} = 0$. Proposition 17.3.4 yields (since $p+r+1 > 0$)$\delta\omega^{p+r,q-r} = -d\omega^{p+r+1,q-r-1}$ for some $\omega^{p+r+1,q-r-1}$. The sequence is obtained.

Now it will be proved that all the $\omega^{p+r,q-r}$ of this sequence can be represented as

(17.3.10) $\qquad \omega^{p+r,q-r} = d\varphi^{p+r,q-r-1} + \delta\varphi^{p+n-1,q-r}$

for some forms $\{\varphi^{p+r,q-r-1}\}$. We have $\delta\omega^{p+q,0} = d\omega^{p-q+1,-1} = 0$. By virtue of the proposition 17.3.2 a form $\varphi^{p+q-1,0}$ exists such that $\omega^{p+q,0} = \delta\varphi^{p+q-1,0}$. Let (17.3.10) be proved for $r > r_0$. We shall prove this for r_0. We have

$$\omega^{p+r_0+1,q-r_0-1} = d\varphi^{p+r_0+1,q-r_0-2} + \delta\varphi^{p+r_0,q-r_0-1}.$$

Further, $d\omega^{p+r_0+1,q-r_0-1} = -\delta\omega^{p+r_0,q-r_0}$ which yields $d\delta\varphi^{p+r_0,q-r_0-1} = -\delta\omega^{p+r_0,q-r_0}$, i.e., $\delta(d\varphi^{p+r_0,q-r_0-1} - \omega^{p+r_0,q-r_0}) = 0$. According to 17.3.2 a form $\varphi^{p+r_0-1,q-r_0}$ exists such that $d\varphi^{p+r_0,q-r_0-1} - \omega^{p+r_0,q-r_0} = -\delta\varphi^{p+r_0-1,q-r_0}$. The sequence is constructed by induction. In particular, the required assertion is proved. □

17.3.11. Proposition (Tulczyjev). All the sequences of the bi-complex, both vertical and horizontal, are exact.

Proof: It remains to prove this for the bottom row. Let $d\omega^{0,q} = 0, q < n$. Then $\delta d\omega^{0,q} = 0$ and $d\delta\omega^{0,q} = 0$. Proposition 17.3.4 yields that a form $\omega^{1,q-1}$ exists for which $\delta\omega^{0,q} = d\omega^{1,q-1}$. Then, according to 17.3.9 $\omega^{0,q} = d\varphi^{0,q-1}$ for some $\varphi^{0,q-1}$. □

17.3.12. Proposition. The associated complex $\{\mathcal{A}^{(r)}\}$ is exact with respect to the differential $d + \delta$.

Proof: Let $\omega^r = \sum_{p+q=r} \omega^{p,q}$ be a closed form: $(d+\delta)\omega^r = 0$, i.e. $d\omega^{p,q} + \delta\omega^{p-1,q+1} = 0$. There are such forms (see 17.3.9) $\{\varphi^{p-1,q}\}$ that $\omega^{p,q} = d\varphi^{p,q-1} + \delta\varphi^{p-1,q}$ i.e. $\omega^r = (d+\delta) \sum_{p+q=r-1} \varphi^{p,q}$. □

17.4. Variational derivative.

17.4.1. We consider now the last column of the bi-complex. The problem is to find the image of the mapping $d : \mathcal{A}^{(p,n-1)} \to \mathcal{A}^{(p,n)}$. Let $E_1^{(p,n)} = \mathcal{A}^{(p,n)}/d\mathcal{A}^{(p,n-1)}$ (the notation is taken from the theory of spectral sequences, see e.g. [100], 3.5). We amplify the bi-complex with one more column

$$
\begin{array}{ccccccc}
\delta\uparrow & & \delta\uparrow & & \delta\uparrow & \\
0 \to a^{(1,0)} & \xrightarrow{d} & \cdots & \xrightarrow{d} & a^{(1,n)} & \xrightarrow{\pi_1} & E_1^{(1,n)} \to 0 \\
\delta\uparrow & & \delta\uparrow & & \delta\uparrow & \\
0 \to \overset{\circ}{a}{}^{(0,0)} & \xrightarrow{d} & \cdots & \xrightarrow{d} & \overset{\circ}{a}{}^{(0,n)} & \xrightarrow{\pi_0} & E_1^{(0,n)} \to 0 \\
\uparrow & & \uparrow & & \uparrow & \\
0 & & 0 & & 0 &
\end{array}
$$

(17.4.2)

The anticommutativity of d and δ implies that δ can be considered in the quotient space $E_1^{p,n}$.

17.4.3. Proposition. The amplified bi-complex (17.4.2) remains exact.

We denote the elements of $E_1^{(p,n)}$ by $\omega_*^{p,n}$; $\omega^{p,n}$ will be a representative of the class $\omega_*^{p,n}$. Let $\delta\omega_*^{p,n} = 0$. For a representative $\omega^{p,n}$, this means that $\delta\omega^{p,n} = -d\omega^{p,n-1}$ for some $\omega^{p,n-1} \in \mathcal{A}^{(p,n-1)}$. Owing to 17.3.9 forms $\varphi^{p,n-1}$ and $\varphi^{p-1,n}$ exist such that $\omega^{p,n} = d\varphi^{p,n-1} + \delta\varphi^{p-1,n}$ (if $p = 0$ the second form is absent and $\omega^{0,n} = d\varphi^{0,n-1}$). This means that $\omega_*^{p,n} = \delta\varphi_*^{p-1,n}$, where $\varphi_*^{p-1,n}$ is the class of the form $\varphi^{p-1,n}$. □

17.4.4. Proposition. If $\omega^{0,n} = f dx_1 \wedge \ldots \wedge dx_n \in \mathcal{A}^{(0,n)}$ then $\delta\omega^{0,n}$ can uniquely represented as

$$\delta\omega^{0,n} = \sum_{k=1}^{n} A_k \delta u_k \wedge dx_1 \wedge \ldots \wedge dx_n + d\omega^{1,n-1}$$

where $\omega^{1,n-1} \in \mathcal{A}^{(1,n-1)}/d\mathcal{A}^{(1,n-2)}$. The coefficients A_k will be denoted as $\delta f/\delta u_k$ and called variational derivatives with respect to u_k. Let $\delta f/\delta u = \{\delta f/\delta u_k\}$.

Proof: Let us transform the expression

$$\delta f \wedge dx_1 \wedge \ldots \wedge dx_n = \sum \partial f/\partial u_k^{(i)} \delta u_k^{(i)} \wedge dx_1 \wedge \ldots \wedge dx_n$$

by means of repeated integration by parts

$$\sum \partial f/\partial u_k^{(i)} \delta u_k^{(i)} = \sum (-1)^{|i|} (\partial^{(i)} \partial f/\partial u_k^{(i)}) \cdot \delta u_k + \sum \partial_\alpha B_\alpha,$$

where B_α are some forms. It remains to put

(17.4.5) $$\delta f/\delta u_k = \sum (-1)^{|i|} \partial^{(i)} \partial f/\partial u_k^{(i)}$$

and $\omega^{1,n-1} = \sum (-1)^\alpha B_\alpha dx_1 \wedge \ldots \wedge \widehat{dx_\alpha} \wedge \ldots \wedge dx_n$.

The uniqueness can be proved thus. If D is the Tulczyjev operator then $Dd\omega^{1,n-1} + dD\omega^{1,n-1} = \omega^{1,n-1}$ i.e. $\omega^{1,n-1} = d(D\omega^{1,n-1}) - D\sum A_k \delta u_k \wedge dx_1 \wedge \ldots$ According to 17.3.8 the last term vanishes, and $\omega^{1,n-1} = d(\ )$ whence $\sum A_k \delta u_k \wedge dx_1 \wedge \ldots = 0$ and $\{A_k = 0\}$. □

17.4.6. Proposition. The variational derivative vanishes, $\delta f/\delta u = 0$ if and only if the form $\omega^{0,n} = f dx_1 \wedge \ldots \wedge dx_n$ is exact, $\omega^{0,n} = d\varphi^{0,n-1}$.

Proof: The equation $\delta f/\delta u = 0$ is equivalent to $\delta \omega^{0,n} = d\omega^{1,n-1}$ for some $\omega^{1,n-1}$. This is the case if $\omega^{0,n} = d\varphi^{0,n-1}$. Conversely, if $\delta\omega^{0,n} = d\omega^{1,n-1}$ then $\omega^{0,n} = d\varphi^{0,n-1}$, according to 17.3.9. □

Thus, the operation of the variational derivative can be transferred to $E_1^{(0,n)} = \mathcal{A}^{(0,n)}/d\mathcal{A}^{(0,n-1)}$. It can be considered as the mapping $\delta : E_1^{(0,n)} \to E_1^{(1,n)}$. In this sense the variational derivative can be generalized to all $E_1^{(p,n)}$.

17.4.7. Proposition. A set $\{A_k\}, k = 1, \ldots, m$, is a variational derivative of some $f \in \mathcal{A}$ if and only if $\sum \delta A_k \wedge \delta u_k \wedge dx_1 \wedge \ldots \wedge dx_n$ is a d-differential of a form:

$$\sum \delta A_k \wedge \delta u_k \wedge dx_1 \wedge \ldots \wedge dx_n = -d\omega^{2,n-1}.$$

Proof: $\{A_k\}$ is a variational derivative if and only if $\sum A_k \delta u_k$ can be represented as $\delta f \wedge dx_1 \wedge \ldots \wedge dx_n - d\omega^{1,n-1}$ i.e. this form, as an element of $E_1^{(1,n)}$, belongs to $\delta E_0^{(1,n)}$. According to 17.4.3, this is equivalent to the fact that $\delta \sum A_k \delta u_k \wedge dx_1 \wedge = \sum \delta A_k \wedge \delta u_k \wedge dx_1 \wedge \ldots$ vanishes as an element of $E_1^{(2,n)}$, i.e. is a d-differential of a form. □

We shall give another condition for this. Let $g = \{g_k\}, k = 1, \ldots, m$, and $h = \{h_k\}, k = 1, \ldots, m$. The expression

$$\partial_g h = \{\partial_g h_k\} = \left\{ \sum_{(i),l} g_l^{(i)} \partial h_k/\partial u_l^{(i)} \right\}$$

is linear with respect to both h and g. This expression can be considered as the result of application of a matrix differential operator to the vector $g : Q_h g = \left\{ \sum_{(i),l} \partial h_k/\partial u_l^{(i)} \partial^{(i)} g_l \right\}, k = 1, \ldots, m$. The operator Q_h is called the Frechet derivative. Thus, $\partial_g h = Q_h g$. In the vector space $\mathcal{A}^m$ there is a scalar product

(17.4.8) $$\langle h, g \rangle = \sum h_k g_k dx_1 \wedge \ldots \wedge dx_n \in E_1^{(0,n)}.$$

(The fact that the scalar product is considered as an element of $E_1^{(0,n)}$, i.e. up to an exact differential $d\omega^{0,n-1}$, in analytical terms means that the scalar product is the integral $\int \sum h_k g_k dx_1 \ldots dx_n$.)

17.4.9. Proposition. A set $g = \{g_k\}, k = 1, \ldots, m$ is a variational derivative $g_k = \delta f/\delta u_k, k = 1, \ldots, m$ if and only if Q_g is a self-adjoint operator with respect to the scalar product (17.4.8). (This theorem for $n = 1$ was proved by Gelfand, Manin and Shubin [101] in one direction, and by Dorfman [102] in the opposite direction.)

Proof: According to 17.4.7, the set g is a variational derivative if and only if a form $\omega^{2,n-1}$ exists such that $\sum \delta u_k \wedge \delta g_k \wedge dx_1 \wedge \ldots \wedge dx_n = d\omega^{2,n-1}$. This is, in its turn, equivalent to the fact that after the substitution of two arbitrary vector fields ∂_{h_1} and ∂_{h_2} an exact differential will be obtained, i.e. zero in $E_1^{0,n}$:

$$d\omega^{0,n-1} = \sum \{(\partial_{h_1} u_k)(\partial_{h_2} g_k) - (\partial_{h_2} u_k)(\partial_{h_1} g_k)\}$$
$$dx_1 \wedge \ldots \wedge dx_n = \langle \partial_{h_1} u, \partial_{h_2} g \rangle - \langle \partial_{h_2} u, \partial_{h_1} g \rangle$$
$$= \langle h_1, \partial_{h_2} g \rangle - \langle h_2, \partial_{h_1} g \rangle = \langle h_1, Q_g h_2 \rangle - \langle h_2, Q_g h_1 \rangle.$$

This equality expresses the fact that Q_g is a self-adjoint operator. □

The simplest way to reconstruct a function f knowing its variational derivative is by way of the following proposition.

17.4.10. Proposition. If the condition 17.4.9 is fulfilled and $\{g_k\}$ are homogeneous polynomials in $\{u_k^{(i)}\}$ of degree r then

$$f = (r+1)^{-1} \sum u_k g_k$$

satisfies the equations $\delta f / \delta u_k = g_k, k = 1, \ldots, m$.

Proof: We find (all the computations in $E_1^{(0,n)}$):

$$\forall h \in \mathcal{A}^m, \partial_h f dx_1 \wedge \ldots \wedge dx_n = (r+1)^{-1} \sum h_k g_k dx_1 \wedge \ldots \wedge dx_n$$
$$+ (r+1)^{-1} \sum u_k (\partial_h g_k) dx_1 \wedge \ldots \wedge dx_n = (r+1)^{-1}$$
$$\sum h_k g_k dx_1 \wedge \ldots \wedge dx_n + (r+1)^{-1} \langle Q_g h, u \rangle = (r+1)^{-1}$$
$$\left(\sum h_k g_k + \sum h_k \partial g_k / \partial u_l^{(i)} \cdot u_l^{(i)} \right) dx_1 \wedge \ldots \wedge dx_n$$

Applying Euler's theorem about homogeneous functions we obtain

$$\partial_h f dx_1 \wedge \ldots \wedge dx_n = \sum (r+1)^{-1}(1+r) h_k g_k dx_1 \wedge \ldots \wedge dx_n = \langle h, g \rangle.$$

On the other hand

$$\partial_h f dx_1 \wedge \ldots \wedge dx_n = \sum \partial f / \partial u_k^{(i)} h_k^{(i)} dx_1 \wedge \ldots \wedge dx_n$$
$$= \sum \delta f / \delta u_k \cdot h_k dx_1 \wedge \ldots \wedge dx_n = \langle \delta f / \delta u, h \rangle.$$

where h is arbitrary which implies $g = \delta f/\delta u$. □

17.4.11. Remark. It is possible to do this without separation of homogeneous parts of $\{g_k\}$. The same result will be obtained substituting $pu_l^{(i)}$ for $u_l^{(i)}$ into $\{g_k\}$ an integrating:

$$f = \sum_k \int_0^1 u_k g_k(pu) dp.$$

17.5. Lagrangian-Hamiltonian formalism.

17.5.1. An arbitrary element of $E_1^{(0,n)} = \mathcal{A}^{(0,n)}/d\mathcal{A}^{(0,n-1)}$ can be taken as a Lagrangian (or action, to be more precise). Let $\mathbf{\Lambda} = \Lambda dx_1 \wedge \ldots \wedge dx_n$ be a representative of the class. Then (see 17.4.4) a form $\omega^{1,n-1}$ exists such that $\delta\mathbf{\Lambda} = \sum \delta u_k \wedge \delta\mathbf{\Lambda}/\delta u_k + d\omega^{1,n-1}$; $\delta\mathbf{\Lambda}/\delta u_k = \delta\Lambda/\delta u_k dx_1 \wedge \ldots \wedge dx_n$. The form $\omega^{1,n-1}$ is especially important and therefore has a special notation $\Omega^{(1)}$. Thus,

(17.5.2) $$\delta\mathbf{\Lambda} = \sum \delta u_k \wedge \delta\mathbf{\Lambda}/\delta u_k - d\Omega^{(1)}.$$

The form $\Omega^{(1)}$ is determined up to a form of the type $\delta\omega^{0,n-1} + d\omega^{1,n-2}$. Let $\Omega = \delta\Omega^{(1)} \in \mathcal{A}^{(0,n-1)}$ (this form is determined up to a form $d\delta\omega^{1,n-2}$). The form will be called symplectic corresponding to the given Lagrangian. Evidently

(17.5.3) $$d\Omega = \sum \delta u_k \wedge \delta(\delta\mathbf{\Lambda}/\delta u_k).$$

Put

(17.5.4) $$T_j = -i(\partial_j)\mathbf{\Lambda} + i(\tilde{\partial}_j)\Omega^{(1)}, j = 1, \ldots, n.$$

The set of these forms is called the energy-momentum tensor.

17.5.5. Proposition. The relation

$$dT_j = -\sum_k (\partial_j u_k) \cdot \delta\mathbf{\Lambda}/\delta u_k - \partial_j' \mathbf{\Lambda}$$

holds (recall that ∂_j' is the partial derivative arising on account of explicit dependence of the Lagrangian on time).

Proof:

$$\begin{aligned}dT_j &= -di(\partial_j)\mathbf{\Lambda} + di(\tilde{\partial}_j)\Omega^{(1)} = -\partial_j\mathbf{\Lambda} + di(\tilde{\partial}_j)\Omega^{(i)}\\ &= -\partial'_j\mathbf{\Lambda} - \tilde{\partial}_j\mathbf{\Lambda} + di(\tilde{\partial}_j)\Omega^{(1)} = \partial'_j\mathbf{\Lambda} - i(\tilde{\partial}_j)\delta\mathbf{\Lambda} + di(\tilde{\partial}_j)\Omega^{(1)}\\ &= -\partial'_j\mathbf{\Lambda} - i(\tilde{\partial}_j)\sum \delta u_k \wedge \delta\mathbf{\Lambda}/\delta u_k = -\partial'_j\mathbf{\Lambda} - \sum(\partial_j u_k)\delta\mathbf{\Lambda}/\delta u_k \;.\square\end{aligned}$$

The following proposition expresses, as we shall show below, the main property of the energy-momentum tensor.

17.5.6. Proposition. The relation

$$\delta T_j = -i(\tilde{\partial}_j)\Omega + i(\partial_j)\sum \delta u_k \wedge \delta\mathbf{\Lambda}/\delta u_k + di(\partial_j)\Omega^{(1)} - \partial'_j\Omega^{(1)}$$

holds.

Proof: Owing to 17.2.10

$$\begin{aligned}\delta T_j =&\, i(\partial_j)\delta\mathbf{\Lambda} + \delta i(\tilde{\partial}_j)\Omega^{(1)} = i(\partial_j)(\sum \delta u_k \wedge \delta\mathbf{\Lambda}/\delta u_k - d\Omega^{(1)})\\ &+ \delta i(\tilde{\partial}_j)\Omega^{(1)} = i(\partial_j)\sum \delta u_k \wedge \delta\mathbf{\Lambda}/\delta u_k - (i(\partial_j)d + di(\partial_j))\Omega^{(1)}\\ &+ di(\partial_j)\Omega^{(1)} + (\delta i(\tilde{\partial}_j) + i(\tilde{\partial}_j)\delta)\Omega^{(1)} - i(\tilde{\partial}_j)\delta\Omega^{(1)} = i(\partial_j)\\ &\sum \delta u_k \wedge \delta\mathbf{\Lambda}/\delta u_k - \partial_j\Omega^{(1)} + \tilde{\partial}_j\Omega^{(1)} + di(\partial_j)\Omega^{(1)} - i(\tilde{\partial}_j)\Omega \;.\square\end{aligned}$$

17.5.7. Corollary. If the Lagrangian does not depend on time explicitly then

$$\delta T_j = -i(\tilde{\partial}_j)\Omega$$

modulo a d-differential and by virtue of the system of equations $\{\delta\mathbf{\Lambda}/\delta u_k = 0\}$.

This proposition generalizes the main relation of the usual (single-time) mechanics: $\delta\mathcal{H} = -i(\partial_t)\Omega$ is equivalent to the variational system $\delta\mathbf{\Lambda}/\delta u_k = 0$ where $\mathcal{H} = -\mathbf{\Lambda} + i(\partial_t)\Omega^{(1)}$.

17.5.8. Definition. The field Hamiltonian is the form

$$\mathcal{H} = \sum dx_j \wedge T_j + (n-1)\mathbf{\Lambda}$$

17.5.9. Lemma. It can also be written

$$\mathcal{H} = -\mathbf{\Lambda} + \sum_{j=1}^{n} dx_j \wedge i(\tilde{\partial}_j)\Omega^{(1)}\;.$$

Proof: This immediately follows from (17.5.4) if the equality $dx_j \wedge i(\partial_j)\mathbf{\Lambda} = \mathbf{\Lambda}$ is taken into account which holds for any $\mathbf{\Lambda} \in \mathcal{A}^{(0,n)}$.

17.5.10. Proposition.

$$\delta\mathcal{H} = \sum_{j=1}^{n} dx_j \wedge i(\tilde{\partial}_j)\Omega - \sum_{k=1}^{m} \delta u_k \wedge \delta\mathbf{\Lambda}/\delta u_k + \sum_{j=1}^{n} dx_j \wedge \partial'_j \Omega^{(1)}.$$

Proof:

$$\delta\mathcal{H} = -\sum_{j=1}^{n} dx_j \wedge \delta T_j + (n-1)\delta\mathbf{\Lambda}$$

$$= \sum dx_j \wedge i(\tilde{\partial}_j)\Omega - \sum dx_j \wedge i(\partial_j) \sum \delta u_k \wedge \delta\mathbf{\Lambda}/\delta u_k$$

$$- \sum dx_j \wedge di(\partial_j)\Omega^{(1)} + \sum dx_j \wedge \partial'_j \Omega^{(1)} + (n-1)\delta\mathbf{\Lambda}$$

$$= \sum dx_j \wedge i(\tilde{\partial}_j)\Omega - n\sum \delta u_k \wedge \delta\mathbf{\Lambda}/\delta u_k - \sum dx_j \wedge (di(\partial_j)$$

$$+ i(\partial_j)d)\Omega^{(1)} + \sum dx_j \wedge i(\partial_j)d\Omega^{(1)} + \sum dx_j \wedge \partial'_j \Omega^{(1)}$$

$$+ (n-1)\delta\mathbf{\Lambda} = \sum dx_j \wedge i(\tilde{\partial}_j)\Omega - \sum \delta u_k \wedge \delta\mathbf{\Lambda}/\delta u_k$$

$$- (n-1)d\Omega^{(1)} - \sum dx_j \wedge \partial_j \Omega^{(1)} + nd\Omega^{(1)} + \sum dx_j \wedge \partial'_j \Omega^{(1)}$$

$$= \sum dx_j \wedge i(\tilde{\partial}_j)\Omega - \sum \delta u_k \wedge \delta\mathbf{\Lambda}/\delta u_k + \sum dx_j \wedge \partial'_j \Omega^{(1)}. \quad \square$$

17.5.11. Corollary. If the Lagrangian does not depend on times explicitly then the equation $\{\delta\mathbf{\Lambda}/\delta u_k = 0\}$ is equivalent to

$$\delta\mathcal{H} = \sum dx_j \wedge i(\tilde{\partial}_j)\Omega.$$

This equation we call the Hamilton equation.

17.5.12. Remark. Below we shall meet an example where the Lagrangian depends on times but the additional term $\sum dx_j \wedge \partial'_j \Omega^{(1)}$ in 17.5.10 vanishes and the equation has the Hamilton form.

17.5.13. *The "coordinate-momentum" variables.* We shall find the canonical variables. Let $c_{(i)} = |i|!/i_1! \ldots i_n!$ (if one of i_k is negative then $c_{(i)} = 0$). It is easy to verify that

$$c_{(i)} = \sum_{j=1}^{n} c_{(i)-e_j}$$

(Pascal's triangle). Let

$$\delta/\delta u_k^{(i)} = \sum_{(m)} (-1)^{|m|} c_{(m)}/c_{(i)+(m)} \partial^{(m)} \partial/\partial u_k^{(i)+(m)}$$

If $(i) = 0$ this coincides with $\delta/\delta u_k$. Expressions

$$P_{k,(i)} = \delta \Lambda / \delta u_k^{(i)} \in \mathcal{A}^{(0,n)}$$

are called momenta. The variables $\{u_k^{(i)}\}$ are "coordinates".

17.5.14. Lemma. The equation

$$P_{k,(i)} = c_{(i)}^{-1} \partial \Lambda / \partial u_k^{(i)} - \sum_{j=1}^{n} \partial_j P_{k,(i)+e_j}$$

holds.

Proof:

$$P_{k,(i)} = c_{(i)}^{-1} \partial \Lambda / \partial u_k^{(i)}$$
$$+ \sum_{(m)\neq 0} \sum_{j=1}^{n} (-1)^{|m|} c_{(m)-e_j}/c_{(i)+(m)} \partial^{(m)} \partial \Lambda / \partial u_k^{(i)+(m)}$$
$$= c_{(i)}^{-1} \partial \Lambda / \partial u_k^{(i)} -$$
$$\sum_{(m)\neq 0} \sum_{j=1}^{n} (-1)^{|m|-1} c_{(m)-e_j}/c_{(i)+e_j+(m)-e_j} \partial^{(m)-e_j} \partial_j \partial \Lambda / \partial u_k^{(i)+e_j+(m)-e_j}$$
$$= c_{(i)}^{-1} \partial \Lambda / \partial u_k^{(i)} -$$
$$\sum_{(m)\neq 0} \sum_{j=1}^{n} (-1)^{|m|} \partial_j c_{(m)}/c_{(i)+e_j+(m)} \partial^{(m)} \partial \Lambda / \partial u_k^{(i)+e_j+(m)}$$
$$= c_{(i)}^{-1} \partial \Lambda / \partial u_k^{(i)} - \sum_{j=1}^{n} \partial_j P_{k,(i)+e_j} . \quad \square$$

17.5.15. Proposition. The form $\Omega^{(1)}$ is

$$\Omega^{(1)} = \sum_{k=1}^{m} \sum_{j=1}^{n} \sum_{(i)} c_{(i)} \delta u_k^{(i)} \wedge i(\partial_j) P_{k,(i)+e_j}$$

Proof: One must verify that Eq. (17.5.2) holds for this form:

$$d\Omega^{(1)} = -\sum_{k,j,(i)} c_{(i)}\partial_j(\delta u_k^{(i)} \wedge P_{k,(i)+e_j}) = -\sum_{k,(i)} \delta u_k^{(i)} \wedge$$

$$(\partial \mathbf{\Lambda}/\partial u_k^{(i)} - c_{(i)}P_{k,(i)}) - \sum_{k,(i),j} \delta u_k^{(i)+e_j} \wedge c_{(i)}P_{k,(i)+e_j}$$

$$= -\delta \mathbf{\Lambda} + \sum \delta u_k \wedge \delta \mathbf{\Lambda}/\delta u_k. \square$$

17.5.16. *Corollary.* In coordinate-momentum variables:

$$\Omega = \sum_{k=1}^{m}\sum_{j=1}^{n}\sum_{(i)} c_{(i)}\delta u_k^{(i)} \wedge i(\partial_j)\delta P_{k,(i)+e_j},$$

$$T_j = \sum_{k,(i),l} c_{(i)} u_k^{(i)+e_j} i(\partial_l) P_{k,(i)+e_l},$$

$$\mathcal{H} = \sum_{k,(i),j} c_{(i)} u_k^{(i)+e_j} P_{k,(i)+e_j} - \mathbf{\Lambda}.$$

17.5.17. *Proposition.* The Hamilton equation 17.5.11 is now

$$\delta \mathcal{H} = \sum c_{(i)} u_k^{(i)+e_j} \delta P_{k,(i)+e_j} - \sum c_{(i)}\delta u_k^{(i)} \wedge \partial_j P_{k,(i)+e_j}$$

Proof: The expression obtained for Ω must be substituted into 17.5.11. $\square$

If some $c_{(i)} u_k^{(i)}$ and the corresponding $P_{k,(i)+e_j}$ can be taken as independent variables in the phase space, then the last equation is equivalent to the system

(17.5.18) $\quad \partial_j c_{(i)} u_k^{(i)} = \delta h/\delta p_{k,(i)+e_j}, \sum_j \partial_j p_{k,(i)+e_j} = -\delta h/\delta c_{(i)} u_k^{(i)}$

where $\mathcal{H} = h dx_1 \wedge \ldots \wedge dx_n$ and $P_{k,(i)} = p_{k,(i)} dx_1 \wedge \ldots \wedge dx_n$ having the canonical form (17.1.3).

17.6. *Variational bi-complex of a differential equation. First integrals.*

17.6.1. Let

(17.6.2) $\qquad Q_k = 0, \, k = 1, \ldots, m, Q_k \in \mathcal{A}$

be a system of differential equations. The elements Q_k generate in $\mathcal{A}$ a differential ideal J_Q. Let $\mathcal{A}_Q = \mathcal{A}/J_Q$. The system of equations (17.6.2) is called a system of pseudo Cauchy-Kovalevski type (pCK) (see [90]) if

$$Q_k = u_k^{(i)_k} + Q_k^*$$

where $(i)_k$ are multi-indices depending on k, and Q_k^* do not depend on $\{u_l^{(i)}\}$ with $(i) \geq (i)_l$ (we say that $(i) \geq (l)$ if all $i_j \geq l_j$).

If a system is of the pCK type, then the elements of the two kinds below

(17.6.3)   (a) $\partial^{(i)} Q_k$ $\forall_{(i),k}$
               (b) $\{u_k^{(i)}\}$, $(i) \not\geq (i)_k$

can be taken as generators of $\mathcal{A}$. An element $f \in \mathcal{A}$ belongs to J_Q if and only if each term in its representation in terms of the generators (17.6.3) contains at least one factor of the type (a).

Now we develop a theory of characteristics quite similar to that in Sec. 11.2. Let $f \in J_Q$ then

$$f = \sum_{k,(i)} a_{k,(i)} \tilde{\partial}^{(i)} Q_k, \quad a_{k,(i)} \in \mathcal{A}.$$

Put

$$\chi_{f,k} = \left(\sum (-1)^{|i|} \partial^{(i)} a_{k,(i)} \right)_Q, \quad k = 1, \ldots, m.$$

The subscript Q denotes the natural projection $\mathcal{A} \to \mathcal{A}_Q$: The set $\chi_f = (\chi_{f,1}, \ldots \chi_{f,m}) \in \mathcal{A}_Q^m$ is called the characteristic of the element f.

17.6.4. Lemma. If $\sum a_{k,(i)} \partial^{(i)} Q_k = 0$ then all $a_{k,(i)} \in J_Q$. The proof is the same as in 11.2.3. □

17.6.5. Lemma. If $f = \partial_j g, g \in J_Q$ then $\chi_f = 0$.
Proof: See 11.2.5. □

17.6.6. Lemma. The characteristic is well defined.
Proof: See 11.2.4. □

17.6.7. A vector field ξ is called a tangent to $\mathcal{A}_Q$ if $\xi J_Q \subset J_Q$.

A tangent vector field is defined in $\mathcal{A}_Q$. Two tangent vector fields are equivalent (modulo the equation) if $(\xi - \eta)\mathcal{A} \subset J_Q$. Then the coefficients $\xi_{(k),j}$ and ξ_j^* of these vector fields are equal as elements of $\mathcal{A}_Q$. The linear space of the equivalence classes is denoted as $T\mathcal{A}_Q$.

It will be given two notions of equivalence of forms, a strong one and a weak one.

17.6.8. Definition. A form ω is equivalent to zero modulo the equation in the strong sense, $\omega \stackrel{QQ}{=} 0$, if all its coefficients belong to J_Q. A form is equivalent to zero in the weak sense $\omega \stackrel{Q}{=} 0$ if the result of substitution of tangent vector fields into this form is zero in $\mathcal{A}_Q$. It is easy to see that this condition can be expressed thus: if the form is written in terms of generators (17.6.3) each term contains a generator of type (a) either in the coefficient or under the sign δ.

Example. $\delta Q_k \stackrel{Q}{=} 0$ but $\delta Q_k \stackrel{QQ}{\neq} 0$.

If $\omega \stackrel{QQ}{=} 0$ then $i(\xi)\omega \stackrel{QQ}{=} 0$ for each $\xi \in T\mathcal{A}$. If $\omega \stackrel{Q}{=} 0$ then $i(\xi)\omega \stackrel{Q}{=} 0$ for $\xi \in T\mathcal{A}_Q$.

The classes of equivalence of forms make complexes $\mathcal{A}_Q^{(\cdot,\cdot)} = \left\{\mathcal{A}_Q^{(p,q)}\right\}$ and $\mathcal{A}_{QQ}^{(\cdot,\cdot)} = \left\{\mathcal{A}_{QQ}^{(p,q)}\right\}$. The complex $\mathcal{A}_{QQ}^{(\cdot,\cdot)}$ has only one differential d.

Let $\tilde{\mathcal{A}}^{(p,q)} = \mathcal{A}^{(p,q)}/d\mathcal{A}^{(p,q-1)}, \tilde{\mathcal{A}}^{(\cdot,\cdot)} = \left\{\tilde{\mathcal{A}}^{(p,q)}\right\}$. There is a differential δ in this complex, and all the sequences are exact with respect to this differential (the proof is the same as in 17.4.3 where $q = n$). The mapping d can be considered as a mapping $\tilde{\mathcal{A}}^{p,q} \to \mathcal{A}^{p,q+1}$. Vector fields ∂_h can be transferred to $\tilde{\mathcal{A}}^{(\cdot,\cdot)}$ since they commute with d, see 17.2.16. Finally, we introduce the complex $\tilde{\mathcal{A}}_Q^{(p,q)} = \mathcal{A}_Q^{(p,q)}/d\mathcal{A}_Q^{(p,q-1)}, \tilde{\mathcal{A}}_Q^{(\cdot,\cdot)} = \left\{\tilde{\mathcal{A}}_Q^{(p,q)}\right\}$.

17.6.9. Proposition. All the sequences

$$\mathcal{A}_{QQ}^{(p,q)} \xrightarrow{d} \mathcal{A}_{QQ}^{(p,q+1)} \xrightarrow{d} \mathcal{A}_{QQ}^{(p,q+2)}$$

are exact if $p > 0$.

Proof: The proof of 17.3.4 remains valid since the operator D has the property: if all the coefficients of a form ω belong to J_Q then so do the coefficients of $D\omega$, i.e. the operator D can be transferred to $\mathcal{A}_{QQ}^{(\cdot,\cdot)}$. □

Note that if $p = 0$ then the sequence is not exact (otherwise there would be no first integrals). The proof of exactness by diagram chasing cannot be carried out in this case since there is no differential δ in the complex $\mathcal{A}_{QQ}^{(\cdot,\cdot)}$.

17.6.10. Definition. First integral is an element $\tilde{F} \in \tilde{\mathcal{A}}_Q^{(0,n-1)}$ such that $d\tilde{F} = 0$ in $\mathcal{A}_Q^{(0,n)}$.

If F is a representative of the class then $dF = adx_1 \wedge \ldots \wedge dx_n$ where $A \in J_Q$.

17.6.11. Definition. The characteristic of the element a of the ideal is called the characteristic of the first integral $\tilde{F}$.

According to the lemma 17.6.5 the characteristic of a first integral is independent of the choice of a representative.

17.6.12. Lemma. If a representative of the class $\tilde{F}$ is taken in the reduced form i.e. contains only generators of the type (17.6.3b) then a has the form $\sum a_k Q_k$ and $\chi_F = \{a_k\}$.
The proof is obvious. □

17.6.13. Proposition. Let $Q_k = \delta\Lambda/\delta u_k$ i.e. let the equation have a variational Euler-Lagrange form. Let $\tilde{F}$ be a first integral with characteristic $\chi_F = \{\chi_{F,k}\}$. Then the vector field ∂_{χ_F} has the property $\partial_{\chi_F}\Lambda = 0$ in $\tilde{\mathcal{A}}^{(0,n)}$.

Proof:
$$\partial_{\chi_F}\Lambda = \sum_{k,(i)} \chi_{F,k}^{(i)} \partial\Lambda/\partial u_k^{(i)} = \sum_k \chi_{F,k} \delta\Lambda/\delta u_k \text{ in } \tilde{\mathcal{A}}^{(0,n)}.$$

On the other hand,
$$0 = dF = \sum a_{k,(i)} \partial^{(i)} \delta\Lambda/\delta u_k = \sum_k \chi_{F,k} \delta\Lambda/\delta u_k \text{ in } \tilde{\mathcal{A}}^{(0,n)}.$$

Therefore $\partial_{\chi_F}\Lambda = 0$ in $\tilde{\mathcal{A}}^{(0,n)}$ as required. □

17.6.14. Lemma. If $\sum \delta u_k \wedge \varphi_k + d\omega_1^{1,n-1} \stackrel{QQ}{=} 0, \varphi_k \in \mathcal{A}_{QQ}^{(0,n)}, \omega_1^{1,n-1} \in \mathcal{A}_{QQ}^{(1,n-1)}$ then $\varphi_k \stackrel{QQ}{=} 0$ and a form $\omega^{1,n-2}$ exists such that $\omega_1^{1,n-1} \stackrel{QQ}{=} d\omega^{1,n-2}$.

Proof: The proof of uniqueness in 17.4.4 with the aid of Tulczyjev's operator D remains valid since this operator acts also in $\mathcal{A}_{QQ}$.

17.6.15. Proposition. The vector field ∂_{χ_F} preserves the ideal J_Q i.e. it belongs to $T\mathcal{A}_Q$.

Proof: Let us act with this field on the equality $\delta\Lambda = \sum \delta u_k \wedge \delta\Lambda/\delta u_k - d\Omega^{(1)}$:

$$\delta\partial_{\chi_F}\Lambda = \sum \delta u_k \wedge \partial_{\chi_F} \delta\Lambda/\delta u_k + \sum (\partial_{\chi_F} \delta u_k) \wedge \delta\Lambda/\delta u_k + d(\ ).$$

The proposition 17.6.13 yields $\delta\partial_{\chi_F}\Lambda = 0$ in $\tilde{\mathcal{A}}$. Hence $\sum \delta u_k \wedge \partial_{\chi_F} \delta\Lambda/\delta u_k + \sum (\partial_{\chi_F} \delta u_k) \wedge \delta\Lambda/\delta u_k = 0$ in $\tilde{\mathcal{A}}$ and

$$\sum \delta u_k \wedge \partial_{\chi_F} \delta\Lambda/\delta u_k \stackrel{QQ}{=} d(\ ).$$

Then lemma 17.6.14 gives $\partial_{\chi_F}\delta\Lambda/\delta u_k \stackrel{QQ}{=} 0$ i.e. $\partial_{\chi_F}\delta\Lambda/\delta u_k \subset J_Q$ and $\partial_{\chi_F} J_Q \subset J_Q$. □

17.6.16. *The main proposition about first integrals.* The relation

$$\delta \tilde{F} \stackrel{Q}{=} i(\partial_{\chi_F})\Omega + d\omega^{1,n-2}$$

(where $\omega^{1,n-2} \in \mathcal{A}_Q^{1,n-2}$ is a form) holds.

Proof: We have $\tilde{F} \in \tilde{\mathcal{A}}_Q^{(0,n-1)}$ and $d\tilde{F} \stackrel{Q}{=} 0$. Let F be a representative of the class having the reduced form. Then $dF = \sum \chi_{F,k} \delta\Lambda/\delta u_k$ (see 17.6.12). We have

$$di(\partial_{\chi_F})\Omega = -i(\partial_{\chi_F})d\Omega = -i(\partial_{\chi_F})\sum \delta u_k \wedge \delta(\delta\Lambda/\delta u_k) = -\sum \chi_{F,k}$$
$$\delta(\delta\Lambda/\delta u_k) + \sum \delta u_k \partial_{\chi_F}\delta\Lambda/\delta u_k \stackrel{QQ}{=} -\sum \chi_{F,k}\delta(\delta\Lambda/\delta u_k).$$

(17.6.5 is used). On the other hand

$$d\delta F = -\delta dF = -\delta \sum \chi_{F,k}\delta\Lambda/\delta u_k \stackrel{QQ}{=} -\sum \chi_{F,k}\delta(\delta\Lambda/\delta u_k),$$

whence $d(\delta F - i(\partial_{\chi_F})\Omega) \stackrel{QQ}{=} 0$. It remains to apply the lemma 17.6.14.

17.6.17. *Example.* The energy-momentum tensor is a set of first integrals T_j if the Lagrangian does not depend on the times explicitly. The proposition 17.5.5 shows that the characteristic of the first integral T_j is $\chi_{T_j} = \{-\partial_j u_k\}, k = 1, \ldots, m$. Hence $\partial_{\chi_{T_j}} = -\sum u_k^{(i)+e_j} \partial/\partial u_k^{(i)} = -\tilde{\partial}_j$. The same was obtained earlier in 17.5.7.

17.7. Poisson bracket.

17.7.1. Let $V \subset \tilde{\mathcal{A}}_Q^{(0,n-1)}$ be the subspace of first integrals. As it was proven, a vector field $\xi_F = -\partial_{\chi_F} \in T\mathcal{A}_Q$ corresponds to each $\tilde{F} \in V$ such that $\delta \tilde{F} = -i(\xi_F)\Omega$ in $\tilde{\mathcal{A}}_Q^{(1,n-1)}$.

For two elements $\tilde{F}, \tilde{G} \in V$ we define the Poisson bracket

(17.7.2) $$\{\tilde{F}, \tilde{G}\} = \xi_F \tilde{G}.$$

17.7.3. *Proposition.* The space V is a Lie algebra with respect to the Poisson bracket and

$$\xi_{\{\tilde{F},\tilde{G}\}} = [\xi_F, \xi_G].$$

Proof: From $d\tilde{G} \stackrel{Q}{=} 0$ we obtain $\xi_F d\tilde{G} \stackrel{G}{=} 0$ since $\xi_F \in T\mathcal{A}_Q$, and $d\xi_F \tilde{G} \stackrel{Q}{=} 0$ since ξ_F commutes with d. This proves that $\{\tilde{F}, \tilde{G}\} \in V$. Applying the operator $i(\xi_G)$ to both sides of $\delta \tilde{F} = -i(\xi_F)\Omega$ (in $\tilde{\mathcal{A}}_Q^{(\cdot\cdot)}$) which can be done by the same reason as above we get

$$\xi_G \tilde{F} = -\Omega(\xi_F, \xi_G) \; (\text{in } \tilde{\mathcal{A}}_Q^{(\cdot\cdot)})$$

which implies the skew symmetry of $\{\,,\,\}$. Now, using the proposition 17.2.17 and the fact that the form Ω is closed, $\delta\Omega = 0$, we have (calculations in $\tilde{\mathcal{A}}_Q^{(\cdot\cdot)}$)

$$\begin{aligned}
0 &= i(\xi_F)i(\xi_G)\delta\Omega = i(\xi_F)\xi_G\Omega - i(\xi_F)\delta i(\xi_G)\Omega \\
&= \xi_G i(\xi_F)\Omega - i([\xi_G, \xi_F])\Omega - i(\xi_F)\delta i(\xi_G)\Omega \\
&= -\xi_G \delta \tilde{F} - i([\xi_G, \xi_F])\Omega + i(\xi_F)\delta\delta\tilde{G} \\
&= -\delta\xi_G \tilde{F} - i([\xi_G, \xi_F])\Omega = \delta\{\tilde{F}, \tilde{G}\} + i([\xi_F, \xi_G])\Omega\,.
\end{aligned}$$

This is equivalent to the required equality. At the same time this proves the Jacobi identity

$$\begin{aligned}
\{\{\tilde{F}, \tilde{G}\}, \tilde{H}\} &= \xi_{\{\tilde{F}, \tilde{G}\}}\tilde{H} = [\xi_F, \xi_G]\tilde{H} = \xi_F\{\tilde{G}, \tilde{H}\} \\
&\quad - \xi_G\{\tilde{F}, \tilde{H}\} = \{\tilde{F}, \{\tilde{G}, \tilde{H}\}\} - \{\tilde{G}, \{\tilde{F}, \tilde{H}\}\} \quad \square
\end{aligned}$$

Two elements of V are said to be in involution if their Poisson bracket vanishes.

It is not clear how to extend the notion of the Poisson bracket to more general objects than the first integrals.

17.8. Connection with the single-time formalism.

17.8.1. Let

$$\Omega^{(1)} = \sum_{r=1}^{n} \Omega_r^{(1)} \wedge i(\partial_r)dx_1 \wedge \ldots \wedge dx_n,$$

$$\Omega = \sum_{r=1}^{n} \Omega_r \wedge i(\partial_r)dx_1 \wedge \ldots \wedge dx_n,$$

$$T_j = \sum_{r=1}^{n} T_{jr} i(\partial_r)dx_1 \wedge \ldots \wedge dx_n$$

be the forms written in coordinates. Let one of variables $x_1, \ldots, x_n$ be taken as a time, say $t = x_n$. The term with $i(\partial_n)dx_1 \wedge \ldots \wedge dx_n = (-1)^{n-1}dx_1 \wedge \ldots \wedge dx_{n-1}$ in Eq. 17.5.6 for $j = n$ is

$$\delta T_{nn} \wedge dx_1 \wedge \ldots \wedge dx_{n-1} \stackrel{Q}{=} -i(\tilde{\partial}_n)\Omega_n \wedge dx_1 \wedge \ldots \wedge dx_{n-1} + d^{(n-1)}\omega$$

where $d^{(n-1)}$ means $\sum_{1}^{n-1} dx_i \partial_i$, a differential with respect to $n-1$ independent variables, and ω is a $(1, n-2)$-form with respect to these variables. As usual, $\stackrel{Q}{=}$ means that the non-phase variables must be eliminated with the aid of the equation $\{\delta\Lambda/\delta u_k = 0\}$. Then the differentiation $\tilde{\partial}_n$ is the vector field ξ associated with the equation. This equality can also be written as

$$\delta \int \ldots \int T_{nn} dx_1 \wedge \ldots \wedge dx_{n-1} = -i(\xi) \int \ldots \int \Omega_n dx_1 \wedge \ldots \wedge dx_{n-1}$$

Thus, $\int \ldots \int T_{nn} dx_1 \wedge \ldots \wedge dx_{n-1}$ is the Hamiltonian, and $\int \ldots \int \Omega_n \wedge dx_1 \wedge \ldots \wedge dx_{n-1}$ the symplectic form of the single-time formalism.

17.8.2. Conversely, if one knows a Hamiltonian and a symplectic form in the single-time formalism then the Lagrangian and all components of the multi-time formalism can be restored. It follows from 17.5.4:
(17.8.3)
$$\Lambda = -dx_j \wedge T_j + dx_j \wedge i(\tilde{\partial}_j)\Omega^{(1)} = -\left(T_{jj} - i(\tilde{\partial}_j)\Omega_j^{(1)}\right) dx_1 \wedge \ldots \wedge dx_n.$$

If $x_j = t$ is the chosen time variable, T_{jj} and $\Omega_j^{(1)}$ are supposed to be known up to $\partial_j(\ )$. Then Λ is determined up to $d(\ )$ as it must be the case.

17.8.4. *The first example: the KdV equation* $\dot{u} = 6uu' + u'''$. Recall that the forms $\Omega^{(1)}$ and Ω are defined on the vector fields $\partial_{a'}$, and $\Omega^{(1)}(\partial_{a'}) = \frac{1}{2}\int uadx$, $\Omega(\partial_{a'}, \partial_{b'}) = \frac{1}{2}\int a'b\,dx$. In order to write these forms in terms of the basic differentials we use a trick, which will be systematically discussed in the next chapter, namely, we make a substitution:

(17.8.5) $$u = \varphi'$$

which embeds the differential algebra $\mathcal{A} = \mathcal{A}_u$ into $\mathcal{A}_\varphi$. This is the so-called "potential representation", see Witham [109]. The equation takes the form $\dot{\varphi}' = 6\varphi'\varphi'' + \varphi^{IV}$. Vector fields $\partial_{a'}$ can be written as

$$\partial_{a'} = \sum_{0}^{\infty} a^{(i+1)} \partial/\partial u^{(i)} = \sum_{0}^{\infty} a^{(i+1)} \partial/\partial \varphi^{(i+1)} = \sum_{1}^{\infty} a^{(i)} \partial/\partial \varphi^{(i)}.$$

These fields, originally defined on $\mathcal{A}_u$, can be extended to the whole of $\mathcal{A}_\varphi$ by

$$\tilde{\partial}_a = \sum_0^\infty a^{(i)} \partial/\partial\varphi^{(i)}.$$

17.8.6. Lemma. The forms $\Omega^{(1)}$ and Ω can be written as

$$\Omega^{(1)} = \frac{1}{2}\int \varphi'\delta\varphi \wedge dx, \Omega = \frac{1}{2}\int \delta\varphi' \wedge \delta\varphi \wedge dx.$$

Proof:

$$\Omega^{(1)}(\partial_{a'}) = i(\tilde{\partial}_a)\frac{1}{2}\int \varphi'\delta\varphi \wedge dx = \frac{1}{2}\int \varphi' a\, dx = \frac{1}{2}\int u a\, dx,$$

$$\Omega(\partial_{a'}, \partial_{b'}) = i(\tilde{\partial}_b)i(\tilde{\partial}_a)\frac{1}{2}\int \delta\varphi' \wedge \delta\varphi \wedge dx = \frac{1}{2}\int (a'b - ab')dx = \int a'b\, dx$$

as required. □

The Hamiltonian of the KdV-equation is $H = \int(u^3 - (u')^2/2)dx = \int((\varphi')^3 - (\varphi')^2/2)dx$.

Now we pass to the multi-time formalism. $\Omega^{(1)}$ and Ω will now denote forms in this formalism. Let $x_1 = x$ and $x_2 = t$. We know the following components of the forms $\Omega^{(1)}$ and $\{T_j\}$: $\Omega_t^{(1)} = \Omega_2^{(1)} = \varphi'\delta\varphi/2$ and $T_{tt} = T_{22} = -\varphi'^3 + \varphi''^2/2$. Equation (17.8.3) yields

$$\mathbf{\Lambda} = -(T_{tt} - i(\tilde{\partial}_t)\Omega_t^{(1)})dx \wedge dt = -(-\varphi'^3 + \frac{1}{2}\varphi''^2 + \frac{1}{2}\varphi'\dot{\varphi})dx \wedge dt$$

Now all components of the forms $\Omega^{(1)}$ and Ω, and of the energy-momentum tensor $\{T_j\}$ can be found. We have

$$\delta\mathbf{\Lambda} = -(6\varphi'\varphi'' + \varphi^{IV} - \varphi')\delta\varphi \wedge dx \wedge dt$$
$$- d\left\{\frac{1}{2}\varphi'\delta\varphi \wedge dx + (3\varphi'^2\delta\varphi + \varphi'''\delta\varphi - \varphi''\delta\varphi' - \frac{1}{2}\dot{\varphi}\delta\varphi) \wedge dt\right\},$$

whence

(17.8.7)
$$\Omega^{(1)} = \Omega_t^{(1)} \wedge dx - \Omega_x^{(1)} \wedge dt = \frac{1}{2}\varphi'\delta\varphi \wedge dx +$$
$$(3\varphi'^2\delta\varphi + \varphi'''\delta\varphi - \varphi''\delta\varphi' - \frac{1}{2}\dot{\varphi}\delta\varphi) \wedge dt.$$

Further, $-\delta\mathbf{\Lambda}/\delta\varphi = 6\varphi'\varphi'' + \varphi^{IV} - \dot\varphi'$, i.e. the fact that $\mathbf{\Lambda}$ is indeed the Lagrangian is verified. Then

$$\begin{aligned}
T_t &= -i(\partial_t)\mathbf{\Lambda} + i(\tilde\partial_t)\Omega^{(1)} = -(-\varphi'^3 + \frac{1}{2}\varphi''^2 + \frac{1}{2}\varphi'\dot\varphi)dx \\
&\quad + \frac{1}{2}\varphi'\dot\varphi dx + (3\varphi'^2\dot\varphi + \varphi'''\dot\varphi - \varphi''\dot\varphi' - \frac{1}{2}\dot\varphi^2)dt \\
&= (\varphi'^3 - \frac{1}{2}\varphi''^2)dx + (3\varphi'^2\dot\varphi + \varphi'''\dot\varphi - \varphi''\dot\varphi' - \frac{1}{2}\dot\varphi^2)dt \\
&= -T_{tt}dx + T_{tx}dt, \\
T_x &= -i(\partial_x)\mathbf{\Lambda} + i(\tilde\partial_x)\Omega^{(1)} = (-\varphi'^3 + \frac{1}{2}\varphi''^2 + \frac{1}{2}\varphi'\dot\varphi)dt \\
&\quad + \frac{1}{2}\varphi'^2 dx + (3\varphi'^3 + \varphi'''\varphi' - \varphi''^2 - \frac{1}{2}\dot\varphi\varphi')dt \\
&= \frac{1}{2}\varphi'^2 dx + (2\varphi'^3 + \varphi'''\varphi' - \frac{1}{2}\varphi''^2)dt = -T_{xt}dx + T_{xx}dt.
\end{aligned}$$

The field Hamiltonian is

$$\mathcal{H} = dx \wedge T_x + dt \wedge T_t + \mathbf{\Lambda} = (2\varphi'^3 - \frac{1}{2}\varphi''^2 + \varphi'\varphi''' - \frac{1}{2}\varphi'\dot\varphi)dx \wedge dt$$

and the Hamiltonian form of the equation is

$$\delta\mathcal{H} = (dx \wedge i(\tilde\partial_x) + dt \wedge i(\tilde\partial_t))\Omega$$

(see 17.5.11).

17.8.8. Exercise. Using the obtained expressions for Ω and $\mathcal{H}$ in terms of φ check that this equation is, indeed, equivalent to the KdV equation (recall that $i(\tilde\partial_x)\delta\varphi = \varphi', i(\tilde\partial_x)\delta\varphi' = \varphi'', i(\tilde\partial_t)\delta\varphi = \dot\varphi, i(\tilde\partial_t)\delta\varphi' = \dot\varphi'$ etc).

The main examples will be given in the next chapter. In particular we shall see that the substitution (17.8.5) is none other than the dressing of the operator L.

17.8.9. Example. The expression

$$\begin{aligned}
F = &(10\varphi'^3 + 10\varphi'\varphi''' + 5\varphi''^2 + \varphi^V)dx + (\varphi^{VII} + 24\varphi^{IV}\varphi'' + 23\varphi'''^2 \\
&60\varphi'\varphi''^2 + 16\varphi'\varphi^V + 90\varphi'^2\varphi''' + 45\varphi'^4)dt
\end{aligned}$$

is a first integral of the KdV-equation since

$$-dF = (30\varphi'^2 + 10\varphi''')\delta\mathbf{\Lambda}/\delta\varphi + 10\varphi''(\delta\mathbf{\Lambda}/\delta\varphi)' + 10\varphi'(\delta\mathbf{\Lambda}/\delta\varphi)'' + (\delta\mathbf{\Lambda}/\delta\varphi)^{IV}$$

(check this!). Whence

$$-\chi_F = 30\varphi'^2 + 10\varphi'''.$$

It is also instructive to verify the relation

$$\delta F = i(\partial_{30\varphi'^2+10\varphi'''})\Omega + d(\;)$$

directly. The momenta (see 17.5.3) can be found:

$$P_{1,0} = \delta\Lambda/\delta\varphi' = (3\varphi'^2 + \varphi''' - \frac{1}{2}\dot\varphi)dx \wedge dt,$$

$$P_{0,1} = \delta\Lambda/\delta\dot\varphi = -\frac{1}{2}\varphi', \; P_{2,0} = \delta\Lambda/\delta\varphi'' = 0.$$

and the proposition 17.5.13 can be verified. However, no one set of $c_{(i)}\varphi^{(i)}$ and the corresponding $P_{(i)+e_j}$ can be chosen as independent variables, thus the representation (17.5.18) has no sense.

CHAPTER 18. Further Examples and Applications.

18.1. KP-hierarchy.

18.1.1. At first we discuss the KP-equation in the proper sense:

$$-3u_{yy} + (4\dot{u} - u''' - 6uu')' = 0.$$

Letting $u = \varphi'$ we get

(18.1.2) $$-3\varphi_{yy} + 4\dot{\varphi}' - \varphi^{IV} - 6\varphi'\varphi'' = 0.$$

In order to obtain the Lagrangian of this equation, we multiply the equation by φ and divide every term by its degree in φ and its derivatives (see 17.4.10):

(18.1.3) $$\mathbf{\Lambda} = -\left(-\frac{3}{2}\varphi_y^2 + 2\dot{\varphi}\varphi' + \frac{1}{2}\varphi''^2 - \varphi'^3\right) dx \wedge dy \wedge dt$$

(we have used our right to add to $\mathbf{\Lambda}$ a complete differential). The variation of this Lagrangian is

$$\delta\mathbf{\Lambda} = -(3\varphi_{yy} - 4\dot{\varphi}' + \varphi^{IV} + 6\varphi'\varphi'')\delta\varphi \wedge dx \wedge dy \wedge dt - d\Omega^{(1)},$$

where

$$\Omega^{(1)} = -3\varphi_y\delta\varphi \wedge dx \wedge dt - 2\varphi'\delta\varphi \wedge dx \wedge dy + (-2\dot{\varphi}\delta\varphi$$
$$- \varphi''\delta\varphi' + \varphi'''\delta\varphi + 3\varphi'^2\delta\varphi) \wedge dy \wedge dt.$$

This yields
$$\delta\Lambda/\delta\varphi = -3\varphi_{yy} + 4\dot\varphi' - \varphi^{IV} - 6\varphi'\varphi''$$
confirming the fact that Λ is the Lagrangian, and
$$\Omega = -3\delta\varphi_y \wedge \delta\varphi \wedge dx \wedge dt - 2\delta\varphi' \wedge \delta\varphi \wedge dx \wedge dy$$
$$+ (-2\delta\dot\varphi \wedge \delta\varphi - \delta\varphi'' \wedge \delta\varphi' + \delta\varphi''' \wedge \delta\varphi + 6\varphi'\delta\varphi' \wedge \delta\varphi) \wedge dy \wedge dt.$$

The field Hamiltonian is
$$\mathcal{H} = -\Lambda + \left\{ dx \wedge i(\tilde\partial) + dy \wedge i(\tilde\partial_y) + dt \wedge i(\tilde\partial_t) \right\} \Omega^{(1)}$$
$$= \left(\frac{3}{2}\varphi_y^2 - 2\dot\varphi\varphi' - \frac{1}{2}\varphi''^2 + 2\varphi'^3 + \varphi'\varphi''' \right) dx \wedge dy \wedge dt,$$
and the Hamiltonian form of the equation is
$$\delta\mathcal{H} = \left\{ dx \wedge i(\tilde\partial) + dy \wedge i(\tilde\partial_y) + dt \wedge i(\tilde\partial_t) \right\} \Omega$$
(check that this equation is, indeed, equivalent to (18.1.2)). The energy-momentum tensor is the set of three forms
$$T_x = -i(\partial)\Lambda + i(\tilde\partial)\Omega^{(1)} = -3\varphi_y\varphi' dx \wedge dt - 2\varphi'^2 dx \wedge dy$$
$$+ \left(-\frac{1}{2}\varphi''^2 + 2\varphi'^3 + \varphi'\varphi''' + \frac{3}{2}\varphi_y^2 \right) dy \wedge dt,$$
$$T_y = -i(\partial_y)\Lambda + i(\tilde\partial_y)\Omega^{(1)} = -2\varphi'\varphi_y dx \wedge dy - \left(\frac{3}{2}\varphi_y^2 + 2\dot\varphi\varphi' \right.$$
$$\left. + \frac{1}{2}\varphi''^2 - \varphi'^3 \right) dx \wedge dt - (2\varphi\varphi_y + \varphi''\varphi_y'$$
$$- \varphi'''\varphi_y - 3\varphi'^2\varphi_y) dy \wedge dt,$$
$$T_t = -i(\partial_t)\Lambda + i(\tilde\partial_t)\Omega^{(1)} = -\left(\frac{3}{2}\varphi_y^2 - \frac{1}{2}\varphi''^2 + \varphi'^3 \right) dx \wedge dy$$
$$- 3\varphi_y\varphi_t dx \wedge dt - \left(2\dot\varphi^2 + \varphi''\dot\varphi' - \varphi'''\dot\varphi - 3\varphi'^2\dot\varphi \right) dy \wedge dt.$$

The Hamiltonian of the single-time formalism is
$$h = \int_{t=\mathrm{const}} T_{tt} dx \wedge dy = \iint \left(-\frac{3}{2}\varphi_y^2 + \frac{1}{2}\varphi''^2 - \varphi'^3 \right) dx \wedge dy$$

(which coincides with the one usually used, see e.g. Case [103]). The symplectic form of the single-time theory is

$$\omega = \int_{t=\text{const}} \int \Omega_t dx \wedge dy = 2 \int \int \delta\varphi \wedge \delta\varphi' \wedge dx \wedge dy.$$

18.1.4. Now we pass on to the general case. As it became clear earlier one must use a new set of variables $\{\varphi_i\}$ instead of variables $\{u_i\}$ generalizing the substitution $u = \varphi'$. This is the dressing substitution (see (7.1.2)):

$$L = \phi \partial \phi^{-1},$$

where $\phi = 1 + \sum_0^\infty \varphi_i \partial^{-i-1}$.

As we have seen, an operation here is of great importance: dividing all the terms of a differential polynomial in $\{\varphi_0\}$ by their degrees in $\{\varphi_i^{(j)}\}$. This operation can be performed thus: $\int_0^1 p^{-1} f(p\varphi) dp$, cf. 17.4.11. Therefore we introduce the notation $\phi_p = 1 + p \sum_0^\infty \varphi_i \partial^{-i-1}$.

18.1.5. *Proposition.* Equation (5.2.4)

$$\partial_m L_+^n - \partial_n L_+^m = [L_+^m, L_+^n]$$

can be obtained from the Lagrangian

$$\Lambda = \text{res}\left\{ -\int_0^1 p^{-1} \left[(\phi_p \partial^m \phi_p^{-1})_+, (\phi_p \partial^n \phi_p^{-1})_+\right] \phi_p^{-1} dp \right.$$

$$\left. + \partial^n \phi^{-1} \cdot \partial\phi/\partial x_m - \partial^m \phi^{-1} \cdot \partial\phi/\partial x_n \right\} dx \wedge dx_m \wedge dx_n.$$

Proof: First the formula

(18.1.6) $$\delta \text{res} \int_0^p p^{-1} \left[(\phi_p \partial^m \phi_p^{-1})_+, (\phi_p \partial^n \phi_p^{-1})_+\right] \phi_p^{-1} dp$$

$$= -\text{res}\left[(\phi_p \partial^m \phi_p^{-1})_+, (\phi_p \partial^n \phi_p^{-1})_+\right] \delta\phi_p \cdot \phi_p^{-1} + \partial(\;)$$

will be verified. It is sufficient to verify the equation obtained by differentiation with respect to p:

$$\delta \operatorname{res} \left[(\phi_p \partial^m \phi_p^{-1})_+ , (\phi_p \partial^n \phi_p^{-1})_+\right] \phi_p^{-1}$$
$$= -\operatorname{res}[(\phi_p \partial^m \phi_p^{-1})_+ , (\phi_p \partial^n \phi_p^{-1})_+]\delta\phi_p \cdot \phi_p^{-2}$$
$$- \operatorname{res}\left(p\frac{\partial}{\partial p}[(\phi_p \partial^m \phi_p^{-1})_+ , (\phi_p \partial^n \phi_p^{-1})_+]\right)\delta\phi_p \cdot \phi_p^{-1} + \partial(\ )$$

(it was taken into account that $p\partial\phi_p/\partial p = \phi_p - 1$). We transform

$$p\frac{\partial}{\partial p}(\phi_p \partial^m \phi_p^{-1})_+ = ((\phi_p - 1)\partial^m \phi_p^{-1} - \phi_p \partial^m \phi_p^{-1}(\phi_p - 1)\phi_p^{-1})_+$$
$$= -[\phi_p^{-1}, (\phi_p \partial^m \phi_p^{-1})_+]_+ ,$$
$$p\frac{\partial}{\partial p}(\phi_p \partial^n \phi_p^{-1})_+ = -[\phi_p^{-1}, (\phi_p \partial^n \phi_p^{-1})_+]_+ ,$$
$$\delta(\phi_p \partial^m \phi_p^{-1})_+ = [\delta\phi_p \cdot \phi_p^{-1}, (\phi_p \partial^m \phi_p^{-1})_+]_+ ,$$
$$\delta(\phi_p \partial^n \phi_p^{-1})_+ = [\delta\phi_p \cdot \phi_p^{-1}, (\phi_p \partial^n \phi_p^{-1})_+]_+ .$$

Let $S = \phi_p^{-1}, T = \delta\phi_p \cdot \phi_p^{-1}, U = (\phi_p \partial^m \phi_p^{-1})_+, V = (\phi_p \partial^n \phi_p^{-1})_+$. Now we must prove that

$$\operatorname{res}\{([[T,U]_+,V] + [U,[T,V]_+])S + [U,V]TS - [U,V]ST$$
$$- [[S,U]_+,V]T - [U,[S,V]_+]T\} = \partial(\ ).$$

Two of the terms are (we use the general property $\operatorname{res}[A, B] = \partial(\ )$)

$$\operatorname{res}\{[U,[T,V]_+]S - [[S,U]_+,V]T\} = \operatorname{res}\{[S,U] \cdot [T,V]_+$$
$$+ [T,V] \cdot [S,U]_+\} + \partial(\ ) = \operatorname{res}\{[T,V] \cdot [S,U]\} + \partial(\ ).$$

Similarly

$$\operatorname{res}\{[[T,U]_+,V]S - [U,[S,V]_+]T\} = \operatorname{res}\{[T,U] \cdot [V,S]\} + \partial(\ ).$$

The expression takes the form

$$\operatorname{res}\{-[T,U] \cdot [S,V] + [T,V] \cdot [S,U] + [U,V] \cdot [T,S]\} + \partial(\ )$$
$$= \operatorname{res} T\{[U,[V,S]] + [V,[S,U]] + [S,[U,V]]\} + \partial(\ ) = \partial(\ ),$$

which proves Eq. (18.1.6). Put $p = 1$, then

$$\delta\mathrm{res}\int_0^1 p^{-1}[(\phi_p\partial^m\phi_p^{-1})_+, (\phi_p\partial^n\phi_p^{-1})_+]\phi_p^{-1}dp$$
$$= -\mathrm{res}\phi^{-1}[(\phi\partial^m\phi^{-1})_+, (\phi\partial^n\phi^{-1})_+]\delta\phi - \partial\omega_1,$$

where ω_1 is a form. Then we calculate the rest of the variation:

$$\delta\,\mathrm{res}(\partial^n\phi^{-1}\partial\phi/\partial x_m - \partial^n\phi^{-1}\partial\phi/\partial x_n)$$
$$= (\partial/\partial x_m)\mathrm{res}\,(\partial^n\phi^{-1}\delta\phi) - (\partial/\partial x_n)\mathrm{res}(\partial^m\phi^{-1}\delta\phi)$$
$$+ \mathrm{res}(\partial^n\phi^{-1}(\partial\phi/\partial x_m)\phi^{-1}\delta\phi - \partial^m\phi^{-1}(\partial\phi/\partial x_n)\phi^{-1}\delta\phi$$
$$- \phi^{-1}(\partial\phi/\partial x_m)\partial^n\phi^{-1}\delta\phi + \phi^{-1}(\partial\phi/\partial x_n)\partial^m\phi^{-1}\delta\phi) + \partial\omega_2$$
$$= (\partial/\partial x_m)\mathrm{res}(\partial^n\phi^{-1}\delta\phi) - (\partial/\partial x_n)\mathrm{res}(\partial^m\phi^{-1}\delta\phi)$$
$$+ \mathrm{res}\phi^{-1}(\partial L_+^m/\partial x_n - \partial L_+^n/\partial x_m)\delta\phi + \partial\omega_2,$$

where ω_2 is another form. Thus
(18.1.7)
$$\delta\mathbf{\Lambda} = \mathrm{res}\{\phi^{-1}(-\partial L_+^n/\partial x_m + \partial L_+^m/\partial x_n + [L_+^m, L_-^n]) \cdot \delta\phi\}$$
$$\wedge dx \wedge dx_m \wedge dx_n + d\{-\omega \wedge dx_m \wedge dx_n + \mathrm{res}(\partial^n\phi^{-1}\delta\phi \wedge dx \wedge dx_n$$
$$+ \partial^m\phi^{-1}\delta\phi \wedge dx \wedge dx_m),$$

where $\omega = \omega_1 + \omega_2$. This implies the expression for the variational derivative:

$$\delta\Lambda/\delta\phi = \{\phi^{-1}(-\partial L_+^n/\partial x_m + \partial L_+^m/\partial x_n + [L_+^m, L_+^n])\}_+.$$

Equating this to zero one gets an equation which is equivalent to Eq. (5.2.4). This completes the proof. $\square$

18.1.8. Proposition. The 1-form corresponding to the Lagrangian is

$$\Omega^{(1)} = \omega \wedge dx_m \wedge dx_n - \mathrm{res}(\partial^n\phi^{-1}\delta\phi \wedge dx \wedge dx_n + \partial^m\phi^{-1}\delta\phi \wedge dx \wedge dx_m),$$

and the 2-form is

$$\Omega = \delta\Omega^{(1)} = \delta\omega \wedge dx_m \wedge dx_n + \mathrm{res}(\partial^n\phi^{-1}\delta\phi \wedge \phi^{-1}\delta\varphi \wedge dx \wedge dx_n$$
$$+ \partial^m\phi^{-1}\delta\phi \wedge \phi^{-1}\delta\phi \wedge dx \wedge dx_m).$$

The field Hamiltonian is

$$\mathcal{H} = -\Lambda + (dx \wedge i(\tilde{\partial}) + dx_m \wedge i(\tilde{\partial}_m) + dx_n \wedge i(\tilde{\partial}_n))\Omega^{(1)} = (i(\tilde{\partial})\omega$$
$$+ \operatorname{res} \int_0^1 p^{-1}[(\phi_p \partial^m \phi_p^{-1})_+ , (\phi_p \partial^n \phi_p^{-1})_+]\phi_p^{-1} dp) dx \wedge dx_m \wedge dx_n .$$

The Hamiltonian form of the equation is, as usual,

$$\delta \mathcal{H} = (dx \wedge i(\tilde{\partial}) + dx_m \wedge i(\tilde{\partial}_m) + dx_n \wedge i(\tilde{\partial}_n))\Omega .$$

Proof: All this immediately follows from (18.1.7). □

18.2. Polynomial matrix family.

18.2.1. We return to the matrix differential operators polynomially depending on a parameter, studied earlier in Chap. 16. Now we intend to make both the independent variables, x and t equal in rights. We have a system

(18.2.2) $$\dot{U}_+ + V'_+ = [U_+, V_+],$$

where

$$U_+ = \sum_0^{m+1} U_i \varsigma^i , \ V_+ = \sum_0^{m+1} V_i \varsigma^i .$$

Equation (18.2.2) differs from that in Sec. 16.2 first by that the notations U_+ and V_+ stand for U and V there. Secondly, the degrees of the polynomials U and V in 16.2 were, in general, distinct, being $m+1$ and $m+p$ respectively, now we make them equal. Without difficulty we could make them distinct here also. However, from the point of view of possibility to perform linear transformations in x and t, the case of distinct degrees seems to be degenerate. Equation (18.2.2) must be satisfied identically in ς. The matrices U_i and V_i are of order n: $U_{m+1} = A$ and $V_{m+1} = B$, where A and B are constant and diagonal with distinct diagonal elements.

Let $\mathfrak{M}_{\Lambda,M}$ be a manifold of matrix bundles U_+ and V_+ which can be expressed in terms of a single bundle $\phi = 1 + \sum_1^\infty \varphi_i \varsigma^{-i}$, diag $\varphi_i = 0$ as

(18.2.3) $$U_+ = (\phi^{-1} \Lambda \phi)_+ , V = (\phi^{-1} M \phi)_+ ,$$

where the subscript $+$ means taking terms with non-negative powers of ς and omitting those with negative powers, Λ and M are series of diagonal matrices

$$(18.2.4) \qquad \Lambda = A\varsigma^{m+1} + \sum_{-m}^{\infty} \lambda_k \varsigma^{-k}, M = B\varsigma^{m+1} + \sum_{-m}^{\infty} \mu_k \varsigma^{-k}$$

which are considered as given functions of x and t under a unique condition

$$(18.2.5) \qquad \dot{\Lambda} + M' = 0.$$

This condition is equivalent to the existence of a bundle of diagonal matrices $T = \sum_{-m}^{\infty} \tau_k \varsigma^{-k}$ for which $\Lambda = A\varsigma^{m+1} - T'$ and $M = B\varsigma^{m+1} + \dot{T}$. Note that expressions (18.2.3) involve, in fact, only φ_i with $i \leq m+1$ and λ_k, μ_k with $k \leq 0$.

18.2.6. Proposition. Equation (18.2.2) on the manifold $\mathfrak{M}_{\Lambda,M}$ turns into a consistent system for matrices $\varphi_i, i \leq m+1$.

Proof. Let $U = \phi^{-1}\Lambda\phi$ and $V = \phi^{-1}M\phi$ (thus $U_+ = (U)_+$ and $V_+ = (V)_+$). Let $U_- = U - U_+$ and $V_- = V - V_+$, φ_i being arbitrary when $i > n+1$. Obviously, $[U, V] = 0$ and hence Eq. (18.2.2) takes the form

$$(18.2.7) \qquad \dot{U}_+ + V'_+ = -[U_+, V_-]_+ - [U_-, V_+]_+.$$

This is a system of $m+1$ matrix equations for matrices $\varphi_i, i = 1, \ldots, m+1$. However the number of equations is here greater than that of variables since diag $\varphi_i = 0$. Let us replace the system (18.2.7) by an equivalent one:

$$(18.2.8) \qquad (\phi(\dot{U}_+ + V'_+ + [U_+, V_-]_+ + [U_-, V_+]_+)\phi^{-1})_+ = 0$$

(the equivalence follows from the fact that, as it can be easily seen, the systems are related to each other by a triangular transformation). Now the subscript $+$ inside the parenthesis can be omitted:

$$(\phi(\dot{U} + V' + [U, V_-] + [U_-, V])\phi^{-1})_+ = 0.$$

Expanding this equation in powers of ς we obtain for the term with ς^k the relation

$$[A, \dot{\varphi}_{m+1-k}] + [B, \varphi'_{m+1-k}] = F_k, k = m, m-1, \ldots, 0,$$

where F_k contains only φ_i with $i < m+1-k$. Considering non-diagonal parts of these equations we obtain a system of the Cauchy-Kovalevski type (both with respect to x and t). As to the diagonal parts of these equations, they vanish identically. Indeed,

$$(\phi \dot{U} \phi^{-1})_+ = \dot{\Lambda}_+ + (\Lambda \dot{\phi} \phi^{-1} - \dot{\phi}\phi^{-1}\Lambda)_+ = \dot{\Lambda}_+ + [\Lambda, \dot{\phi}\phi^{-1}].$$

The diagonal part is $\dot{\Lambda}_+$. Similarly, the diagonal part of $(\phi V' \phi^{-1})_+$ is M'_+. Further,

$$(\phi[U, V_-]\phi^{-1})_+ = [\Lambda, \phi V_- \phi^{-1}]_+.$$

The diagonal part is zero. This is true also for $(\phi[U_-, V]\phi^{-1})_+$. It remains to take into account (18.2.5) and then arrive at the desired assertion. □

18.2.9. *Proposition.* Any solution U_+, V_+ of Eq. (18.2.2) belongs to one of the manifolds $\mathfrak{M}_{\Lambda,M}$ for some Λ and M.

Proof: Let U_+ and V_+ be a solution of Eq. (18.2.2). Let us find ϕ and Λ satisfying the first of the equations (18.2.3). For the term with ς^k we have

$$U_k - [A, \varphi_{m+1-k}] - \lambda_{-k} = F_k$$

with F_k depending only on $\varphi_i, i < m+1-k$ and $\lambda_{-i}, i > k$. Taking $k = m, m-1, \ldots, 0$ we determine in succession $\varphi_1, \ldots, \varphi_{m+1}$ and diagonal matrices $\lambda_{-m}, \ldots, \lambda_0$. Then we put $M = \phi V_+ \phi^{-1}$, i.e. $V_+ = \phi^{-1} M \phi$ and prove that M is diagonal. By virtue of (18.2.2) the commutator $[U_+, V_+]$ does not contain the powers ς^i with $i \geq m+1$. Thus $[\Lambda_+, M_+]$ does not contain these powers either. This yields in succession $[A, \mu_m] + [\lambda_m, B] = 0$, i.e. $[A, \mu_m] = 0, [A, \mu_{m-1}] + [\lambda_m, \mu_m] + [\lambda_{m-1}, B] = 0$, i.e. $[A, \mu_{m-1}] = 0$, etc., in other words all the matrices $\mu_i, i = m, \ldots, 0$ are diagonal. It remains to add that the proof of the foregoing proposition showed that Eq. (18.2.2), or the equivalent equation (18.2.8), for the powers $\varsigma^i, i = 0, \ldots, m$ are equivalent to Eq. (18.2.5). The proposition is proved. □

18.2.10. *Remark.* In 16.2.4 we have already noted that the Zakharov-Shabat system is underdetermined, there remaining the freedom of choosing $m+1$ diagonal matrices. We now use this freedom to choose matrices Λ and M with an only restraint (18.2.5).

18.2.11. *Lagrangian.* Denote res $\sum a_k \varsigma^{-k} = a_1$. Let $\phi_p = 1 + \sum_{1}^{\infty} p\varphi_i \varsigma^{-i}$,

$$U_p = (\phi_p^{-1} \Lambda \phi_p)_+, V_p = (\phi_p^{-1} M \phi_p)_+,$$

where p is a parameter. Put

$$\Lambda = \operatorname{tr}\operatorname{res}\left\{\int_0^1 p^{-1}[U_{(p)},V_p]\phi_p^{-1}dp - \phi^{-1}M\phi' - \phi^{-1}\Lambda\dot\phi\right\} dx \wedge dt.$$

We shall prove that this is the Lagrangian of the system (18.2.2) on the manifold $\mathfrak{M}_{\Lambda,M}$.

18.2.12. Proposition.

$$\delta\Lambda = -\operatorname{tr}\operatorname{res}\{([U_+,V_+]-\dot U_+ - V'_+)\phi^{-1}\delta\phi\} \wedge dx \wedge dt \\ + d\operatorname{tr}\operatorname{res}\{V_+\phi^{-1}\delta\phi \wedge dt - U_+\phi^{-1}\delta\phi \wedge dx\}.$$

Proof: It does not differ from the proof of Eq. (18.1.7). $\square$
There is no additional term with ω now because the equality $\operatorname{tr}\operatorname{res}AB = \operatorname{tr}\operatorname{res}BA$ is exact, without the term $d\omega$.

18.2.13. Corollary.

$$\delta\Lambda/\delta\phi = -(([U_+,V_+]-\dot U_+ - V'_+)\phi^{-1})_+,$$

where $\delta\Lambda/\delta\phi = \sum_1^{n+1} \delta\Lambda/\delta\varphi_i\varsigma^{i-1}$, $(\delta\Lambda/\delta\varphi_i)_{kl} = \delta\Lambda/\delta(\varphi_i)_{lk}$.

18.2.14. Corollary. The form $\Omega^{(1)}$ is

$$\Omega^{(1)} = \operatorname{tr}\operatorname{res}\phi^{-1}(\Lambda\delta\phi \wedge dx - M\delta\phi\Lambda dt).$$

The variational equation $\delta\Lambda/\delta\varphi = 0$ is equivalent to Eq. (18.2.2). The other elements of the Hamiltonian formalism are the energy-momentum tensor

(18.2.15)
$$T_x = \operatorname{tr}\operatorname{res}\left\{-\int_0^1 p^{-1}[U_p,V_p]\phi_p^{-1}dpdt + \phi'^{-1} \wedge d\phi\right\}$$

$$T_t = \operatorname{tr}\operatorname{res}\left\{\int_0^1 p^{-1}[U_p,V_p]\phi_p^{-1}dpdt + \phi^{-1}Md\phi\right\}$$

and the field Hamiltonian

(18.2.16)
$$\mathcal{H} = \operatorname{tr}\operatorname{res}\left\{-\int_0^1 p^{-1}[U_p,V_p]\phi_p^{-1}dp\right\} dx \wedge dt.$$

The Hamiltonian form of Eq. (18.2.2) is

$$\delta \mathcal{H} = (dx \wedge i(\tilde{\partial}) + dt \wedge i(\tilde{\partial}_t))\Omega.$$

18.2.17. Remark. In our case the coefficients of the system of equations depend on x and t explicitly (through Λ and M). However, this does not prevent the equation from being Hamiltonian since $(dx \wedge \partial' + dt \wedge \partial'_t)\Omega^{(1)} = 0$ (see remark 17.5.12). Indeed,

$$(dx \wedge \partial'_x + dtr\, \partial'_t)\Omega^{(1)} = \operatorname{tr} \operatorname{res} \phi^{-1}(\dot{\Lambda} + M')\delta\phi \wedge dx \wedge dt = 0.$$

In what follows we drop the subscript $+$ in U_+ and V_+ and simply write U and V.

18.2.18. Resolvents and first integrals. A resolvent is a common solution of two equations

(18.2.19)   a) $\ -R' = [R, U]$, b) $\ \dot{R} = [R, V]$

having the form of a formal series $R = \sum_0^\infty R_k \varsigma^{-k}$, the equations being satisfied by virtue of Eq. (18.2.2).

The construction of a resolvent rests on three lemmas.

18.2.20. Lemma. Equation (18.2.19a) has a solution of the required form where R_0 is an arbitrary constant diagonal matrix and the elements of all matrices R_k are differential polynomials in elements of U (with differentiations with respect to x only), these polynomials do not contain free terms (i.e. constants) when $k > 0$. The same holds for Eq. (18.2.19b) with substitution of V for U and of t for x. The solution is uniquely determined by the matrix R_0.

Proof: In Chap. 16 there were two proofs of this lemma: in 16.1.3 a direct one and in 16.5.7 with the aid of a dressing matrix ϕ. □

18.2.21. Lemma. If $F' \stackrel{Q}{=} 0$ or $\dot{F} \stackrel{Q}{=} 0$ (recalling that $\stackrel{Q}{=}$ denotes equality by virtue of Eq. (18.2.2)) then $F = \text{const} + F_1$ where $F_1 \stackrel{Q}{=} 0$. In other words a first integral $F = F_{(x)}dx + F_{(t)}dt$ necessarily contains both non-trivial components $F_{(x)}$ and $F_{(t)}$.

Proof: We have already seen (proof of proposition 8.2.6) that on the manifold $\mathfrak{M}_{\Lambda,M}$ Eq. (18.2.2) turns out to be a Cauchy-Kowalevski equation (in variables φ). This makes it possible to write F in a reduced form, i.e.

containing only derivatives $\varphi', \varphi'', \ldots$. Then $F' \stackrel{Q}{=} 0$ implies $F' = 0$ and $F = $ const. □

18.2.22. Lemma. If $R = \sum\limits_0^\infty R_k \varsigma^{-k}$ satisfies Eq. (18.2.19a) by virtue of Eq. (18.2.2), $-R' \stackrel{Q}{=} [R, U]$, and all R_k do not contain constants, then $R \stackrel{Q}{=} 0$. (Similarly, (18.1.19b).)

Proof. This can be proved in the same way as for the proposition 16.1.3 (uniqueness) using the previous lemma. □

18.2.23. Proposition. The solution of (18.2.17a) constructed in lemma 2.7.18 satisfies also (18.2.17b) by virtue of Eq. (18.2.2).

Proof: Let $S = \dot{R} - [R, V]$, $-R' = [R, U]$. Then

$$S' + [S, U] = -[R, \dot{U} + V'] + [[R, U], V] - [[R, V], U]$$
$$= [\dot{U} + V' - [U, V], R] \stackrel{Q}{=} 0.$$

Since S does not contain constants, which is obvious, we have $S \stackrel{Q}{=} 0$, i.e. $\dot{R} \stackrel{Q}{=} [U, V]$ as required. □

All the resolvents are linear combinations of n basic resolvents $R^{[i]}$ with $(R_0^{[i]})_{\alpha\beta} = \delta_{\alpha i}\delta_{\beta i}$; the coefficients of linear combinations are series with constant coefficients.

18.2.24. Proposition. If R is a resolvent then

$$F = \text{tr}(-U_\varsigma dx + V_\varsigma dt)R = \sum_{-m}^\infty F_k \varsigma^{-k}$$

($U_\varsigma = \partial U/\partial \varsigma$ and $V_\varsigma = \partial V/\partial \varsigma$) is a generator of first integrals, i.e. F_k are first integrals.

Proof:

$$dF = \text{tr}(\dot{U}_\varsigma R + V'_\varsigma R + U_\varsigma \dot{R} + V_\varsigma R')dx \wedge dt$$
$$\stackrel{Q}{=} \text{tr}([U, V]_\varsigma R + U_\varsigma [R, V] - V_\varsigma [R, U])dx \wedge dt = 0. □$$

18.2.25. Proposition. If R is a resolvent with $R_0 = $ const and with other coefficients R_k non-constant, then

$$\delta F = \text{tr}\{(-\delta U \wedge dx + \delta V \wedge dt)R\}_\varsigma + d(\ \ )$$

holds.
Proof: We have (see (16.5.8))

$$\delta \operatorname{tr} U_\varsigma R \stackrel{Q}{=} \operatorname{tr}(\delta U \cdot R)_\varsigma + \partial(\ ).$$

Similarly,

$$\delta \operatorname{tr} V_\varsigma R \stackrel{Q}{=} \operatorname{tr}(\delta V \cdot R)_\varsigma + \partial(\ ).$$

These two relations are equivalent to the required one. □

Now we look for vector fields corresponding to first integrals. A vector field has the form

$$\partial_a = \operatorname{tr} \sum_{i=0}^{m} \sum_{(k)} a_i^{(k)} \partial/\partial \varphi_i^{(k)} = \operatorname{tr} \operatorname{res} \sum_{(k)} a^{(k)} \partial/\partial \varphi^{(k)},$$

where a_i are matrices and $(k) = (k_1, k_2)$ is a multi-index indicating k_1-fold differentiation with respect to x and k_2-fold differentiation with respect to t, $\partial/\partial \varphi = \sum_{1}^{m+1} \varsigma^{i-1} \partial/\partial \varphi_i$ and $a = \sum_{1}^{m+1} a_i \varsigma^{-i}$. A vector field corresponds to a first integral F if

$$\delta F \stackrel{Q}{=} -i(\partial_a)\Omega + d(\ ).$$

18.2.26. Proposition. The vector field corresponding to the first integrals F_{k+1} are $\partial_{a_{k+1}}$ where

$$a_{k+1} = k\phi(R\varsigma^{k-1})_-$$

(the subscript $-$ means that only terms with negative powers of ς are taken). The above formula involves only the terms with $\varsigma^{-1}, \ldots, \varsigma^{-m-1}$ in the expansion of a, the other terms being unimportant.
Proof: We have

$$\begin{aligned}
i(\partial_{a_{k+1}})\Omega &= k \operatorname{tr} \operatorname{res} \phi(R\varsigma^{k-1})_- \phi^{-1}\delta\phi\phi^{-1}(\Lambda dx - M dt) \\
&\quad - k \operatorname{tr} \operatorname{res} \delta\phi(R\varsigma^{k-1})_- \phi^{-1}(\Lambda dx - M dt) \\
&= -k \operatorname{tr} \operatorname{res}(R\varsigma^{k-1})_- \{\phi^{-1}(\Lambda dx - M dt)\delta\phi \\
&\quad - \phi^{-1}\delta\phi\phi^{-1}(\Lambda dx - M dt)\phi\} \\
&= -k \operatorname{tr} \operatorname{res}(R\varsigma^{k-1})_- \cdot \delta(\phi^{-1}(\Lambda dx - M dt)\phi).
\end{aligned}$$

Now we note that for any two series A and B in the relation, $\operatorname{res} A_- B_+ = \operatorname{res} A_- B = \operatorname{res} AB_+$ holds. Therefore

$$\begin{aligned}
i(\partial_{a_{k+1}})\Omega &= -k \operatorname{tr} \operatorname{res} R\varsigma^{k-1}\delta(\phi^{-1}(\Lambda dx - M dt)\phi)_+ \\
&= -k \operatorname{tr} \operatorname{res}(R\delta(U dx - V dt)).
\end{aligned}$$

On the other hand, from 18.2.23 we have

$$\delta F_{k+1} = -k \operatorname{tr} \operatorname{res} R\delta(-U\,dx + V\,dt) + d(\quad)$$

which yields

$$\delta F_{k+1} = -i(\partial_{a_{k+1}})\Omega + d(\quad). \quad\square$$

Recall the notation S_p from 16.1.6: $S_p = \sum\limits_{k=0}^{m} R_{p+k}\varsigma^{-k-1}$. Then we can write

(18.2.27) $$a_{k+1} = k\Phi S_k$$

(the rest of the terms are not involved in any formula).

The Poisson bracket of two first integrals is

$$\begin{aligned}\{F^{(1)}, F^{(2)}\} &= \Omega(\partial_{a^{(1)}}, \partial_{a^{(2)}}) = \operatorname{const},\operatorname{tr}\operatorname{res}\{\phi S^{(1)}_{k_1} S^{(2)}_{k_2}\\&\quad \cdot \phi^{-1}(\Lambda\,dx - M\,dt) - \phi S^{(2)}_{k_2} S^{(1)}_{k_1}\phi^{-1}(\Lambda\,dx - M\,dt)\}\\&= \operatorname{const}\operatorname{tr}\operatorname{res}[S^{(1)}_{k_1}, U\,dx - V\,dt]S^{(2)}_{k_2}.\end{aligned}$$

18.2.28. Proposition. Any two of the first integrals are in involution.
Proof: We use the scheme suggested in 9.3.13. For any series $X = \sum X_i \varsigma^i$, let us denote $X_{(0,m)} = \sum\limits_0^m X_i \varsigma^i$. Multiply both sides of (18.2.17) by ς^{m+k} and apply the operator $(0,m)$ (cf. 16.1.7):

$$\begin{aligned}(\varsigma^{m+1}[-\partial + U, S_k])_{(0,m)} + ([-\partial + U, S_{k+m+1}])_{(0,m)} &= 0,\\(\varsigma^{m+1}[\partial_t + V, S_k])_{(0,m)} + ([\partial_t + V, S_{k+m+1}])_{(0,m)} &= 0.\end{aligned}$$

It is clear that $([\partial, S_k])_{(0,m)} = ([\partial_t, S_k])_{(0,m)} = 0$. Taking into account $\operatorname{res} S_k X = \operatorname{res} S_k X_{(0,m)}$ we have

$$\begin{aligned}&\operatorname{tr}\operatorname{res}[S^{(1)}_{k_1}, U\,dx - V\,dt]S^{(2)}_{k_2}\\&= \operatorname{tr}\operatorname{res}\{\varsigma^{m+1}[-\partial + U, S^{(1)}_{k_1-m-1}]dx - \varsigma^{m+1}[\partial_t + V, S^{(1)}_{k_1-m-1}]dt\}S^{(2)}_{k_2}\\&= \operatorname{tr}\operatorname{res}\varsigma^{m+1}\{-[-\partial + U, S^{(2)}_{k_2}]dx + [\partial_t + V, S^{(2)}_{k_2}]dt\}S^{(1)}_{k_1-m-1}\\&= \operatorname{tr}\operatorname{res}\{-[U, S^{(2)}_{k_2+m+1}]dx + [V, S^{(2)}_{k_2+m+1}]dt\}S^{(1)}_{k_1-m-1}\\&= -\operatorname{tr}\operatorname{res}[S^{(1)}_{k_1-m-1}, U\,dx - V\,dt]S^{(1)}_{k_2+m+1}.\end{aligned}$$

We have managed to increase the subscript k_2 by $m+1$ and decrease k_1 by $m+1$. This process can be repeated until one of the subscripts becomes less than $-m$. □

18.2.29. As has already been said, the degree of polynomials U and V can be different:
$$U = \sum_0^{m+1} U_i \varsigma^i, \quad V = \sum_0^{p+1} V_i \varsigma^i.$$

Then
$$\Lambda = A\varsigma^{m+1} + \sum_{-m}^{\infty} \lambda_k \varsigma^{-k}, \quad M = B\varsigma^{p+1} + \sum_{-p}^{\infty} \mu_k \varsigma^{-k}.$$

The definition of S_k must be slightly changed in this case. Namely, $S_k = \sum_0^s R_{k+i} \varsigma^{-i-1}$, where s is an arbitrary integer greater than or equal to the maximum of m and p. Then everything remains valid.

18.2.30. *Example.* Let $m = 0, p = 1$ and the matrices A and B be equal, $A = B$. The system takes the form
$$[U_0, A] + [A, V_1] = 0$$
$$[U_0, V_1] + [A, V_0] = V_1'$$
$$[U_0, V_0] = \dot{U}_0 + V_0'.$$

All the matrices are supposed to be complex and skew symmetric. Let
$$A = i \begin{pmatrix} 1 & 0 \\ 0 & -1 \end{pmatrix}, \quad U_0 = \begin{pmatrix} 0 & u \\ -\bar{u} & 0 \end{pmatrix}.$$

We find
$$V_1 = U_0, \quad V_0 = \frac{1}{2i} \begin{pmatrix} |u|^2 & u' \\ \bar{u}' & -|u|^2 \end{pmatrix}.$$

For u we obtain the equation
$$-2i\dot{u} = u'' + 2|u|^2 u.$$

This is the non-linear Schrödinger equation. The representation (18.2.3) where
$$\Lambda = A\varsigma, \quad M = A\varsigma^2, \quad \phi = I + \varphi_0 \varsigma^{-1} + \varphi_1 \varsigma^{-2}$$
$$= I + \frac{1}{2i} \begin{pmatrix} 0 & u \\ \bar{u} & 0 \end{pmatrix} \varsigma^{-1} + \frac{1}{4} \begin{pmatrix} 0 & u' \\ -\bar{u}' & 0 \end{pmatrix} \varsigma^{-2}$$

can be obtained. The Lagrangian 18.2.9 becomes

$$\Lambda = (-|u|^4 + |u'|^2 + i(u\dot{\bar{u}} - \bar{u}\dot{u}))dx \wedge dt.$$

This can also be verified directly:

$$\delta\Lambda = -(2|u|^2\bar{u} + \bar{u}'' - 2i\dot{\bar{u}})\delta u \wedge dx \wedge dt - (2|u|^2 u + u'' + 2i\dot{u})$$
$$\times \delta\bar{u} \wedge dx \wedge dt + d\{-(\bar{u}'\delta u + u'\delta\bar{u}) \wedge dt + i(-\bar{u}\delta u + u\delta\bar{u})dx\}.$$

Thus the variational equation $\delta\Lambda/\delta\bar{u} = 0$ coincides with the non-linear Schrödinger equation, and the forms are

$$\Omega^{(1)} = -2\operatorname{Im}\bar{u}\delta u \wedge dx + 2\operatorname{Re}\bar{u}'\delta u \wedge dt$$
$$\Omega = -2\operatorname{Im}\delta\bar{u} \wedge \delta u \wedge dx + 2\operatorname{Re}\delta\bar{u}' \wedge \delta u \wedge dt.$$

The energy-momentum tensor is

$$T_x = -2\operatorname{Im}\bar{u}u'dx - (-|u|^4 - |u'|^2 + 2\operatorname{Im}\bar{u}\dot{u})dt,$$
$$T_t = -(|u|^4 - |u'|^2)dx + 2\operatorname{Re}\bar{u}'\dot{u}dt$$

and, finally, the field Hamiltonian is

$$\mathcal{H} = (|u|^4 + |u'|^2)dx \wedge dt.$$

The Hamiltonian form of the equation is, as usual,

$$\delta\mathcal{H} = (dxi(\tilde{\partial}_x) + dti(\tilde{\partial}_t))\Omega.$$

First integrals can be obtained by a restriction of the first integrals of the general system to the submanifold of matrices having the above special form.

18.3. Principal chiral field.

18.3.1. The theory of the principal chiral field (see 16.7) will be presented here from the point of view of multi-time formalism. One of the Eqs. (16.7.3) $M_\eta - N_\xi = [N, M]$ is equivalent to the possibility of finding a matrix $g(\xi, \eta)$ such that

$$M = g_\xi g^{-1}, N = g_\eta g^{-1}.$$

Now take
$$\mathbf{\Lambda} = \operatorname{tr} g_\xi g^{-1} g_\eta g^{-1} d\xi \wedge d\eta$$
as a Lagrangian. We have
$$\delta\mathbf{\Lambda} = -\operatorname{tr}(M_\eta + N_\xi)\delta g \cdot g^{-1} \wedge d\xi \wedge d\eta + d\operatorname{tr}(-Md\xi + Nd\eta) \wedge \delta g \cdot g^{-1}$$
whence
$$\delta\mathbf{\Lambda}/\delta g = -g^{-1}(M_\eta + N_\xi),$$
where $\delta\mathbf{\Lambda}/\delta g$ is as usual, the matrix $(\delta\mathbf{\Lambda}/\delta g)_{ij} = \delta\mathbf{\Lambda}/\delta g_{ji}$. The equation $\delta\mathbf{\Lambda}/\delta g = 0$ is therefore the remaining equation $M_\eta + N_\xi = 0$.

According to the general rule we find
$$\Omega^{(1)} = -\operatorname{tr}(-Md\xi + Nd\eta) \wedge \delta g \cdot g^{-1}$$
$$\Omega = \delta\Omega^{(1)} = -\operatorname{tr}\{(-\delta M \wedge d\xi + \delta N \wedge d\eta) \wedge \delta g \cdot g^{-1} - (-Md\xi + Nd\eta)$$
$$\wedge g \cdot g^{-1} \wedge \delta g \cdot g^{-1}\} = \operatorname{tr}(-\delta g_\eta d\eta + \delta g_\xi d\xi)g^{-1} \wedge \delta g.$$

The energy-momentum tensor is the set of two forms
$$T_\xi = -\operatorname{tr} M^2 d\xi, \; T_\eta = -\operatorname{tr} N^2 d\eta.$$

The field Hamiltonian is
$$\mathcal{H} = \mathbf{\Lambda} = \operatorname{tr} MN d\xi \wedge d\eta.$$

The Hamiltonian form of the chiral equation is
$$\delta\mathcal{H} = (d\xi \wedge i(\tilde{\partial}_\xi) + d\eta \wedge i(\tilde{\partial}_\eta))\Omega$$

(check this directly!).

In this case the coordinate-momentum variables can be used. We choose the elements of the matrix g as coordinate, and the elements of the matrices $p_\xi = \delta\Lambda/\delta g_\xi = g^{-1}N$ and $p_\eta = \delta\Lambda/\delta g_\eta = g^{-1}M$ as momenta. Then
$$\Omega^{(1)} = -\operatorname{tr}(p_\xi d\eta - p_\eta d\xi)\delta g, \; \Omega = -\operatorname{tr} \delta g(\delta p_\xi d\eta - \delta p_\eta d\xi),$$
$$\mathcal{H} = \operatorname{tr} gp_\eta gp_\xi d\xi \wedge d\eta,$$
and the Hamiltonian equations have the form
$$\partial_\xi g = \delta\mathcal{H}/\delta p_\xi, \; \partial_\eta g = \delta\mathcal{H}/\delta p_\eta,$$
$$\partial_\xi p_\xi + \partial_\eta p_\eta = -\delta\mathcal{H}/\delta g.$$

18.3.2. Resolvents and first integrals. The system (16.7.3) can be written as a zero curvature equation $[\partial_\xi + U, \partial_\eta + V] = 0$, i.e.

(18.3.3) $$U_\eta - V_\xi = [U, V],$$

where
$$U = -\frac{1}{2}(1-z)M\,,\, V = -\frac{1}{2}(1-z^{-1})N$$

(Eq. (18.3.3) being satisfied identically in the parameter z).

A resolvent is a series in z^{-1} or in z:

$$\text{(a)} \quad R = \sum_0^\infty R_k z^{-k}\,, \quad \text{(b)} \quad R = \sum_0^\infty R_k z^k$$

which is a common solution of the equations

$$R_\xi = [R, U]\,,\, R_\eta = [R, V]$$

and the matrices U and V are assumed to satisfy Eq. (18.3.3).

We present construction of resolvents. Suppose a resolvent of the first type, (a) is to be constructed. The zero curvature equation (18.3.3) admits of a group of gauge transformations

$$\tilde{U} = h^{-1}Uh + h^{-1}h_\xi\,,\, \tilde{V} = h^{-1}Vh + h^{-1}h_\eta$$

with arbitrary h. This matrix can be chosen in such a way that $\tilde{U}$ and $\tilde{V}$ have the form $\tilde{U} = U_1 + zA, V = z^{-1}U_0$ with some matrices A, U_0 and U_1. In more detail this transformation is

(18.3.4)
$$\text{(a)} \quad -\frac{1}{2}h^{-1}Mh + h^{-1}h_\xi = U_1$$
$$\text{(b)} \quad \frac{1}{2}h^{-1}Mh = A$$
$$\text{(c)} \quad -\frac{1}{2}h^{-1}Nh + h^{-1}h_\eta = 0$$
$$\text{(d)} \quad \frac{1}{2}h^{-1}Nh = U_0$$

(the matrix h can be found from (c) and then the rest of the equations determine U_1, A and U_0). The zero curvature equation for $\tilde{U}$ and $\tilde{V}$ now has the form

(18.3.5) $$A_\eta = 0\,,\, U_{1\eta} = [A, U_0]\,,\, U_{0\xi} = [U_0, U_1].$$

(we have already met this in (16.7.4), where A was supposed to be constant).

Note that the above procedure admits of some arbitrariness in the determination of h, namely, it can be multiplied on the right by an arbitrary matrix h_1 depending on ξ solely. This implies that the system (18.3.5) still admits of a group of gauge transformations
(18.3.6)
$$\tilde{A} = h_1^{-1} A h_1, \tilde{U}_0 = h_1^{-1} U_0 h_1, \tilde{U}_1 = h_1^{-1} U_1 h_1 + h_1^{-1} h_{1\xi}, h_1 = h_1(\xi).$$

The first of Eqs. (18.3.5) means that A depends only on ξ. Therefore a matrix h_1 can be found such that $\tilde{A}$ is diagonal. After that some freedom is still left. The transformations (18.3.6) can be performed with diagonal matrices h_1 depending on ξ. We fix this transformation by requiring that $\tilde{U}_0$ should have a zero diagonal.

Thus, we have the system (18.3.5) where A is a diagonal matrix and U has zero diagonal. Now put $R = hSh^{-1}$ where h is now uniquely determined. We have

$$S_\xi = [S, U_1 + zA], S_\eta = [S, z^{-1}U_0].$$

Exactly as in Sec. 9.1, we can find a solution $S = \sum_0^\infty S_k z^{-k}$ of the first of the equations and, as in 18.2.21, prove that this solution satisfies also the second equation by virtue of the system (18.3.35).

18.3.7. *Exercise.* Writing several first terms of expansion of S in z^{-1} (see 9.1.9 where U' must be replaced by $-U_{1\xi}$), verify that S satisfies the second equation by virtue of (18.3.35).

Let us find first integrals of the chiral field. Let R be a resolvent. Put

(18.3.8) $\qquad F = \operatorname{tr}(U_z d\xi + V_z d\eta)R, U_z = \partial U/\partial z, V_z = \partial V/\partial z.$

18.3.9. *Proposition.* By virtue of Eq. (18.3.3) we have $dF = 0$.
Proof: Indeed,

$$dF = \operatorname{tr}(-U_{z\eta} + V_{z\xi})R d\xi \wedge d\eta + \operatorname{tr}(-U_z R_\eta + V_z R_\xi) d\xi \wedge d\eta$$
$$\stackrel{Q}{=} \operatorname{tr}(-[U,V]_z - U_z[R,V] + V_z[R,U]) d\xi \wedge d\eta = 0 . \square$$

The expression F is a generator of first integrals, i.e. the coefficients of its expansion in powers of z (or z^{-1}) are first integrals.

18.3.10. Proposition. The relation

$$\delta F = \partial_z \,\mathrm{tr}(\delta U \wedge d\xi + \delta V \wedge d\eta) R + d(\ )$$

holds.

Proof: It can be proved as in 18.2.23. □

18.3.11. Proposition. The characteristic of the first integral (18.3.8) is

(18.3.12) $$a = \frac{1}{4}\partial_z\left(z - \frac{1}{z}\right)Rg$$

(this is understood in the sense that the coefficients in the expansion of (18.3.12) correspond to those of (18.3.8)).

Proof: The following equality should be verified:

$$\delta F \stackrel{Q}{=} i(\partial_a)\Omega + d(\ )$$

(as usual, $\stackrel{Q}{=}$ means equality by virtue of the equation, this time of Eq. (18.3.3)). Let $b = Rg$, $a = \frac{1}{4}\partial_z(z - \frac{1}{z})b$. We have

$$i(\partial_b)\Omega = -\mathrm{tr}\{(Rg)_\xi g^{-1}\delta g g^{-1} \wedge d\xi - \delta g_\xi g^{-1} R \wedge d\xi - (\xi \leftrightarrow \eta)\}.$$

The expression $\mathrm{tr}(Rg)_\xi g^{-1}\delta g g^{-1}$ can be transformed in two ways. On the one hand,

$$\mathrm{tr}(Rg)_\xi g^{-1}\delta g g^{-1}$$
$$= \mathrm{tr}(\partial_\xi(R\delta g g^{-1}) + Rg_\xi g^{-1}\delta g g^{-1} - R\delta g_\xi g^{-1} + R\delta g g^{-1}g_\xi g^{-1})$$
$$= \mathrm{tr}(\partial_\xi(R\delta g g^{-1}) - g_\xi g^{-1}[\delta g g^{-1}, R] + 2\delta g g^{-1}Rg_\xi g^{-1} - R\delta g_\xi g^{-1}).$$

On the other hand,

$$\mathrm{tr}(Rg)_\xi g^{-1}\delta g g^{-1} \stackrel{Q}{=} \mathrm{tr}([R, U]\delta g g^{-1} + Rg_\xi g^{-1}\delta g g^{-1})$$
$$= \mathrm{tr}\left(\frac{1}{2}(1-z)[g_\xi g^{-1}, R]\delta g \cdot g^{-1} + Rg_\xi g^{-1}\delta g g^{-1}\right)$$
$$= \mathrm{tr}\left(\frac{1}{2}(1-z)[R, \delta g \cdot g^{-1}]g_\xi g^{-1} + Rg_\xi g^{-1}\delta g g^{-1}\right).$$

Equating one expression to the other we have

$$\mathrm{tr}\frac{1}{2}(1+z)g_\xi g^{-1}[R, \delta g g^{-1}] \stackrel{Q}{=} \mathrm{tr}(-\partial_\xi(R\delta g g^{-1}) - Rg_\xi g^{-1}\delta g g^{-1} + R\delta g_\xi g^{-1}).$$

Substituting this into the expression for $i(\partial_b)\Omega$ we obtain

$$i(\partial_b)\Omega \stackrel{Q}{=} -\operatorname{tr}\{\partial_\xi(R\delta gg^{-1}) + \frac{2}{1+z}(-\partial_\xi(R\delta gg^{-1}) - Rg_\xi g^{-1}\delta gg^{-1}$$
$$+ R\delta g_\xi g^{-1}) + 2\delta gg^{-1}Rg_\xi g^{-1} - 2R\delta g_\xi g^{-1}\} \wedge d\xi - (\xi \leftrightarrow \eta, z \leftrightarrow z^{-1})$$
$$= -\operatorname{tr}\left\{-\frac{1-z}{1+z}\partial_\xi(R\delta gg^{-1}) - \frac{2z}{1+z}(-Rg_\xi g^{-1}\delta gg^{-1} + R\delta g_\xi g^{-1})\right\}d\xi$$
$$- (\xi \leftrightarrow \eta, z \leftrightarrow z^{-1})$$

whence

$$i(\partial_a)\Omega = \frac{1}{4}\partial_z(z - \frac{1}{z})i(\partial_b)\Omega$$
$$= \partial_z \operatorname{tr}\left\{\frac{1}{2}(z-1)(-Rg_\xi g^{-1}\delta gg^{-1} + R\delta g_\xi g^{-1}) \wedge d\xi\right.$$
$$+ \frac{1}{2}(\frac{1}{z} - 1)(-Rg_\eta g^{-1}\delta gg^{-1} + R\delta g_\eta g^{-1}) \wedge d\eta + d(\ ).$$

Now, according to 18.3.10, we have

$$\delta F = \partial_z \operatorname{tr}(\delta U \wedge d\xi + \delta V \wedge d\eta)R + d(\ )$$
$$= \partial_z \operatorname{tr}\left\{-\frac{1}{2}(1-z)\delta(g_\xi g^{-1})R \wedge d\xi - \frac{1}{2}(1-\frac{1}{z})\delta(g_\eta g^{-1})R \wedge d\eta\right\}$$
$$+ d(\ ) = \partial_z \operatorname{tr}\left\{\frac{1}{2}(1-z)(-\delta g_\xi g^{-1}R + g_\xi g^{-1}\delta gg^{-1}R) \wedge d\xi\right.$$
$$+ \frac{1}{2}(1-\frac{1}{z})(-\delta g_\eta g^{-1}R + g_\eta g^{-1}\delta gg^{-1}R) \wedge d\eta.$$

Thus, $i(\partial_a)\Omega = \delta F + d(\ )$ as required. □

18.3.13. Proposition. Let F_k and $F_l^{(1)}$ be two first integrals constructed with the aid of resolvents R and $R^{(1)}$. Then $\{F_k, F_l^{(1)}\} = 0$, i.o. the first integrals are in involution.

Proof: The generators of first integrals are

$$F = \operatorname{tr}(U_z d\xi + V_z d\eta)R, \quad F^{(1)} = \operatorname{tr}(U_{z_1} d\xi + V_{z_1} d\eta)R^{(1)}$$

(in the expression for $F^{(1)}$ the parameter z_1 is sustituted for z). It is convenient here to consider, instead of F and $F^{(1)}$, two other generators, $\hat{F}$ and $\hat{F}^{(1)}$, such that $F = \partial_z \hat{F}$ and $F^{(1)} = \partial_z \hat{F}^{(1)}$ and to prove their involutiveness. Their characteristics are $\hat{a} = \frac{1}{4}(z - \frac{1}{z})Rg$ and $\hat{a}^{(1)} = \frac{1}{4}(z^{(1)} - \frac{1}{z^{(1)}})Rg^{(1)}$. For them

$$\delta \hat{F} = \text{tr}(\delta U \wedge d\xi + \delta V \wedge d\eta)R + d(\ )\,,$$
$$\delta \hat{F}^{(1)} = \text{tr}(\delta U^{(1)} \wedge d\xi + \delta V^{(1)} \wedge d\eta)R^{(1)} + d(\ )\,.$$

Now we calculate

$$\begin{aligned}\{\hat{F}, \hat{F}^{(1)}\} &= \partial_{\hat{a}} F^{(1)} = i(\partial_{\hat{a}})\delta F^{(1)} = i(\partial_{\hat{a}})\text{tr}(\delta U^{(1)} \wedge d\xi + dV^{(1)} \wedge d\eta)R^{(1)} \\
&= i(\partial_{\hat{a}})\text{tr}(-\frac{1}{2}(1 - z_1)(\delta g_\xi g^{-1} - g_\xi g^{-1}\delta g g^{-1}) \wedge d\xi \\
&\quad - \frac{1}{2}(1 - z_1^{-1})(\delta g_\eta g^{-1} - g_\eta g^{-1}\delta g g^{-1}) \wedge d\eta)R^{(1)} \\
&= \frac{1}{4}(z - \frac{1}{z})\text{tr}(-\frac{1}{2}(1 - z_1)((Rg)_\xi g^{-1} - g_\xi g^{-1}R)d\xi \\
&\quad - \frac{1}{2}(1 - z_1^{-1})((Rg)\eta g^{-1} - g_\eta g^{-1}R)d\eta)R^{(1)} \\
&= \frac{1}{4}(z - \frac{1}{z})\text{tr}(-\frac{1}{2}(1 - z_1)([R, U] + [R, g_\xi g^{-1}])d\xi \\
&\quad - \frac{1}{2}(1 - z_1^{-1})([R, V] + [R, g_\eta g^{-1}])d\eta)R^{(1)} \\
&= \frac{1}{4}(z - \frac{1}{z})\text{tr}\{(\frac{1}{4}(1 - z_1)(1 - z)[R, g_\xi g^{-1}]R^{(1)} - R_\xi^{(1)}R)d\xi \\
&\quad + (\frac{1}{4}(1 - z_1^{-1})(1 - z^{-1})[R, g_\eta g^{-1}]R^{(1)} - R_\eta^{(1)}R)d\eta\}\end{aligned}$$

(all equalities are understood to be up to complete differentials $d(\ )$). The expression within the braces is skew symmetric with respect to interchange of R and $R^{(1)}, z$ and z_1. The multiplier $z - 1/z$ violates this symmetry; however, the Poisson bracket must be skew symmetric. This is possible only if the expression is zero. □

18.3.14. *An additional set of first integrals.* Besides the first integrals related to resolvents we can produce another set of first integrals. Namely,

$$G_k = \frac{2}{k+2}f(\xi)\text{tr}\, M^{k+2}d\xi\,,\ H_k = \frac{2}{k+2}f(\eta)\text{tr}\, N^{k+2}d\eta\,,$$

where f is an arbitrary function. Let us check this:

$$\begin{aligned}dG_k &= -2f(\xi)\text{tr}M^{k+1}M_\eta d\xi \wedge d\eta = -f(\xi)\text{tr}M^{k+1}(M_\eta - N_\xi)d\xi \wedge d\eta \\
&\quad - f(\xi)\text{tr}\, M^{k+1}(M_\eta + N_\xi)d\xi \wedge d\eta \\
&= -f(\xi)\text{tr}\, M^{k+1}[N, M]d\xi \wedge d\eta + f(\xi)\text{tr}\, M^{k+1}g\delta\Lambda/\delta g \\
&= f(\xi)\text{tr}\, M^{k+1}g\delta\Lambda/\delta g\end{aligned}$$

(because $M_\eta + N_\xi = -g\delta\Lambda/\delta g$). Thus $dG_k \stackrel{Q}{=} 0$. Similarly $dH_k = -f(\eta)\mathrm{tr}\, N^{k+1}g\delta\Lambda/\delta g$. At the same time we have obtained characteristics of these first integrals:

$$\chi_{G_k} = f(\xi)M^{k+1}g, \quad \chi_{H_k} = -f(\eta)N^{k+1}g.$$

18.3.15. Proposition. The Poisson brackets are

$$\{G_{k_1}, G_{k_2}\} = \frac{2}{k_1 + k_2 + 2}\{(k_2+1)f_1'f_2 - (k_1+1)f_1f_2'\}$$
$$\times \mathrm{tr}\, M^{k_1+k_2+2}d\xi,$$
$$\{H_{k_1}, H_{k_2}\} = -\frac{2}{k_1 + k_2 + 2}\{(k_2+1)f_1'f_2 - (k_1+1)f_1f_2'\}$$
$$\times \mathrm{tr}\, N^{k_1+k_2+1}d\eta,$$
$$\{G_{k_1}, H_{k_2}\} = \{G_k, F\} = \{H_k, F\} = 0,$$

where F is our old first integral.

Proof: We calculate:

$$\{G_{k_1}, G_{k_2}\} = \partial_{G_{k_1}} G_{k_2} = i(\partial_{G_{k_1}})\delta G_{k_2}$$
$$= i(\partial_{G_{k_1}})\mathrm{tr}\, f_2(\xi) \cdot (k_2+2)M^{k_2+1}(\delta g_\xi \cdot g^{-1} - g_\xi g^{-1}\delta g g^{-1}) \cdot \frac{2}{k_2+2}d\xi$$
$$= 2\mathrm{tr}\, f_2 M^{k_2+1}\{(f_1 M^{k_1+1}g)_\xi g^{-1} - g_\xi g^{-1}(f_1 \cdot M^{k_1+1}g)g^{-1}\}d\xi$$
$$= 2\mathrm{tr}\, f_2 M^{k_2+1}\{f_1' M^{k_1+1} + (k_1+1)f_1 M^{k_1} M_\xi\}d\xi.$$

The expression $2\mathrm{tr}\, f_2 M^{k_1+k_2+1}f_1(k_1+1)M_\eta d\eta$ can be added since $\mathrm{tr}\, M^{k_1+k_2+1}M_\eta \stackrel{Q}{=} \mathrm{tr}\, M^{k_1+k_2+1}\frac{1}{2}[N, M] = 0$. Then

$$\{G_{k_1}, G_{k_2}\} = 2\mathrm{tr}\{f_1'f_2 M^{k_1+k_2+2}\}d\xi + 2\mathrm{tr}\, f_1 f_2 \cdot \frac{k_1+1}{k_1+k_2+2} \cdot dM^{k_1+k_2+2}$$
$$= 2\mathrm{tr}\{f_1'f_2 M^{k_1+k_2+2} - \frac{k_1+1}{k_1+k_2+2}M^{k_1+k_2+2}(f_1 f_2)_\xi\}d\xi$$
$$= \frac{2}{k_1+k_2+2}\mathrm{tr}\{(k_2+1)f_1'f_2 - (k_1+1)f_1 f_2'\}M^{k_1+k_2+2}d\xi.$$
$$\{F, G_k\} = \partial_F G_k = i(\partial_F)\delta G_k = \partial_z(z - z^{-1})i(\partial_{Rg})\delta G_k$$
$$= \partial_z(z - z^{-1})2\mathrm{tr}\, M^{k+1}f \cdot \{(Rg)_\xi g^{-1} - g_\xi g^{-1}Rg \cdot g^{-1}\}d\xi$$
$$= \partial_z(z - z^{-1})2\mathrm{tr}\, M^{k+1}f\{R_\xi + RM - MR\}d\xi$$
$$\stackrel{Q}{=} \partial_z(z - z^{-1}) \cdot 2 \cdot (-\frac{1}{2}(1-z))\mathrm{tr}\, fM^{k+1}[R, M]d\xi = 0,$$

etc. □

Thus, the first integrals G_k (and H_k) form a graded Lie algebra $\mathfrak{E} = \oplus \, \mathfrak{E}_k$ (respectively, $\mathfrak{H}$). The simplest way to describe it is the following. The elements of $\{\mathfrak{E}_k\}$ are pairs (k, f), where f are functions, and

$$(k, \lambda_1 f_1 + \lambda_2 f_2) = \lambda_1(k, f_1) + \lambda_2(k, f_2)$$
$$[(k_1, f_1), (k_2, f_2)] = (k_1 + k_2, (k_2 + 1) f_1' f_2 - (k_1 + 1) f_1 f_2')$$

Commutators in $\mathfrak{H}$ differ by the sign.

18.3.16. *Other Lagrangians*. There are also other possibilities to introduce Lagrangians for the chiral field equations. Zakharov and Mikhailov [104] have suggested Lagrangians for a more general scheme. Applied to the chiral field, their method can be described as follows. A gauge transformation (18.3.4) was introduced above which reduced the matrix M to a diagonal form, and the system took the form (18.3.5). Since matrices M and N are equal in rights, we can reduce the matrix N to the diagonal form by a gauge transformation

(18.3.17)
$$-\frac{1}{2} g^{-1} N g + g^{-1} g_\eta = V_1$$
$$\frac{1}{2} g^{-1} N g = B$$
$$-\frac{1}{2} g^{-1} M g + g^{-1} g_\xi = 0$$
$$\frac{1}{2} g^{-1} M g = V_0$$

which yields the system

$$B_\xi = 0, \; V_{1\xi} = [B, V_0], \; V_{0\eta} = [V_0, V_1].$$

Now we use both the transformations with matrices h and g. Let $A(\xi)$ and $B(\eta)$ be given diagonal matrices. Put

$$\mathbf{\Lambda} = \mathrm{tr}(h^{-1} h_\eta A - g^{-1} g_\xi B + h A h^{-1} g B g^{-1}) d\xi \wedge d\eta \, .$$

It is easy to calculate

$$\begin{aligned}\delta \mathbf{\Lambda} =& \mathrm{tr}\{(-h^{-1} h_\eta A h^{-1} + A h^{-1} h_\eta h^{-1} + A h^{-1} g B g^{-1} \\ & - h^{-1} g B g^{-1} h A h^{-1}) \delta h + (+ g^{-1} g_\xi B g^{-1} - B g^{-1} g_\xi g^{-1} \\ & - B g^{-1} h A h^{-1} + g^{-1} h A h^{-1} g B g^{-1}) \delta g \} \wedge d\xi \wedge d\eta \\ & + d \mathrm{tr}\{A h^{-1} \delta h \wedge d\xi + B g^{-1} \delta g \wedge d\eta\}\end{aligned}$$

whence
$$\delta\Lambda/\delta h = h^{-1}\{-(hAh^{-1})_\eta + [hAh^{-1}, gBg^{-1}]\}$$
$$\delta\Lambda/\delta g = g^{-1}\{(gBg^{-1})_\xi - [gBg^{-1}, hAh^{-1}]\}$$

and
$$\Omega^{(1)} = -\operatorname{tr}\{Ah^{-1}\delta h \wedge d\xi + Bg^{-1}\delta g \wedge d\eta\}.$$

If we put $M = -2hAh^{-1}$ and $N = -2gBg^{-1}$, then the set of variational equations $\delta\Lambda/\delta g = \delta\Lambda/\delta h = 0$ coincides with (16.7.3). The energy-momentum tensor is

$$T_\xi = -\operatorname{tr}\{Ah^{-1}(h_\xi d\xi + h_\eta d\eta) + hAh^{-1}gBg^{-1}d\eta\}$$
$$T_\eta = -\operatorname{tr}\{Bg^{-1}(g_\xi d\xi + g_\eta d\eta) - hAh^{-1}gBg^{-1}d\xi\}$$

and the Hamiltonian is

$$\mathcal{H} = -\operatorname{tr} hAh^{-1}gBg^{-1}d\xi \wedge d\eta.$$

The Hamiltonian form of the equation is, as usual,

$$\delta\mathcal{H} = (d\xi \wedge i(\tilde{\partial}_\xi) + d\eta \wedge i(\tilde{\partial}_\eta))\Omega.$$

Other Lagrangians can also be written.

18.4. Integrable systems related to nth-order linear differential operators.
18.4.1. This section generalizes the examples of 17.8.4. Our aim is to show that symplectic forms of the non-stationary and of the stationary theories can be combined as two components of one united form, that Hamiltonians of these theories are components of the energy-momentum tensor, and that first integrals are also components of one-form. This gives the most natural representation of the theories by Bogoyavlenskii and Novikov [67] and Gelfand and Dickey [25], see 12.2.11.

We return to Chap. 1 and consider the KdV-hierarchies. Let

(18.4.2) $$L = (\phi^{-1}\partial^n \phi)_+, \phi = 1 + \sum_0^\infty \phi_k \partial^{-k-1}$$

be the dressing substitution expressing $\{u_k\}$ in terms of $\phi_k, k = 0, \ldots, n-2$ (the rest of ϕ_k does not play any role and $\phi - 1$ can be considered as an element of $R_-/R_{(-\infty,-n)}$). This substitution embeds the differential algebra

a_u into a_ϕ. Let $T = \sum_0^\infty T_k \partial^{-k-1} = \sum_0^\infty \sum_{r_k}^\infty T_{k,(r)} z^{-r} \partial^{-k-1} = \sum_{-\infty}^\infty T_{(r)} z^{-r}$ be a resolvent. As we know, it satisfies the Adler equation

$$((L - z^n)T)_+(L - z^n) - (L - z^n)(T(L - z^n))_+ = 0.$$

Recall that basic resolvents satisfy also the variational relation 3.5.3; we shall assume our resolvent to be a basic one.

The KdV-hierarchy comprises equations

(18.4.3) $$\dot{L} = [L, \delta \operatorname{res} T_{(r)}/\delta L].$$

18.4.4. Proposition. *The form*

$$\mathbf{\Lambda}_r = (-\operatorname{res} T_{(r)} + \operatorname{res} L\phi^{-1}\dot\phi)dx \wedge dt = \operatorname{res}(-T_{(r)} + \phi^{-1}\partial^n \dot\phi)dx \wedge dt$$

can be taken as a Lagrangian of the equation (18.4.3).
Proof: We have

(18.4.5) $$\delta \operatorname{res} T_{(r)} = \operatorname{res}\{(\delta \operatorname{res} T_{(r)}/\delta L)\delta L\} + \partial \omega^*_{(r)},$$

where $\omega^*_{(r)}$ is a form. Then

$$\begin{aligned}\delta\mathbf{\Lambda}_{(r)} =& \operatorname{res}\{-(\delta \operatorname{res} T_{(r)}/\delta L)\delta L + \phi^{-1}\partial^n \delta\dot\phi - \phi^{-1}\delta\phi\phi^{-1}\partial^n \dot\phi\}dx \wedge dt \\ &+ d(\omega^*_{(r)} \wedge dt) = \operatorname{res}\{-(\delta \operatorname{res} T_{(r)}/\delta L)(\phi^{-1}\partial^n \delta\phi - \phi^{-1}\delta\phi\phi^{-1}\partial^n \phi) \\ &+ \phi^{-1}\dot\phi\delta^{-1}\partial^n \delta\phi - \phi^{-1}\delta\phi\phi^{-1}\partial^n \dot\phi\}dx \wedge dt \\ &+ d(\omega^*_{(r)} \wedge dt + \operatorname{res} \phi^{-1}\partial^n \delta\phi \wedge dx) \\ =& \operatorname{res}\{-[\delta \operatorname{res} T_{(r)}/\delta L, (\phi^{-1}\partial^n \phi)_+] - (\phi^{-1}\partial^n \phi)^\bullet_+\}\phi^{-1}\delta\phi \\ &+ d\{(\omega^*_{(r)} + \omega^{**}_{(r)}) \wedge dt + \operatorname{res} \phi^{-1}\partial^n \delta\phi \wedge dx\},\end{aligned}$$

where $\omega^{**}_{(r)}$ should be determined from

(18.4.6) $$\begin{aligned}\partial \omega^{**}_{(r)} =& \operatorname{res}[(\delta \operatorname{res} T_{(r)}/\delta L)\phi^{-1}\delta\phi, \phi^{-1}\partial^n \phi] \\ &+ \operatorname{res}[\phi^{-1}\partial^n \dot\phi, \phi^{-1}\delta\phi].\end{aligned}$$

Let $\omega^*_{(r)} + \omega^{**}_{(r)} = \omega_{(r)}$. As a corollary we get

$$\delta\mathbf{\Lambda}_{(r)}/\delta L = ([L, \delta \operatorname{res} T_{(r)}/\delta L] - \dot{L})\phi^{-1}\delta\phi$$

whence the required assertion immediately follows. □

Moreover, the 1-form is

(18.4.7) $$\Omega^{(1)} = -\omega_{(r)} \wedge dt - \operatorname{res} \phi^{-1} \partial^n \delta\phi \wedge dx$$

and

$$\Omega = \operatorname{res} \phi^{-1}\delta\phi\phi^{-1}\partial^n \wedge \delta\phi \wedge dx - \delta\omega_{(r)} \wedge dt = -\Omega_{(t)}dx + \Omega_{(x)}dt.$$

The energy-momentum tensor is

$$T_{(x)} = -\operatorname{res} \phi^{-1}\partial^n \phi' dx - (\operatorname{res} \phi^{-1}\partial^n \dot\phi + i(\tilde\partial)\omega_{(r)} - \operatorname{res} T_{(r)})dt$$
$$= -T_{(xt)}dx + T_{(xx)}dt$$
$$T_{(t)} = -\operatorname{res} T_{(r)}dx - i(\tilde\partial_t)\omega_{(r)}dt = -T_{(tt)}dx + T_{(tx)}dt,$$

and the Hamiltonian is

$$\mathcal{H} = (\operatorname{res} T_{(r)} - i(\tilde\partial)\omega_{(r)})dx \wedge dt.$$

18.4.8. Now we compare three theories: (a) the present field theory where x and t are equal in rights, (b) the former single-time theory, and (c) the stationary theory, i.e. the theory of the stationary equation

$$\delta \operatorname{res} T_{(r)}/\delta L = 0.$$

18.4.9. *Proposition.* The form $-\int \Omega_{(t)}dx$ restricted to the algebra $\mathcal{A}_u$ is none other than the symplectic form $\Omega^{(\infty)}$ of the single-time theory b).
Proof: Let X be an element of $R_-/R_{(-\infty,-n-1)}$ and $X \mapsto H, H = [L, X]_+$ be the Hamiltonian mapping.

18.4.10. *Lemma.* The vector field ∂_X can be extended from $\mathcal{A}_u$ to the whole of a_ϕ by its action on the generator:

$$\phi^{-1}\partial_H \phi = X.$$

Proof: Let the vector field ∂_H be defined by this formula. Then

$$\partial_H L = \partial_H(\phi^{-1}\partial^n \phi)_+ = (\phi^{-1}\partial^n \partial_H \phi)_+ - (\phi^{-1}\partial_H \phi \cdot \phi^{-1}\partial^n \phi)_+ = [L, X]_+,$$

i.e. it acts on $\mathcal{A}_u$ exactly as required. □

We go on with the proof of the proposition

$$-\Omega_{(t)}(\partial_{H_1}, \partial_{H_2})$$
$$= -\operatorname{res} \phi^{-1}(\partial_{H_1}\phi)\phi^{-1}\partial^n(\partial_{H_2}\phi) + \operatorname{res} \phi^{-1}(\partial_{H_2}\phi)\phi^{-1}\partial^n(\partial_{H_1}\phi)$$
$$= \operatorname{res}\{-X_1\phi^{-1}\partial^n\phi X_2 + X_2\phi^{-1}\partial^n\phi X_1\} = \operatorname{res}[L, X_1]X_2 + \partial(\ )$$

and $-\int \Omega_{(t)}(\partial_{H_1}, \partial_{H_2})dx = \int \operatorname{res}[L, X_1]X_2 dx$ (see also 17.8.1). □

18.4.11. Proposition. The Hamiltonian in the single-time theory is $-\int T_{(tt)}dx$.

Proof: This is evident. □

18.4.12. Proposition. Let $F = -F_{(t)}dx + F_{(x)}dt$ be a first integral of the theory (a) i.e. $dF \stackrel{Q}{=} 0$. Then $\int F_{(t)}dx$ is a first integral of the theory (b). Conversely, if $\int F_{(t)}dx$ is a first integral of the theory (b) then there exists $F_{(x)}$ such that F is the first integral of the theory (a).

Proof: The equation $dF \stackrel{Q}{=} 0$ is equivalent to $\partial F_{(t)}/\partial t + \partial F_{(x)}/\partial x \stackrel{Q}{=} 0$. The sign $\stackrel{Q}{=}$ means that ∂_t acts as the vector field corresponding to the equation. Then $\partial_t \int F_{(t)}dx = \int \partial_t F_{(t)}dx = -\int \partial_x F_{(x)}dx = 0$. Conversely, if $\partial_t \int F_{(t)}dx = 0$ then $\partial F_{(t)}/\partial t$ is an exact derivative, i.e. $F_{(x)}$ exists such that $\partial F_{(t)}/\partial t = -\partial F_{(x)}/\partial x$. □

Now we compare the theories (a) and (c).

18.4.13. Proposition. The form $\Omega_{(x)}$ is the symplectic form of the stationary equation (all the functions are assumed independent of t and are understood modulo the equation $\delta \operatorname{res} T_{(r)}/\delta L = 0$).

Proof: Equation (18.4.5) shows that $\omega^*_{(r)}$ is the 1-form of the stationary equation. Equation (18.4.6) yields that the form $\omega^{**}_{(r)}$ vanishes under our assumptions. □

18.4.14. Proposition. The Hamiltonian of the stationary equation is $T_{(xx)}$.

Proof: The equation $T_{(xx)} = -\operatorname{res} T_{(r)} + i(\tilde{\partial})\omega_{(r)}$ holds for functions independent of t. If the fact is taken into account that $\operatorname{res} T_{(r)}$ is the Lagrangian and $\omega_{(r)}$ is the 1-form of the stationary equation, then this formula yields the usual expression for a Hamiltonian. □

Finally, the proposition is given below.

18.4.15. Proposition. If $F = -F_{(t)}dx + F_{(x)}dt$ is a first integral then $F_{(x)}$ is the first integral of the stationary equation. If ∂_a is the vector field related to the first integral F then it is also the vector field related to the first integral $F_{(x)}$ of the stationary theory.

Proof: We have $dF \stackrel{Q}{=} 0$, i.e. $\partial F_{(x)}/\partial x + \partial F_{(t)}/\partial t \stackrel{Q}{=} 0$. For solutions independent of t this reduces to $\partial F_{(x)}/\partial x \stackrel{Q}{=} 0$, i.e. $F_{(x)}$ is the first integral. Further, $\delta F \stackrel{Q}{=} -i(\partial_a)\Omega + d(\ )$. For the terms containing dt this means

$$\delta F_{(x)} = -i(\partial_a)\Omega_{(x)} + \partial_t(\ ).$$

The last term vanishes for solutions independent of t. □

REFERENCES

1. Zakharov V. E., Manakov S. V., Novikov S. P., and Pitajevski L. P., ed. Novikov, Theory of solitons, Nauka, 1980 (Russian).
2. Ablowitz M. J., and Segur H., Solitons and inverse scattering transform, SIAM, 1981.
3. Calogero F., and Degasperis A., Spectral transform and solitons, North-Holland, 1982.
4. Takhtajan L. A. and Faddeev L. D., Hamiltonian approach in the soliton theory, Nauka, 1986 (Russian).
5. Newell A. C., Solitons in mathematics and physics, Lectures in Appl. Math. 15, Philadelphia, 1985.
6. Dubrovin B. A., Matveev V. B. and Novikov S. P., Nonlinear equations of the Korteweg-de Vries type, finite-zonal linear operators and Abel manifolds, Uspehi Mat. Nauk, 31, no1, 55–136, 1976 (Russian).
7. Dickey L. A., Korteweg-de Vries and all that, Ann. N. Y. Acad. Sc., 410, 301–316, 1983.
8. Gardner C. S., Green J. M., Kruskal M. D., and Miura R. M., Method for solving the Korteweg-de Vries equation, Phys. Rew. Lett., 19, no19, 1095–97, 1967.
9. Zakharov V. E., and Faddeev L. D., The Korteweg-de Vries equation is a completely integrable Hamiltonian system, Funkz. Anal. Priloz., 5, no 4, 18–27, 1971 (Russian).

10. Gelfand I. M., On identities for eigenvalues of second-order differential operators, Uspehi Mat. Nauk, 11, no1, 191, 1956.
11. Dickey L. A., Trace formulas for Sturm-Liouville's operators, Uspehi Mat. Nauk, 13, no3, 111–143, 1958 (Russian).
12. Kruskal M. D., The birth of the solitons, in "Nonlinear evolution equations solvable by the spectral transform", ed. Calogero F., Research notes in Math, no26, Pitman, 1978.
13. Lax P., Integrals of nonlinear equation of evolution and solitary waves, Comm. Pure Appl. Math., 21, no5, 467–90, 1968.
14. Gardner C. S., Korteweg-de Vries equation and generalizations IV, J. Math. Phys., 12, no8, 1548–51, 1971.
15. Krichever I. M., Algebraic curves and commuting matrix differential operators, Funkz. Anal. Prilozh., 10, no2, 75–77, 1976 (Russian).
16. Gelfand I. M. and Dickey L. A., Fractional powers of operators and Hamiltonian systems, Funkz. Anal. Priloz., 10, no4, 13–29, 1976 (Russian).
17. Adler M., On a trace functional for formal pseudo-differential operators and the symplectic structure of the Korteweg-de Vries equations, Invent. Math., 50, no3, 219–48, 1979.
18. Wilson G., Commuting flows and conservation laws for Lax equations, Math. Proc. Cambridge Phil. Soc., 86, no1, 131–143, 1979.
19. Gelfand I. M., and Dickey L. A., Asymptotics of the resolvent of Sturm-Liouville's equations and algebra of the Korteweg-de Vries equation, Uspehi Mat. Nauk, 30, no5, 67–100, 1975 (Russian).
20. Gelfand I. M. and Dickey L. A., Resolvent and Hamiltonian systems, Funkz. Anal. Priloz., 11, no2, 11–27, 1977 (Russian).
21. Manin Yu. I., Matrix solitons and vector bundles over curves with singularities, Funkz. Anal. Priloz. 12, no4, 53–67, 1978 (Russian).
22. Drinfeld V. G., on commutative subrings of some noncommutative rings, Funkz. Anal. Priloz., 11, no1, 11–14, 1977 (Russian).
23. Arnold V. I., Mathematical methods of classical mechanics, 1974 (Russian), translation:
24. Bishop R. L., Crittenden R. J., Geometry of manifolds, Ac. Press, 1964.
25. Gelfand I. M. and Dickey L. A., Lie algebra structure in the formal variational calculus, Funkz. Anal. Priloz., 10, no1, 18–25, 1976 (Russian).
26. Gelfand I. M., and Dorfman I. Ya., Hamiltonian operators and associated algebraic structures, Funkz. Anal. Priloz., 15, no3, 13–30, 1979 (Russian).
27. Gelfand I. M., and Dorfman I. Ya., Schouten bracket and Hamiltonian operators, Funkz. Anal. Priloz., 14, no3, 71–74, 1980 (Russian).

28. Gelfand I. M., and Dorfman I. Ya., Hamiltonian operators and infinite-dimensional Lie algebras, Funkz. Anal. Priloz., 15, no3, 23–40, 1981 (Russian).
29. Gelfand I. M., and Dickey L. A., Family of Hamiltonian structures connected with integrable nonlinear differential equations, Preprint IPM (Moscow) 136, 1978 (Russian).
30. Lebedev D. R., and Manin Yu. I., Hamiltonian Gelfand-Dickey operator and coadjoint representation of the Volterra group, Funkz. Anal. Priloz., 13, no4, 40–46, 1979.
31. Magri F., A simple model of the integrable Hamiltonian equation, J. Math. Phys., 19, no5, 1156–62, 1978.
32. Khovanova T. G., Lie algebras of Gelfand-Dickey and the Virasoro algebra, Funkz. Anal. Priloz., 20, no4, 89–90, 1986 (Russian).
33. Focas A. S., and Fuchssteiner B., On the structure of symplectic operators and hereditary symmetries, Lettere al Nuovo Cemento 30, no17, 539–44, 1981.
34. Adler M., On the Bäcklund transformation for the Gelfand-Dikii equations, Comm. Math. Phys., 80, no4, 517–527, 1981.
35. Kupershmidt B. A., and Wilson G., Modifying Lax equations and the second Hamiltonian structure, Invent. Math. 62, 403–36, 1981.
36. Dickey L. A., A short proof of a Kupershmidt and Wilson theorem, Comm. Math. Phys., 87, 127–129, 1983.
37. Date E., Jimbo M., Kashiwara M., and Miwa T., Transformation groups for soliton equations, in Jimbo and Miwa (ed.) Non-linear integrable systems — classical theory and quantum theory Proc. RIMS symposium, Singapore, 1983.
38. Hirota R., Direct method of finding exact solutions of nonlinear evolution equations, Lecture Notes in Math., 515, 40, 1976.
39. Segal G., and Wilson G., Loop groops and equations of KdV-type, Publ. Mathem. IHES, 63, 1–64, 1985.
40. Watanabe Y., Hamiltonian structure of Sato's hierarchy of KP equations and a coadjoint orbit of a certain formal Lie group, Lett. Math. Phys. 7, no2, 99–106, 1983.
41. Dickey L. A., On Hamiltonian and Lagrangian formalisms for the KP-hierarchy of integrable equtions, Ann. N. Y. Acad. Sc., 491, 131–148, 1987.

42. Radul A. O., Two series of Hamiltonian structures for the hierarchy of Kadomtsev-Petviashvili equations, in "Applied methods of nonlinear analysis and control", ed. Mironov, Moroz, and Tshernjatin, MGU, 1987 (Russian), pp. 149–157.
43. Faà di Bruno M., Note sur une nouvelle formule de calcul differentiel, Quart. J. Pure and Appl. Math., 1, 359–360, 1857.
44. Chen H. H., Lee Y. C., and Lin J. E., On a new hierarchy of symmetries for the Kadomtsev-Petviashvili Equation, Physica D, 9D, no3, 439–445, 1983.
45. Orlov A. Yu., and Shulman, E. I., Additional symmetries for integrable equations and conformal algebra representations, Lett. Math. Phys. 12, 171–179 (1986).
46. Krichever I. M., Algebraic geometrical methods in the theory of nonlinear equations, Ispehi Mat. Nauk, 32, 183, 1977 (Russian).
47. Springer G., Introduction to Riemann surfaces, Addison-Wesley, 1957.
48. Forster O., Riemannsche Flächen, Springer, 1977.
49. Dubrovin B. A., Riemann surfaces and nonlinear equations I, MGU, Moscow, 1986 (Russian).
50. Its A. R., Conversion of hyperelliptic integrals and integration of nonlinear equations, Vestnik MGU, ser. Math. Mech. Astr., 7, 28–39, 1976 (Russian).
51. Ablowitz M. J., Kaup D. J., Newell A. C., and Segur H., Nonlinear evolution equations of physical significance, Phys. Rev. Lett., 31, 125–7, 1973.
52. Ablowitz M. J., Kaup D. J., Newell A. C., and Segur H., The inverse scattering transform — Fourier analysis for non-linear problems, Stud. Appl. Math., 53, 249–315, 1974.
53. Dubrovin B. A., Completely integrable Hamiltonian systems connected with matrix operators, and Abel manifolds, Funkz. Anal. Priloz., 11, no4, 28–41, 1977.
54. Dickey L. A., Integrable nonlinear equations and Liouville's theorem I, Commun. Math. Phys., 82, 345–360, 1981.
55. Reyman A. G., Semenov-Tjan-Shanski M. A., and Frenkel I. B., Graded Lie algebras and completely integrable dynamic systems Doklady AN SSSR, 247, no4, 802–4, 1979 (Russian).
56. Reyman A. G., and Semenov-Tjan-Shanski M. A., Family of Hamiltonian structures, hierarchy of Hamiltonians, and reduction for matrix first-order differential operators, Funkz. Analys, Priloz., 14, no2, 77–78, 1980 (Russian).
57. Jimbo M., Miwa T., and Ueno K., Monodromy preserving deformation of linear ordinary differential equations with rational coefficients, Preprint RIMS no319, Kyoto, 1980.

58. Its A. R., Liouville's theorem and inverse scattering method, Zapiski nauchn. sem. LOMI, 133, 113–125, 1984 (Russian).
59. Gelfand I. M., and Dickey L. A., Jet calculus and nonlinear Hamiltonian systems, Funkz. Anal. Priloz., 12, no2, 8–23, 1975 (Russian).
60. Drinfeld V. G., and Sokolov V. V., Equations of the Korteweg-de Vries type and simple Lie algebras, Doklady AN SSSR, 258, no1, 11–16, 1981 (Russian).
61. Drinfeld V. G., and Sokolov V. V., Lie algebras and equations of the Korteweg-de Vries type, Itogi nauki i tekhniki, ser. Sovremennyie problemy matematiki, 24, 81–180, 1984.
62. Dickey L. A., Integrable nonlinear equations and Liouville' theorem II, Commun. Math. Phys., 82, 361–75, 1981.
63. Serre J.-P., Algebres de Lie semi-simples complexes, Benjamin, 1966.
64. Yusin B. V., Proof of a variational relation between the coefficients of asymptotics of the resolvent of Sturm-Liouville equation, Uspehi Mat. Nauk, 33, no1, 233–234. 1978.
65. Alber S. I., Study of equtions of Korteweg-de Vries type by the recurrency relations method, preprint I. Ch. Ph., 1976 (Russian).
66. Alber S. I., On stationary problems for equtions of Korteweg-de Vries type, Comm. Pure Appl. Math., 1981, 34, no2, 259–272.
67. Bogoyavlenski O. I., Novikov S. P., On a connection between Hamiltonian formalisms of stationary and nonstationary problems, Funkz. Anal. Priloz., 10, no1, 9–13, 1976.
68. Dubrovin B. A., Periodic problem of Korteweg-de Vries in the class of finite-zonal potentials, Funkz. Anal. Priloz., 9, no3, 41–52, 1975 (Russian).
69. Its A. R., and Matveev V. B., Schrödinger operator with finite-zonal spectrum and N-soliton solutions of the Korteweg-de Vries equation, Teor. Mat. Fiz., 23, no1, 51–68, 1975 (Russian).
70. Arnold V. I., Sur la geometrie differentielle des groupes de Lie de dimension infinite et ses applications a l'hydrodynamique des fluides parfaits, Ann. Inst. Fourier, 16, no1, 319–361, 1966.
71. Arnold V. I., Hamiltonian property of Euler's equations of the rigid body and of the ideal fluid, Uspehi Mat. Nauk, 24, no3, 225–6, 1969.
72. Mishchenko A.S., Integrals of geodesic flows on the Lie groups, Funkz. Anal. Priloz., 4, no3, 73–77, 1970 (Russian).
73. Dickey L. A., A note on Hamiltonian systems connected with the group of rotations, Funkt. Anal. Priloz., 6, no4, 83–84, 1972.

74. Manakov S. V., Note on integrations of Euler's equation of the dynamics of the n-dimensional rigid body, Funkz. Anal. Priloz., 10, no4, 93–4, 1976 (Russian).
75. Veselov A. P., On Hamiltonian formalism for the Novikov-Krichever equations of commutability of two operators, Funkz. Anal. Priloz., 13, no1, 1–7, 1979 (Russian).
76. Dickey L. A., Hamiltonian structures and Lax equations generated by matrix differential operators with polynomial dependence on a parameter, Commun. Math. Phys. 88, 27–42, 1983.
77. Dickey L. A. Symplectic structure, Lagrangian and involutiveness of first integrals of the principal chiral field equation, Comm. Math. Phys., 87, 505–513, 1983.
78. Reyman A. G., and Semenov-Tjan-Shanski M. A., Reduction of Hamiltonian systems, affine Lie algebras and Lax equations I, II, Invent. Math., 54, no1, 81–100, 1979; 63, no3, 423–32, 1981.
79. Adler M., and van Moerbeke P., Completely integrable systems, Euclidian Lie algebras and curves, Adv. in Math., 38, no3, 267–317, 1980.
80. Kac V., Infinite dimensional Lie algebras, Progress in Math., 44, Birhäuser, 1983.
81. Flashka H., Newell A. C., and Ratiu T., Kac-Moody Lie algebras and soliton equations II, III, Physica 9D, no3, 300–323; 324–332, 1983.
82. Caratheodory C., Ueber die Variationsrechnung bei mehrfachen Integralen, Acta Szeged, 4, 193–216, 1929.
83. Weyl H., Geodesic fields, Ann. Math., 36, 607–29, 1935.
84. de Donder Th., Théorie invariantive du calcul des variationes, 1935.
85. Dedecker P., On the generalisation of symplectic geometry to multiple integrals in the calculus of variations, Lecture Notes in Math., 570, 395–456, 1977.
86. Tulczyjev W. M., The Euler-Lagrange resolution, Lecture Notes in Math., 836, 22–48, 1980.
87. Dedecker P., and Tulczyjev W. M., Spectral sequences and the inverse problem of the calculus of variations, Lecture Notes in Math., 836, 498–503, 1980.
88. Vinogradov A. M., Krasilshchik I. S., and Lychagin V. V., Introduction to geometry of nonlinear differential equations, Nauka, 1986 (Russian).
89. Vinogradov A. M., and Kupershmidt B. A., Structure of Hamiltonian mechanics, Uspehi Mat. Nauk, 32, no4, 175–236, 1977 (Russian).
90. Tsujishita T., On variation bicomplexes associated to differential equations, preprint.

91. Goldshmidt H., and Sternberg S., The Hamilton-Cartan formalism in the calculus of variations, Ann. Inst. Fourier, 23, no1, 203–267, 1973.
92. Gabrielov A. M., Gelfand I. M., and Losik M. V., Combinatorial calculation of characteristic classes, IPM preprint 12, 1975 (Russian).
93. Martinez Alonso L., On the Noether map, Lett. Math. Phys. 3, 419–424, 1979.
94. Shadwick W. F., Noether theory and Steudal conserved currents for the sine-Gordon equation, Lett. Math. Phys., 4, no3, 241, 1980.
95. Shadwick W. F., The Hamilton-Cartan formalism for higher-order conserved currents, Reports in Math. Phys., 18, no2, 243–256, 1980.
96. Shadwick W. F., The Hamilton-Cartan formalism for nth order Lagrangian and the integrability of the KdV and modified KdV equation, Lett. Math. Phys., 5, no2, 137–141, 1981.
97. Shadwick W. F., The Hamilton structure associated to evolution-type Lagrangians, Lett. Math. Phys., 6, no4, 271–276, 1982.
98. Shadwick W. F., The Hamilton formulation of regular nth-order Lagrangian field theories, Lett. Math. Phys., 6, no6, 409–16, 1982.
99. Spivak M., Calculus on manifolds, Benjamin, 1965.
100. Griffits Ph., and Harris J., Principles of algebraic geometry, Willey and Sons, 1978.
101. Gelfand I. M., Manin Yu. I., and Shubin M. A., Poisson brackets and the kernel of variational derivatives in the formal calculus of variations, Funkz. Anal. Priloz., 10, no4, 30–34, 1976 (Russian).
102. Dorfman I. Ya., On formal calculus of variations in the algebra of smooth cylindrical functions, Funkz. Anal. Priloz., 12, no2, 32–39, 1978.
103. Case K. M., Symmetries of the higher-order KP-equations, J. Math. Phys., 26, no6, 1158–9, 1985.
104. Zakharov V. E., and Mikhailov A. V., Variational principle for equations integrable by the inverse problem method, Funkz. Anal. Prilozh., 14, no1, 55–6, 1980.
105. Cherednik I. V., Differential equations for the Baker-Akhiezer functions of the algebraic curves, Funkz. Anal. Prilozh., 12, no3, 1978.
106. Gelfand I. M., and Dickey L. A., Integrable nonlinear equations and Liouville's theorem, Funkz. Anal. Prilozh., 13, no1, 8–20, 1979.
107. Leznov A. N., Saveljev M. V., and Smirnov V. G., Theory of group representations and integration of nonlinear dynamic systems, Teor. Mat. Fiz., 48, no1, 565–71, 1981.
108. Manin Yu. I., Algebraic aspects of nonlinear differential equations, Itogi nauki i tekhniki, ser. Sovremennye problemy Mat., 11, 1978.

109. Witham G. B., Linear and non-linear waves, Wiley, 1974.
110. Gervais J.-L., Neveu A., Dual string spectrum in Polyakov's quantisation, Nucl. Phys. B209 (1982), 125–145.
111. Leznov A. N. and Saveljev M. V., Group method of integration of non-linear dynamical systems, Nauka, 1985 (Russian).
112. Wilson G., Infinite-dimensional Lie groups and algebraic geometry in soliton theory, Phil. Trans. Royal Soc. London A315, 393–404, 1985.